BIOCHEMISTRY OF ARACHIDONIC ACID METABOLISM

PROSTAGLANDINS, LEUKOTRIENES, AND CANCER

Series Editors: Kenneth V. Honn and Lawrence J. Marnett
Wayne State University School of Medicine
Detroit, Michigan

W.E.M. Lands, ed.: Biochemistry of Arachidonic Acid Metabolism. 1985. ISBN 0-89838-717-5.
L.J. Marnett, ed.: Arachidonic Acid Metabolism and Tumor Initiation. 1985. ISBN 0-89838-729-9.
S.M. Fischer and T.J. Slaga, eds.: Arachidonic Acid Metabolism and Tumor Promotion. 1985. ISBN 0-89838-724-8.
J.S. Goodwin, ed.: Prostaglandins and Immunity. 1985. ISBN 0-89838-723-X.

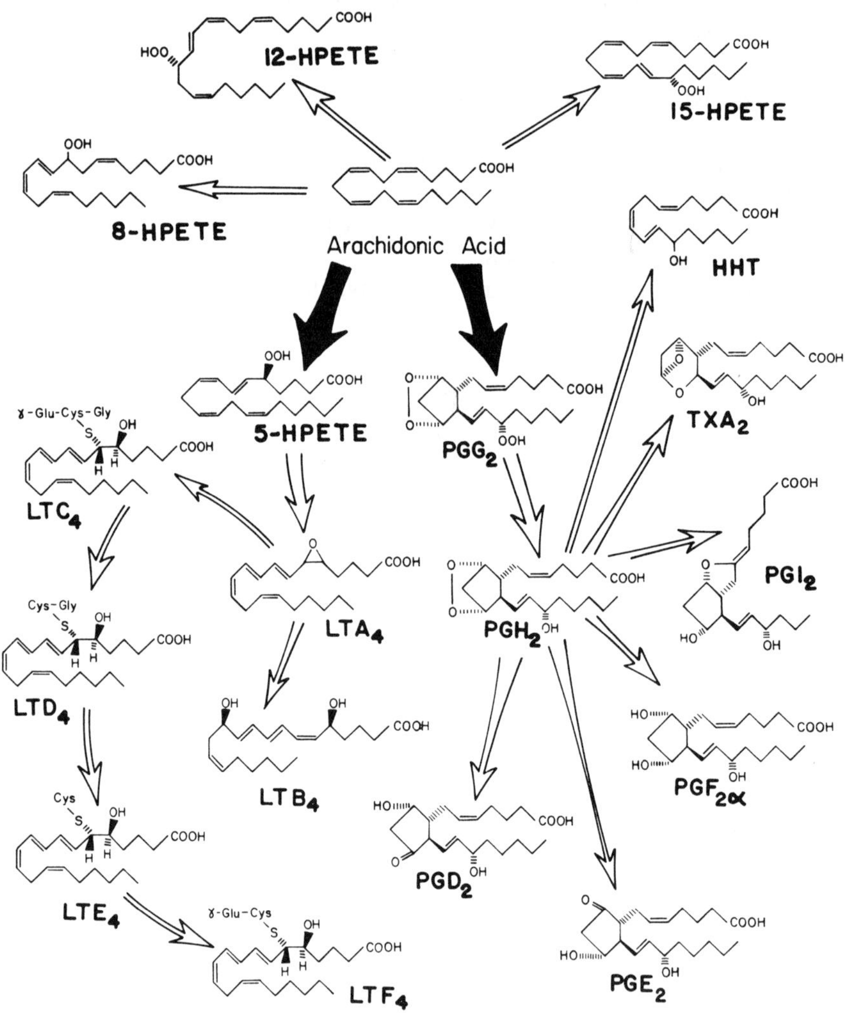

COOH
HOO
12-HPETE
COOH
OOH
15-HPETE
OOH
COOH
COOH
8-HPETE
Arachidonic Acid
COOH
HHT
OH
COOH
OH
TXA2
OOH
COOH
COOH
5-HPETE
COOH
OOH
PGG2
γ-Glu-Cys-Gly
S
OH
COOH
H
H
LTC4
COOH
O
O
COOH
PGI2
HO
OH
COOH
LTA4
COOH
OH
PGH2
Cys-Gly
S
OH
COOH
H
H
LTD4
OH
OH
COOH
LTB4
HO
COOH
HO
OH
PGF2α
Cys
S
OH
COOH
H
H
LTE4
HO
COOH
O
OH
PGD2
O
COOH
HO
OH
PGE2
γ-Glu-Cys
S
OH
COOH
H
H
LTF4

BIOCHEMISTRY OF ARACHIDONIC ACID METABOLISM

edited by

William E.M. Lands
University of Illinois at Chicago
College of Medicine

Martinus Nijhoff Publishing
a member of the Kluwer Academic Publishers Group
Boston/Dordrecht/Lancaster

Distributors for North America:
Kluwer Academic Publishers
190 Old Derby Street
Hingham, MA 02043

Distributors for all other countries:
Kluwer Academic Publishers Group
Distribution Centre
P.O. Box 322
3300 AH Dordrecht
The Netherlands

Library of Congress Cataloging in Publication Data

Main entry under title:

Biochemistry of arachidonic acid metabolism.

(Prostaglandins, leukotrienes, and cancer)
Includes bibliographies and index.
1. Arachidonic acid—Metabolism. 2. Cell transformation. 3. Cancer cells. I. Lands, William E.M., 1930– . II. Series. [DNLM: 1. Arachidonic Acids—biosynthesis. 2. Arachidonic Acids—metabolism. 3. Cell Transformation, Neoplastic. QZ 202 B6147]
QP752.A7B56 1985 599′.09247 85-4853
ISBN 0-89838-717-5

Printed in the United States of America

CONTENTS

CONTRIBUTORS

DR. DOUGLAS ARCHER
Food and Drug Administration
200 C. Street, N.W. HFF231
Washington, D.C. 20204

DR. DONALD CHAMBERS
Periodontal Diseases Research Center
University of Illinois at Chicago
Chicago, IL 60612

DR. RHONNA COHEN
Department of Oral Pathology
University of Illinois at Chicago
Chicago, IL 60612

DR. LARRY W. DANIEL
Department of Biochemistry
Bowman Gray School of Medicine
Wake Forest University
Winston-Salem, NC 27103

DR. BERNARD B. DAVIS
Chief, Medical Service
Veterans Administration Medical Center
St. Louis, MO 63125
and
Professor of Medicine
St. Louis University School of Medicine
St. Louis, MO 63125

DR. R.J. FLOWER
The Wellcome Research Laboratories
Beckenham, Kent BR33BS
ENGLAND

DR. G. HAHN
Department of Biochemistry
Boston University
School of Medicine
80 East Concord Street
Boston, MA 02118

DR. T.W. HARPER
Department of Pharmacology
University of Colorado
School of Medicine
Denver, CO 80262

DR. KENNETH V. HONN
Departments of Radiology and Radiation Oncology
Wayne State University
School of Medicine
5045 Cass Street
Detroit, MI 48202

DR. A.W. FORD-HUTCHINSON
Merck Frosst Canada Inc.
P. O. Box 1005
Pointe Claire-Dorwal
Quebec, Canada H9R 4P8

DR. BARBARA A. JAKSCHIK
Department of Pharmacology
Washington University Medical School
4566 Scott Avenue
St. Louis, MO 63110

DR. RASHIDA A. KARMALI
Memorial Sloan-Kettering Cancer Center
1275 York Avenue
New York, NY 10021

CONTRIBUTORS (CONTINUED)

DR. RICHARD KULMACZ
Department of Biological Chemistry
University of Illinois at Chicago
1853 W. Polk St.
Chicago, IL 60612

DR. C.G. KUO
Department of Pharmacology
Washington University
Medical School
4566 Scott Avenue
St. Louis, MO 63110

DR. WILLIAM E.M. LANDS
Department of Biological Chemistry
University of Illinois at Chicago
1853 W. Polk St.
Chicago, IL 60612

DR. LAWRENCE LEVINE
Department of Biochemistry
Brandeis University
Waltham, MA 02254

DR. EUAN MACINTYRE
Department of Pharmacology
University of Glasgow
Glasgow G12 8QQ
SCOTLAND

DR. JACQUES MACLOUF
Unite 150 Inserm
Hopital Lariboisiere, 6
Rue Guy-Patin, 75475 Paris
Cedex 10
France

DR. PAUL MARSHALL
Department of Biological Chemistry
University of Illinois at Chicago
1853 W. Polk St.
Chicago, IL 60612

DR. ROBERT C. MURPHY
Department of Pharmacology
University of Colorado
School of Medicine
Denver, CO 80262

MR. FRANK PAPATHEOFANIS
Department of Biological Chemistry
University of Illinois at Chicago
1853 W. Polk St.
Chicago, IL 60612

DR. L. PARENTE
The Wellcome Research Laboratories
Beckenham, Kent BR33BS
ENGLAND

DR. PETER POLGAR
Department of Biochemistry
Boston University
School of Medicine
80 East Concord Street
Boston, MA 02118

DR. WILLIAM S. POWELL
Department of Medicine
McGill University, Endocrine Lab
Royal Victoria Hospital
687 Pine Avenue W.
Montreal, Quebec, Canada H3A 1A1

CONTRIBUTORS (CONTINUED)

DR. NORBERTA SCHOENE
U.S. Department of Agriculture
Lipid Nutrition Laboratory
Building 308 111, BARC-East
Beltsville, MD 20705

DR. LINDA L. SLAKEY
Department of Biochemistry
University of Massachusetts
Amherst, MA 01003

DR. BONNIE F. SLOANE
Department of Pharmacology
Wayne State University
School of Medicine
540 E. Canfield
Detroit, MI 48201

DR. WILLIAM L. SMITH
Department of Biochemistry
Michigan State University
East Lansing, MI 48824

DR. L. TAYLOR
Department of Biochemistry
Boston University
School of Medicine
80 East Concord Street
Boston, MA 02118

DR. JACK Y. VANDERHOEK
Department of Biochemistry
George Washington University
School of Medicine
2300 Eye Street, N.W.
Washington, D.C. 20037

DR. JAMES VARANI
Department of Pathology
University of Michigan
Ann Arbor, MI 48109

DR. Y.F. WEI
Department of Pharmacology
Washington University
Medical School
4566 Scott Avenue
St. Louis, MO 63110

DR. JOANN WESS
Food and Drug Administration
Microbiology Division
1090 Tusculum Ave.
Cincinnati, OH 45226

DR. SHOZO YAMAMOTO
Department of Biochemistry
Tokushima University
School of Medicine
Kuramoto-Cho, Tokushima 770
Japan

DR. TERRY V. ZENSER
Research Chemist
GRCC (111G-JB)
Veterans Administration Medical Center
St. Louis, MO 63125
and
Associate Professor of Medicine and Biochemistry
St. Louis University School of Medicine
St. Louis, MO 63125

FOREWORD

Prostaglandins, Leukotrienes, and Cancer is a multi-volume series that will focus on an emerging area of cancer research. In 1968, R.H. Williams first reported that elevated prostaglandin levels are present in human medullary carcinoma. Since that time, the concept that arachidonic acid metabolites may be involved in cancer has expanded to include every aspect of the disease from cell transformation through metastasis.

Prostaglandins and leukotrienes are generic terms used to describe a family of bioactive lipids produced from unsaturated fatty acids (principally from arachidonic acid) via the cyclooxygenase and lipoxygenase pathways, respectively. Cyclooxygenase products consist of diverse products such as prostaglandin E_2 (PGE_2), prostacyclin (PGI_2) and thromboxane A_2 (TXA_2), whereas lipoxygenase products consist of hydroperoxy fatty acids and mono-, di- and tri-hydroxy acids including leukotrienes. The precursor fatty acids for the cyclooxygenase and lipoxygenase pathways are present in cellular phospholipids. This finding established an important control point in their biosynthesis—the release of substrate. This occurs in response to numerous stimuli that act at the cell surface. Dr. Bengt Samuelsson's extensive study of the metabolism of prostaglandins indicated that they are rapidly inactivated on a single pass through pulmonary circulation. Thus, they cannot act as circulating hormones and appear to be made on demand in or in the vicinity of target tissues leading to the concept that prostaglandins are local hormones or autocoids.

Altered production, qualitative and/or quantitative, of prostaglandins and leukotrienes has been implicated in the development of a number of disease states (e.g., atherosclerosis, inflammatory diseases, asthma). Evidence has been accumulating in the literature suggesting that prostaglandins and leukotrienes may stimulate or inhibit various steps in the complex etiology of cancer, i.e., steps in the progression from a transformed cell to a metastatic tumor. The initial volumes in this series will examine the roles of prostaglandins and leukotrienes in tumor initiation, tumor promotion, tumor cell growth and differentiation, tumor immunity, tumor metastasis and cancer therapy. We hope as this field of cancer research develops that this series, Prostaglandins, Leukotrienes, and Cancer, will provide a forum within the framework of current evidence for the synthesis of new hypotheses and discussion of controversial issues.

Kenneth V. Honn
Lawrence J. Marnett

PREFACE

WILLIAM E. M. LANDS

In the past decade, research on cancer has moved to a new level of sophistication. We now understand that the appearance and proliferation of cancer cells can result from a diverse and subtle set of cellular and biochemical events. At this time, the basic biochemical processes that underlie tumor proliferation are not definitely known to involve the eicosanoids, prostaglandins and leukotrienes. Nevertheless, these compounds are steadily becoming recognized as significant modifiers of cellular behavior. In this first volume of our series of books on prostaglandins, leukotrienes and cancer, the authors of the various chapters have joined together to provide their collective insights into some of these basic processes.

The biochemical mechanism of eicosanoid biosynthesis by fatty acid oxygenases led us to develop a central theme in assembling the different chapters of this volume: *the signalling events regulating the appearance and proliferation of cancer cells may be inextricably related through the participation of lipid hydroperoxide.*

Fatty acid oxygenases that catalyze the first committed steps in the biosynthesis of prostaglandins and leukotrienes require a lipid hydroperoxide to initiate the oxygenation reaction, and they produce a lipid hydroperoxide as their first product. The signalling events that lead to the formation of eicosanoids and that are derived from eicosanoids may thus be influenced by the generation and removal of lipid hydroperoxide. The cellular level of the hydroperoxides themselves may be regulated by the rate of eicosanoid biosynthesis leading to an explosive involvement of peroxide with the eicosanoid synthesizing enzymes.

The scheme below illustrates how the fatty acid oxygenases can amplify small amounts of lipid peroxide directly into more cellular hydroperoxide as well as promoting eicosanoid-mediated inflammatory responses that lead to still more oxidant release and peroxide generation. Such a rationale places these eicosanoid-forming enzymes in a pivotal position relative to current concepts of biochemical factors regulating the origins and proliferation of cancer cells.

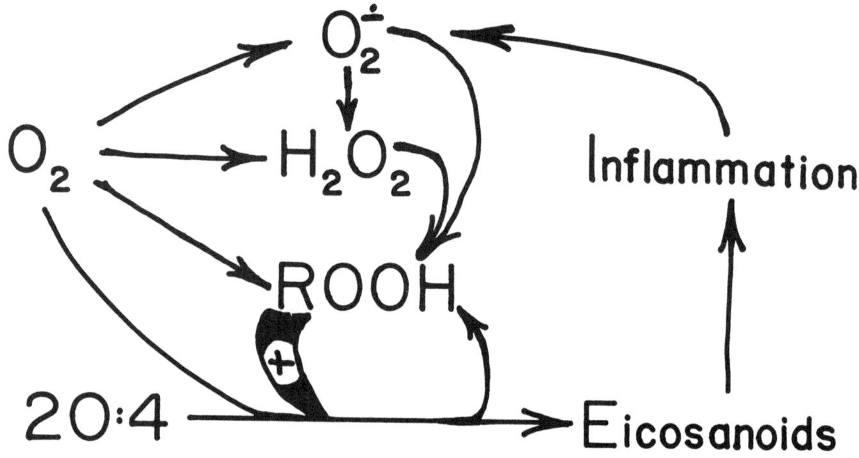

The first five chapters of this book deal with the enzymes of eicosanoid biosynthesis, their behavior, and their subcellular locations. The subsequent four chapters (6-9) deal with factors controlling the generation and removal of lipid hydroperoxides which are necessary factors in eicosanoid biosynthesis. The subsequent three chapters (10-12) deal with the regulation of the fatty acid precursors of eicosanoids through the control of phospholipase action or dietary supply.

Although eicosanoids such as prostaglandins and leukotrienes are not known to be absolutely essential for cell survival, they are becoming widely recognized as important signalling agents in modifying and modulating cellular behavior. The signals that these eicosanoids generate in linking the behavior of one cell to that of another have been extensively studied in relation to inflammatory and thrombotic events and the regulation of immune responses. In all of these areas of cellular behavior, eicosanoids play important modifying roles in signalling the various cells to bring about the overall pathophysiologic responses. Therefore, a major section of this book deals with the nature of signalling events that are associated with prostaglandin and leukotriene formation.

Chapters (13,14) on the effects of hydroxypolyunsaturated fatty acids and chemotaxis are followed by discussions (15,16) of prostaglandin receptors and leukotriene pathology. Five chapters (17-21) introduce the reader to other ways that the integration of cell-cell signals can be maintained with agents other than polyunsaturated fatty acids: cyclic nucleotides, proteases, adenine nucleotides and phorbol esters. These other signalling agents influence the mobilization of prostaglandins and leukotrienes, and therefore play an important additional modulatory role in involving eicosanoids in cell-cell signalling events.

No reliable biochemical information is obtained without careful, controlled measurement of the chemicals involved. In developing sound research on the biochemical aspects of eicosanoids, several analytical methods have proved essential in acquiring the needed information. Without these analytical tools to measure the compounds involved we would know very little about the role of eicosanoids in cellular events. Thus, the final three chapters (22-24) consider the three most valuable methods currently used for studying eicosanoid biochemistry: high pressure liquid chromatography, radioimmunoassay, and gas chromatography-mass spectrometry.

Stimulation of eicosanoid biosynthesis may lead to elevated levels of peroxides and radicals and thereby to damage of DNA, or it may lead to the release of growth promoting materials such as observed in inflammatory and thrombotic events. These possibilities raise the question as to whether peroxides are more important in initiating the damage to DNA or in promoting proliferation of cells more rapidly than DNA damage can be repaired. Cancer research must always deal with this dual problem of whether initiation is more significant than promotion in the clinical appearance of tumors. That concern cannot be answered in this volume. Rather, the following chapters are arranged to help the reader consider whether the lipid peroxides and eicosanoids that are involved in inflammatory events, immune responses, and thrombotic processes may also participate in tumor initiation and in proliferative events involving tissue invasiveness, metastasis and supression of immune surveillance. Although lipid hydroperoxides have often been discussed in terms of initiation of DNA damage, the reader may benefit from examining the hypothesis that lipid hydroperoxides may participate in stimulating eicosanoid-mediated events that are causally linked to proliferation and tumorigenesis. In this way a whole new range of varied biochemical processes may serve as possible mediators of the often postulated action of peroxides in carcinogenesis.

In the coming decade we expect much more sophisticated experiments to provide interpretations of the many cell-cell signalling processes that are associated with cancer cell proliferation. We assembled these twenty four chapters at this time to provide a stronger base from which those necessary future experiments can be designed and interpreted. We invite the readers to join in our attempts to tie these biochemical processes together in understanding the complex phenomena that we call cancer.

BIOCHEMISTRY OF ARACHIDONIC ACID METABOLISM

1

MECHANISMS OF CYCLOOXYGENASE AND PEROXIDASE CATALYSIS BY PROSTAGLANDIN H SYNTHASE

RICHARD J. KULMACZ

1. INTRODUCTION

Eicosanoids, the group of oxygenated arachidonic acid metabolites which includes prostaglandins, thromboxanes, leukotrienes, and various hydroperoxy and hydroxy fatty acids, have come to be implicated at several stages of carcinogenesis and metastasis. The exact roles played by the eicosanoids in the control of cell proliferation are difficult to discern at present, as the literature is filled with conflicting reports of their effects (for a review, see Ref. 1). The resolution of these apparently contradictory effects of eicosanoids will be aided by a close comparison of the concentrations of these eicosanoids which exist _in vivo_ and those concentrations used in many _in vitro_ experiments. An understanding of the details of the biosynthetic mechanisms for the eicosanoids, and the interplay among substrates and effectors in these reactions may also help in unravelling the tangle of conflicting reports.

The first committed step in the biosynthesis of each of the eicosanoids, the incorporation of molecular oxygen into a polyunsaturated fatty acid, is catalyzed by one of a group of enzymes called fatty acid oxygenases. This group includes the cyclooxygenase of PGH synthase, and the various lipoxygenases. One aspect of all fatty acid oxygenases examined thus far, which may be crucial to the regulation of eicosanoid biosynthesis _in vivo_, is that a hydroperoxide activator is required for catalytic activity (2-4). The consequences of this requirement for hydroperoxide are significant: the activities of the cellular fatty acid oxygenases could be modulated by changes in cellular hydroperoxide levels; lipid hydroperoxide produced by one fatty acid oxygenase could activate a fatty acid oxygenase in another part of the cell or in a different cell; and fatty acid oxygenases, given adequate fatty acid

W.E.M. Lands (ed.), BIOCHEMISTRY OF ARACHIDONIC ACID METABOLISM.

substrate, will tend to amplify cellular hydroperoxide levels because of feedback activation from their own products.

The most detailed studies on the biochemical basis for activation of fatty acid oxygenases by hydroperoxide have involved the cyclooxygenase activity of PGH synthase. The synthase is the only mammalian fatty acid oxygenase currently available in pure form, and in milligram qualities (5-7). An examination of the mechanisms of activation and catalysis of the cyclooxygenase activity of the synthase thus may be useful as a model to provide information about the regulation of other mammalian fatty acid oxygenases. The mechanisms may suggest routes for prophylactic or therapeutic intervention in those cases where eicosanoid biosynthesis is deranged.

PGH synthase is an integral membrane protein, and its purification requires it to be solubilized by detergents (5-7). The pure, detergent-solubilized synthase is a dimer composed of two similar 70,000 Da subunits (8,9), although the subunit arrangement in the native membrane is not known.

2. PEROXIDE REQUIREMENTS OF CYCLOOXYGENASE AND PEROXIDASE ACTIVITIES

Besides the cyclooxygenase activity which generates the hydroperoxide, PGG_2, the pure synthase also exhibits a peroxidase activity capable of reducing a wide variety of hydroperoxides to the corresponding alcohols (6,7). Because of the peroxide activator requirement of the cyclooxygenase, the presence of an endogenous peroxide-removing activity in the synthase might be expected to lead to a suppression of cyclooxygenase activity. However, the cyclooxygenase is activated by peroxide concentrations in the range of 10^{-8} M (10). These levels are well below the K_m value of the synthase peroxidase (3 to 5 x 10^{-6} M; Ref. 10) so that the peroxidase is not able to lower peroxide levels sufficiently to impair the cyclooxygenase. PGH synthase thus has the potential to amplify the peroxide concentration from nanomolar to near-micromolar levels.

The conversion of arachidonic acid to PGH_2 by the synthase has many of the characteristics of a free radical reaction (for a review, see Ref. 11). Most attempts to describe a reaction mechanism for the cyclooxygenase activity have been phrased in terms of the participation of enzyme redox centers in the initiation and propagation of a radical chain reaction (12-15).

3. ROLE OF HEME IN CATALYSIS BY THE SYNTHASE

Both of the catalytic activities of the synthase show an absolute requirement for heme (6,7). The peroxidase reaction can be inhibited by cyanide with a K_I value of 0.3 mM, quite similar to the dissociation constant of the synthase-cyanide complex, 0.2 mM (16). A large number of compounds, including phenol, have been found to stimulate peroxidase activity in a manner which suggests that they function as electron-donating cosubstrates (17-19). These observations are consistent with a peroxidase mechanism (20) in which the reduction of hydroperoxide to an alcohol oxidizes the enzyme heme to a formal Fe(V) state (Compound I), which is then reduced by the cosubstrate to the Fe(IV) (Compound II) and finally to the resting Fe(III) state.

Many of the peroxidase cosubstrates also stimulate the cyclooxygenase activity of the synthase (7,21,22). This stimulatory effect has been ascribed to a protective action of the cosubstrate on the cyclooxygenase (23), slowing the self-catalyzed inactivation. However, examination of the cyclooxygenase reaction kinetics indicates that the decay constant for the loss of cyclooxygenase activity due to self-inactivation is not changed by the presence of phenol (15). It seems likely that the stimulatory effect of peroxidase cosubstrates on the cyclooxygenase activity is due to an increase in the catalytic velocity of the enzyme.

The role of heme in cyclooxygenase catalysis appears to be quite complex. The fact that cyclooxygenase activity is not inhibited by carbon monoxide (15,17) has led to a proposal that the cyclooxygenase promotes the conversion of the fatty acid, rather than oxygen, to a reactive state (15). Cyanide, on the other hand, has distinct effects on the cyclooxygenae activity of the synthase (8,15). At concentrations similar to those needed for formation of the synthase-cyanide complex and to inhibit the peroxidase activity, cyanide only slightly reduces the oxygenation velocity while it markedly impairs the activation of the cyclooxygenase (15,16). The coincidence of these effects of cyanide suggests that both activation of the cyclooxygenase and catalysis by peroxidase involve some interaction of hydroperoxide with the heme iron. The large difference in the peroxide level required for activation of the cyclooxygenase and the K_m value of the peroxidase (10) does, however, indicate clearly different roles for the peroxide in the two processes.

Neither activation of the cyclooxygenase by hydroperoxide nor cyclooxygenase catalysis itself appear to require peroxidase catalysis. 15-Hydroperoxyeicosatetraenoic acid was recovered intact after it had activated a cyanide-suppressed cyclooxygenase system (15). In addition, the cyclooxygenase activity of Mn protoporphyrin IX-synthase, which has no peroxidase activty (6) is nevertheless sensitive to inhibition by glutathione peroxidase (15) suggesting that it undergoes an accelerative activation by hydroperoxide.

Conversion of the synthase apoenzyme to the fully functional cyclooxygenase required only one heme per synthase dimer (24). The same stoichiometry of one heme per dimer was found for the peroxidase activity of the synthase (R. Kulmacz, unpublished results). These results indicate that the two subunits of the synthase are functionally distinct, and that the two catalytic activities of the synthase share the same heme cofactor.

4. NON-HEME IRON CONTENT OF PGH SYNTHASE

The crucial role of iron in the mechanism of another fatty acid oxygenase, soybean lipoxygenase (25), has prompted an examination of the synthase for non-heme iron centers (5,7,8, 26-28). Non-heme iron has been repeatedly detected in the synthase, but no consensus has emerged for the stoichiometry of iron/dimer: the reported values range from 0.2 to 1.6 mol iron/mol dimer (5,7,8,26-28). This variable stoichiometry, and the fact that the non-heme iron content of various batches of the synthase did not correlate with their cyclooxygenase specific activity (26,28), are difficult to reconcile with a catalytic role for non-heme iron, but the matter remains to be settled conclusively.

5. CHANGES IN REDOX STATE OF SYNTHASE DURING CATALYSIS

The detailed description of the cyclooxygenase catalytic intermediates in terms of the redox state of the heme (and whatever other redox centers might be present) is still at an early stage, although progress is being made. Spectrophotometric examination of the synthase after addition of PGG_2 showed an increase in absorbance at 430, 525, and 555 nm (29). Absorbance maxima of the ferrous form of the synthase have been reported to be at 424, 530, and 558 nm (8), and so the interaction of the synthase with PGG_2 may result in the conversion of some of the synthase heme from the resting, ferric, state to the ferrous state. This would support the hypothesis of Hemler and Lands (15) that the synthase heme is reduced to the ferrous state in the course of activation of the cyclooxygenase by peroxide. In that hypothetical scheme (15) reduction of the synthase heme by hydroperoxide (ROOH) produces a hydroperoxy radical:

$$\overset{III}{E}{}^{ROOH} \longrightarrow \overset{II}{E}{}^{ROO\cdot} \qquad \underline{1}$$

After binding arachidonic acid (AA), the hydroperoxy radical formed in Reaction 1 abstracts the 13-S-hydrogen from the fatty acid (30):

$$\overset{II}{E}{}^{ROO\cdot}_{AA} \longrightarrow \overset{II}{E}{}^{ROOH}_{AA\cdot} \qquad \underline{2}$$

Reaction with molecular oxygen and cyclization then affords the hydroperoxy radical of PGG_2:

$$\overset{II}{E}{}^{ROOH}_{AA\cdot} + 2\,O_2 \longrightarrow \overset{II}{E}{}^{ROOH}_{PGG_2\cdot} \qquad \underline{3}$$

Finally the PGG_2 radical abstracts a hydrogen from the hydroperoxide to complete the cyclooxygenase catalytic cycle:

$$\overset{II}{E}{}^{ROOH}_{PGG_2\cdot} \longrightarrow \overset{II}{E}{}^{ROO\cdot} + PGG_2 \qquad \underline{4}$$

This model is able to account for most of the known kinetic properties of the synthase, and furnishes a basis for predicting of the response of the cyclooxygenase and peroxidase activities to physiological and pharmacological perturbations, such as changes in tissue peroxide levels or administration of antioxidant drugs (10,31). One difficulty encountered in attempts to observe enzyme intermediates during incubation of the synthase with substrate is the short life span of the enzyme.

Once incubated with arachidonic acid, the cyclooxygenase velocity reaches its maximum within seven seconds, and decays with a time constant of about 6/min (15). Thus, the total concentration of active synthase decreases rapidly, and the enzyme intermediates of interest may make up only a small fraction of the total enzyme forms at any particular time. The kinetics of the cyclooxygenase reaction are considerably slower when the synthase is treated with cyanide and phenol (15) or with indomethacin (32), and such systems may afford a better opportunity to observe and identify enzyme intermediates in the catalytic cycle.

ACKNOWLEDGEMENTS

This investigation was supported in part by Public Health Service grant number GM 30509.

REFERENCES

1. Cornwell DG, Morisaki N: Fatty acid paradoxes in the control of cell proliferation: prostaglandins, lipid peroxides, and cooxidation reactions. In Free radicals in Biology (ed. by W.A. Pryor), Vol. 6, Academic Press, N.Y. 1984, pp. 95-148.
2. Lands WEM, Lee R, Smith W: Factors regulating the biosynthesis of various prostaglandins. Ann. N.Y. Acad. Sci. (180): 107-122, 1971.
3. Egan RW, Tischler AN, Baptista EM, Ham EA, Soderman DD, Gale PH: In Advances in Prostaglandin Thromboxane and Leukotriene Research (ed. by B. Samuelsson, R. Paoletti, and P. Ramwell) Vol. 11, Raven Press, N.Y., 1983, pp. 151-157.
4. Yokoyama C, Mizuno K, Mitachi H, Yoshimoto T, Yamamoto S, Pace-Asciak CR: Partial purification and characterization of arachidonate 12-lipoxygenase from rat lung. Biochim. Biophys. Acta (750): 237-243, 1983.
5. Hemler M, Lands WEM, Smith, WL: Purification of the cyclooxygenase that forms prostaglandins. J. Biol. Chem. (251): 5575-5579, 1976.
6. Miyamoto T, Ogino N, Yamamoto S, Hayaishi O: Purification of prostaglandin endoperoxide synthetase from bovine vesicular gland microsomes. J. Biol. Chem. (251): 2629-2636, 1976.
7. van der Ouderaa FJ, Buytenhek M, Nugteren DH, van Dorp DA: Purification and characterization of prostaglandin endoperoxide synthetase from sheep vesicular glands. Biochim. Biophys. Acta (487): 315-331, 1977.
8. van der Ouderaa FJ, Buytenhek M, Slikkerveer FJ, van Dorp DA: On the haemoprotein character of prostaglandin endoperoxide synthetase. Biochim. Biophys. Acta (572): 2942, 1979.
9. Roth GJ, Siok CJ, Ozols J: Structural characteristics of prostaglandin synthetase from sheep vesicular gland. J. Biol. Chem. (255): 1301-1304, 1980.

10. Kulmacz RJ, Lands WEM.: Requirements for hydroperoxide by the cyclooxygenase and peroxidase activities of prostaglandin H synthase. Prostaglandins (25): 531-540, 1983.
11. Porter NA.: Prostaglandin endoperoxides. In Free Radicals in Biology (ed. by W.A. Pryor), Volume 4, Academic Press, N.Y., 1980, pp. 261-294
12. O'Brien PJ, Rahimtula A: The possible involvement of a peroxidase in prostaglandin biosynthesis. Biochem. Biophys. Res. Commun. (70): 832-838, 1976.
13. O'Brien PJ, Rahimtula AD: Mechanism of oxygen activation involved in the prostaglandin synthetase mechanism. In Advances in Prostaglandin and Thromboxane Research (ed. by B. Samuelsson, P.W. Ramwell, and R. Paoletti) Vol. 6, Raven Press, N.Y., 1980, pp. 145-148.
14. Panganamala RV, Sharma HM, Heikkila RE, Geer JC, Cornwell DG: Role of hydroxyl radical scavengers dimethyl sulfoxide, alcohols and methional in the inhibition of prostaglandin biosynthesis. Prostaglandins (11): 599-607, 1976.
15. Hemler ME, Lands, WEM: Evidence for a peroxide-initiated free radical mechanism of prostaglandin biosynthesis. J. Biol. Chem. (255): 6253-6261, 1980.
16. Kulmacz RJ, Lands WEM: Quantitative similarities in the several actions of cyanide on prostaglandin H synthase. submitted, 1984.
17. Ohki S, Ogino N, Yamamoto S, Hayaishi O: Prostaglandin hydroperoxidase, an integral part of prostaglandin endoperoxide synthetase from bovine vesicular gland microsomes. J. Biol. Chem. (254): 829-836, 1979.
18. Gale PH, Egan RW: Prostaglandin endoperoxide synthase-catalyzed oxidation reactions. In Free Radicals in Biology (ed. by W.A. Pryor) Vol. 6, Academic Press, N.Y., 1984, pp. 1-38.
19. Marnett LJ: Hydroperoxide-dependent oxidations during prostaglandin biosynthesis. In Free Radicals in Biology (ed. by W.A. Pryor) Vol. 6, Academic Press, N.Y., 1984, pp. 63-94.
20. Yamazaki I: Peroxidase. In Molecular Mechanisms of Oxygen Activation (ed. by O. Hayaishi) Academic Press, N.Y., 1974, pp. 535-558,
21. Ogino N, Ohki S, Yamamoto S, Hayaishi O: Prostaglandin endoperoxide synthetase from bovine vesicular gland microsomes. J. Biol. Chem. (253): 5061-5068, 1978.
22. Nugteren DH, Beerthuis RK, van Dorp DA: The enzymic conversion of all-cis 8,11,14-eicosatrienoic acid into prostaglandin E. Recl. Trav. Chim. Pays-Bas (85): 405-419, 1966.
23. Egan RW, Paxton J, Kuehl FA Jr.: Mechanism for irreversible self-deactivation of prostaglandin synthetase. J. Biol. Chem. (251): 7329-7335, 1976.
24. Kulmacz RJ, Lands WEM: Prostaglandin H synthase: stoichiometry of heme cofactor. J. Biol. Chem. (259): 6358-6363, 1984.
25. Pistorius, EK, Axelrod B: Iron, an essential component of lipoxygenase. J. Biol. Chem. (249): 3183-3186, 1974.
26. Hemler ME, Lands WEM: Biosynthesis of prostaglandins. Lipids (12): 591-595, 1977.
27. vanDorp DA, Buytenhek M, Christ-Hazelhof E, Nugteren DH, van der Ouderaa FJ: Isolation and properties of enzymes involved in prostaglandin biosynthesis. Acta Biol. Med. Germ. (37): 691-699, 1978.

28. van der Ouderaa FJ, Buytenhek M, van Dorp DA: Characterization of prostaglandin H synthetase. In Advances in Prostaglandin and Thromboxane Research (ed. by B. Samuelsson, P.W. Ramwell, and R. Paoletti) Vol. 6, Raven Press, N.Y., 1980, pp. 139-144.
29. Nastainczyk W, Ruf HH, Schuhn D: Spectral properties of the enzyme substrate complex of prostaglandin hydroperoxidase and prostaglandin G_2. In Icosanoids and Cancer (ed. by H. Thaler-Dao, A.C. de Paulet, R. Paoletti) Raven Press, N.Y., 1984, pp. 255-258.
30. Hamberg M, Samuelsson B: On the mechanism of the biosynthesis of prostaglandins E_1 and $F_{1\alpha}$. J. Biol Chem. (242): 5336-5343, 1967.
31. Hanel AM, Lands WEM: Modification of anti-inflammatory drug effectiveness by ambient lipid peroxides. Biochem. Pharmacol. (31): 3307-3311, 1982.
32. Kulmacz RJ: Properties of prostaglandin biosynthesizing enzymes. In Prostaglandins and Leukotrienes (ed. by J.M. Bailey), Plenum Press, N.Y., in the press, 1984.

2

LIPOXYGENASE MECHANISMS

FRANK J. PAPATHEOFANIS AND WILLIAM E.M. LANDS

1. GENERAL ENZYMOLOGY

Lipoxygenases (EC 1.13.11.12) derived from animal and plant sources catalyze the insertion of oxygen into polyunsaturated fatty acids. The basic cis, cis-nonconjugated diene system is a required structure of all substrate fatty acids. The usual product from the action of animal and plant lipoxygenases upon such substrates is a cis, trans-conjugated hydroperoxy acid (Figure 1). The action of lipoxygenases upon substrate fatty acids is commonly monitored by polarographic measurement of oxygen uptake, by spectrophotometric observation of conjugated diene at A_{234} in the product fatty acid, or by chromatographic resolution of radioactive products formed from radioactive precursor. Many published reports describe the easily handled substrate, 18:2 (linoleic acid; 9,12-cis,cis-octadecadienoic* acid) although recent interest has increased for 20:4 (arachidonic acid; 5,8,11,14-all-cis-eicosatetraenoic* acid), the major precursor of eicosanoids in mammalian tissues. These two acids appear to share a qualitatively similar overall reaction mechanism, but they have quantitatively different kinetic constants. There has been a tendency for researchers to use 18:2 when studying plant lipoxygenases and 20:4 with animal lipoxygenases.

---

*For convenience, all ethylenic bonds described in this chapter will be of the cis-configuration (c) unless specifically designated to be trans-(t).

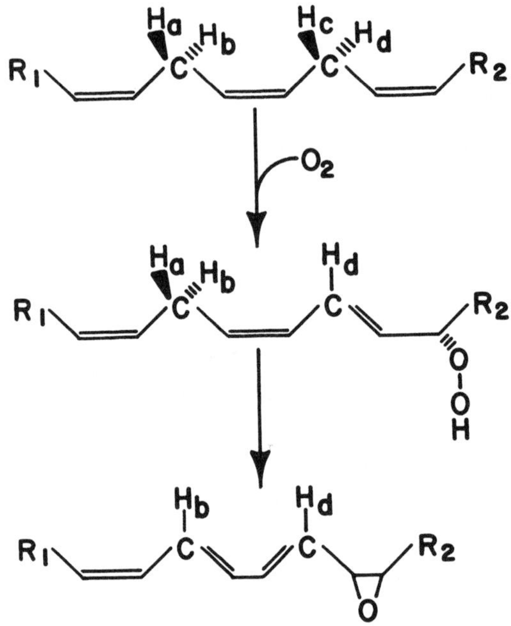

Figure 1. Oxygenation of substrate fatty acid to produce a hydroperoxy acid and subsequent conversion to LTA_4 by lipoxygenase activity.

1.1 Animal Lipoxygenases

1.1.1 Substrates and Positional Specificity

Several lipoxygenase activities have been observed in animal tissues. They are customarily characterized by their positional specificity, and the position(s) of the oxygenation are commonly identified relative to the methyl or carboxyl group of the substrate. In the former convention, the methyl group carbon is denoted as nth carbon and carbons are numbered relative to this group (e.g., n-6). In the latter numbering convention (Geneva), the carboxyl group carbon is designated as 1 and the other carbons are successively numbered from this carbon.

A 5-lipoxygenase activity in rabbit peritoneal neutrophil suspensions was reported by Borgeat et al. (1). Arachidonic acid and dihomo-γ-linolenic acid (8,11,14-eicosatrienoic acid; 20:3) were reported to be converted to 5(L)-hydroxy-6t,8c,11c,14c-eicosatetraenoic acid and 8(L) hydroxy-9t,11c,14c-eicosatrienoic acid, respectively. Most subsequent work has identified the chirality of the products as S, which corresponds to the D-configuration. The rate of formation of the 5-substituted product was 3- to 15-fold faster than that of the 8-derivative, indicating some substrate specificity for the rate-limiting or regulatory enzyme system in leukotriene biosynthesis. The insertion of oxygen at either the 5- or 8-position by the preparation from neutrophils opens the question of whether there are separate enzymes or whether the lipoxygenase has a varied positional selectivity that depends upon the structure of the substrate. Perhaps the selectivity of this enzyme is for the first ethylenic bond nearest to the carboxyl group as suggested for the potato lipoxygenase (2).

The 12-lipoxygenase activity in bovine blood platelets (3) oxygenated various fatty acids relative to 20:4 (100 percent): 81 percent for 8,11,14-eicosatrienoic acid, 60 percent for 5,8,11-eicosatrienoic acid, 60 percent for 5,8,11,14,17-eicosapentaenoic acid, and 26 percent for 8,11-octadecadienoic acid. The 11,14-eicosadienoic isomer was not oxygenated. Apparently the positional specificity of this enzyme is related more to the ethylenic bond distance from the methyl group, and it seems to require a n-9, n-12 diene (4). The use of the methyl-oriented numbering system (n-x) allows recognition of a consistent selectivity that would not be evident with the traditional carboxyl-oriented system. Thus the "12-lipoxygenase" seems to act by removal of the n-11 hydrogen atom and insertion of oxygen at the n-9 position. In the case of 6,9,12-18:3, the product would be a 10-hydroperoxide, and 4,7,10,13, 16,19-22:6 would give a 14-hydroperoxide, both with a n-9 hydroperoxy group. The positional specificity of several lipoxygenase activities are summarized in Table I.

TABLE I. Positional Specificity of Representative Human Lipoxygenases with Respect to Substrate Arachidonic Acid

Position Oxidized	Source	Ref.
5 or n-16	Peripheral leukocytes	(5)
8 or n-13	Platelets	(6)
9 or n-12	Polymorphonuclear leukocytes	(7)
10 or n-11	Platelets	(8)
12 or n-9	Platelets	(5)
15 or n-6	Eosinophils	(5)

1.1.2 Isolation and Physical Properties

The 5-lipoxygenase was found in the cytosol (9) and has been partially purified from the 10,000 x g supernatant fraction of rat basophilic leukemia (RBL-1) cells by Jakschik et al (9). The partial purification of a 12-lipoxygenase from bovine platelets (3) yielded an enzyme specific activity of 1.8 nmol/min/mg protein with arachidonic acid as the substrate. The 15-lipoxygenase has been purified 250-fold from peritoneal polymorphonuclear leukocytes to a specific activity of 93 nmol/min/mg when assayed with arachidonate (10). Although the isolation procedures mentioned above are routinely performed at 0–4°C, the extreme lability of these enzymes contributes to the difficulty in obtaining homogeneous preparations. For example, 30 percent of the original activity of the 15-lipoxygenase is lost after storage at pH 7.0 for 24 hours.

Using gel filtration chromatography and SDS polyacrylamide gel electrophoresis, Parker and Aykent (11) reported the molecular weight of 5-lipoxygenase isolated from RBL-1 cells as about 90,000 Da. Interestingly, addition of 5 mM Ca^{2+} (but not Mn^{2+} or Mg^{2+}) resulted in an increase to 180,000 Da. Significantly, the low molecular weight monomeric form did not exhibit enzymatic activity, whereas the high molecular weight dimeric form was catalytically active. Nugteren (3) reported that the soluble 12-lipoxygenase from bovine platelets behaved as an aggregate with lipid having a molecular weight of >300,000 Da (determined by agarose gel filtration), whereas Ho (12) described a membrane-bound form of this activity in human platelets. The molecular weight of the 15-lipoxygenase from rabbit leukocytes was reported to be 61,000 Da (10). The enzyme had a pH optimum of 6.5, and the apparent K_m for arachidonic acid as substrate was 28 μM. No dependence on divalent cations was observed for the purified enzyme activity.

1.2 Plant Lipoxygenases

1.2.1 Substrates and Positional Specificity

Holman, et al. (13) studied a series of isomeric octadecadienoic acids and reported that linoleic acid (9,12- 18:2) was the best substrate for soybean lipoxygenase-1. The 13,15-18:2 had an oxygenation rate of about 50 percent compared with linoleic acid (13). Conversion of γ-linolenic acid (6,9,12-18:3) was slower than α-linolenic acid (9,12,

15-18:3) and the dihomo analogs (20:3) exhibited a similar difference (14). Among C_{20} fatty acids, the 11,14-20:2, 8,11,14-20:3, and 5,8,11,14-20:4 are all substrates for soybean lipoxygenase-1 (14). With linoleate as substrate, the majority of the lipoxygenases produce a mixture of positional isomers of 9- and 13-hydroperoxylinoleic acid (Table II). The mixture produced varies greatly among soybean isozymes, and it is a function of substrate fatty acid, pH, and other incubation conditions (e.g. Ca^{2+} presence). For example, oxygenation of linoleic acid by soybean lipoxygenase-2 at pH 6.6 yielded isomers in a ratio of 13-L(S):13-D(R):9-L(R):9-D(S) = 12.5:12.5:41:34 whereas at pH 9.0 the ratio was 30:5:51:14 (15). Nevertheless, a thorough study of specificity (16) indicated that the soy enzyme was methyl-oriented with a preference for oxygen insertion at the n-6 position. The high specificity occurred for all polyunsaturated acids except linoleate, the substrate most widely used in studying plant lipoxygenases. Such results bring out the unfortunate level of ignorance that is maintained by investigators not including several other common acids in their studies. When reacting with linoleate (9,12-18:2), the lipoxygenase from potato tubers formed large amounts of the n-10 isomer (see Table II). The product, like that from soybean and corn, was regarded by Corey et al (2) to have the absolute configuration with S chirality at the oxygenated carbon atom. When reacting with arachidonate, however, the mixture of products included a 15% yield of the 5-isomer. Corey and his coworkers inferred that "the strong tendency to attach oxygen at the point nearest to the carboxylic function contrasts, for example, with the much studied soybean lipoxygenase which converts linoleate into 13-(S)-hydroperoxyoctadeca-cis-9, trans-11-dienoic acid and arachidonic acid into 15-(S)-HPETE" (2). The potato lipoxygenase, which has now been highly purified (17,18), may therefore resemble the lipoxygenase activity of neutrophils and be capable of 5-, 8- or 9-lipoxygenase activity depending upon the structure of the substrate provided. In this regard, the data in Table II suggest that the enzmes from corn germ, tomato or Hordeum distichum may also tend to select the ethylenic bond nearest the carboxyl group. Lipoxygenases from other plants are reported to produce different ratios of isomers (see Table II for specific references).

TABLE II. Positional Specificity of Plant Lipoxygenases with Respect to Substrate Linoleic Acid

Enzyme Source	pH	Regiospecificity (13(n-6):9(n-10))	Reference
Soybean-1	6.6	77:23	(19)
Soybean-2	6.6	25:75	(19)
Pea	6.3	42:58	(20)
Potato tuber	5.5	5:95	(21)
Potato tuber	6.3	90%(n-16)*	(18)
Flaxseed	6.5	88:12	(22)
Fusarium oxysporum	9.0	30:70	(23)
Zea mays	6.9	17:83	(24)
Corn germ	6.6	5:95	(25)
Alfalfa	-	50:50	(26)
Peanut	6.5	95:5	(27)
Tomato	5.5	4:96	(28)
Eggplant	6.5	-	(29)
Chlorella pyrenoidosa	7.4	80:20	(30)
Dimorphotheca sinuata	6.9	100%(n-6)	(31)
Thea sinensis	-	84:16	(32)
Wheat flour	-	85:15	(33)
Dutch barley (*carbrinus*)	6.8	10:80	(34)
Hordeum distichum	7.5	90%(n-10)	(35)
Cereal suspensions	-	variable	(36)

*This value is for the case when arachidonate is the substrate.

1.2.2 Physical Properties

The four soybean lipoxygenases have average molecular weights of approximately 100,000 Da. Soybean lipoxygenase-1 has a total of 883 amino acid residues with the ratio of polar:non-polar residues indicating a predominantly hydrophilic protein (37). Spaapen et al. (37) described the presence of five half-cystines in this lipoxygenase, and concluded that the enzyme does not contain cystine residues. Soybean lipoxygenase-1 contains non-heme iron as a prosthetic group. Although iron is bound to the polypeptide backbone, the coordination of iron by the protein is not well-understood (38). The iron atom (Fe^{3+} or Fe^{2+}) appears to be involved in the catalytic function of the enzyme, and is discussed below (section 2.3). The pH optimum for the plant lipoxygenases generally falls in the range of 6-7 for most of the enzymes (See Table II), but varies between 5.5 and 9.0.

2. REACTION MECHANISMS

2.1 General Kinetic Aspects

Most lipoxygenases have not been examined in sufficient detail to permit discussion of their reaction mechanism. Thus in this chapter the general features of the soybean lipoxygenase will be presented as a model for what may occur with the other enzymes. The soy enzyme was also used successfully as a model of the cyclooxygenase activity when purified PGH synthase was not available (39). That approach led to an awareness of some common features of fatty acid oxygenases; a poorly understood class of enzymes prior to 1970. Their unusual mechanism, which is presented in general terms in Figure 2, seems to include activation by product hydroperoxide (K_p) and a self-catalyzed reaction inactivation (k_2) as well as a slower inactivation by lipid hydroperoxides (k_4).

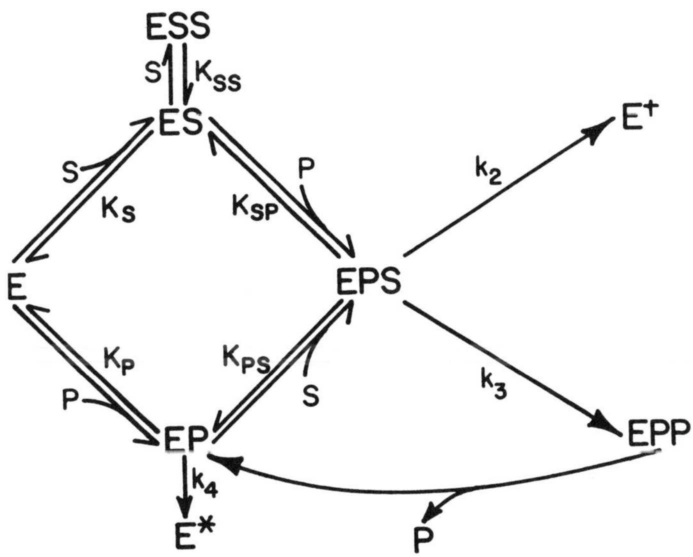

Figure 2. An Early Kinetic Formulation for Lipoxygenase Action (40,44).

2.2 Lipid Hydroperoxide Activator Requirement

Experimental evidence for the hydroperoxide requirement of soybean lipoxygenase was first reported by Haining and Axelrod (42) and confirmed by Lands, et al. (39–41). The early phase of lipoxygenase reactions is characterized by a lag phase and non-linear kinetics. Haining and

Axelrod (42) demonstrated that such lag phases were eliminated with the addition of lipid hydroperoxide. Lands and his colleagues confirmed this phenomenon and used glutathione peroxidase, a hydroperoxide scavenging enzyme, to stop the fatty acid oxygenation reactions (39-41). Thus the two experimental approaches (inhibition by removal of hydroperoxide activator and stimulation by the addition of lipid hydroperoxide) combined to demonstrate the hydroperoxide requirement of soybean lipoxygenase that is the basis for the accelerative lag phase. An argument against a hydroperoxide requirement was introduced following experiments in which sodium borohydride was used to reduce the majority of ambient hydroperoxides to hydroxy acids without the loss of lipoxygenase activity (43). However the results do not unambiquously support this argument since a small remaining amount of non-reduced hydroperoxide may have been sufficient to perpetuate the catalytic activity of the lipoxygenase in a manner analogous to the reaction observed with prostaglandin H synthase. Prostaglandin H synthase, which also requires an activator hydroperoxide for its catalytic activity, converts arachidonic acid to PGG_2 (bearing a hydroperoxide functional group) which remains at a concentration sufficient to sustain the oxygenation reaction, even when PGH_2 is the predominant product.

An important consideration in the requirement for hydroperoxide activation is whether the oxygenase requires the activator only to initiate its catalytic activity or whether the activator must be continuously present during sustained oxygenation of the substrate. Lands _et al_. (39) demonstrated the requirement for a continuous presence of hydroperoxide by adding glutathione peroxidase to remove lipid hydroperoxide activators and observing the cessation of lipoxygenase activity in the experimental system. This direct evidence for the continuous requirement of activator hydroperoxide demonstrates the potential importance of lipid hydroperoxides in regulating lipoxygenase activity in physiological and pathophysiological conditions. In this way, the requirement for hydroperoxide activator has also been demonstrated for lipoxygenase activities isolated from lung (44) and basophilic leukocytes (45).

2.3 Iron Valences States

The requirement for hydroperoxide activator in the mechanism of lipoxygenase seems most probably linked to electron shifts involving the valence state of iron in the enzyme. de Groot et al (46) suggested a conversion of ferrous (Fe^{2+}) to ferric (Fe^{3+}) iron during activation of the soybean fatty acid oxygenation reaction, whereas Lands and Hemler (47,48) proposed that the native oxygenase of PGH synthase may contain Fe^{3+} which is converted to Fe^{2+} upon activation with hydroperoxide. As determined by atomic absorption analysis, Chan (49), Roza and Franke (50), and Pistorius and Axelrod (51) reported that soybean lipoxygenases contain one atom of non-heme iron per mole of protein. Initially, Pistorius and Axelrod used 46 hours of incubation with the Fe^{3+} chelator Tiron (4,5-dihydroxy-m-benzenedisulfonic acid) at 10^{-3} M concentration to inhibit lipoxygenase activity by 74 percent. No inhibition of lipoxygenase was observed after 46 hours of incubation using 10^{-3} M o-phenanthroline, a Fe^{2+} chelator. These results led those authors to propose that the iron in the native lipoxygenase was in the Fe^{3+} state (51). Native lipoxygenase is EPR-silent, but with addition of linoleic acid to an oxygen-saturated reaction mixture, an EPR signal at g=6 was observed (52). The EPR signal around g=6 has been attributed to high spin Fe^{3+} species in axial or rhombic symmetry (53) although high spin Fe^{3+} also gives a strong signal at g=4.3 (46).

Color changes in a lipoxygenase incubation system, which may reflect changes in the iron valence state, were reported to accompany the lipoxygenation of linoleate (53). The major product resulting from lipoxygenation of linoleate at pH 9.0 is 13-L-hydroperoxy-9c,11t-octadecanoic acid. Addition of an equimolar amount of this hydroperoxide to the native enzyme caused the enzyme to change from colorless to yellow and to exhibit a complex EPR signal around g=6. Using this information, de Groot et al. (46) proposed a lipoxygenase activation scheme whereby the native enzyme was in the Fe^{2+} state and existed as an $E\text{-}Fe^{2+}\text{-}O_2$ complex. Addition of hydroperoxide presumably converted the enzyme to an active form, $E\text{-}Fe^{3+}$, and produced an alkoxy radical ($RO^{\cdot}$) and an hydroxyl ion (OH^-). de Groot et al. (46) proposed that upon activation, the system formed superoxide anion ($O_2^{\bullet}$) or excited oxygen (O_2^1) and undesignated reaction products from the hydroperoxide. None of these oxygen-containing species or reaction products were detected or described in those reports. In that proposed reaction scheme, the activated Fe^{3+} enzyme was regarded to react with fatty acid to produce E-Fe^{2+}····R· and H^+. The alkyl free radical (R·) then presumably participated in oxygenation reactions, and the enzyme was left in the E-Fe^{2+} state. The Fe^{2+} enzyme was regarded to form rapidly the Fe^{3+} enzyme spontaneously with O_2 (although the stable native form was considered to be Fe^{2+}). Upon reaction with linoleate, the proposed Fe^{3+} active enzyme intermediate was regarded by de Groot _et al_. to form the Fe^{2+} state (46). This event was complicated by the marked increase in the g=4.3 signal upon aerobic addition of linoleate, which presumably reflected an increased Fe^{3+} state and not a recycling back to the Fe^{2+} form of enzyme.

A different role for iron was proposed from the studies of another fatty acid oxygenase, cyclooxygenase. The native holoenzyme contains heme in the Fe^{3+} state. Lands and Hemler (47,48) reasoned that the requirement for hydroperoxide activator suggested an outer shell electron transfer as a mechanism for generating radical intermediates without cleavage of the hydroperoxide O–O bond. Thus, the initiating event would be a hydroperoxide reduction of Fe^{3+} to Fe^{2+} with the concomitant generation of a hydroperoxide radical. Subsequent substrate activation by H· removal could then occur as a result of the hydroperoxide radical leading to a free radical oxygenation chain reaction with the polyunsaturated fatty acid substrate. Application of this concept to lipoxygenase action (scheme B below) would entail valence changes during activation opposite to those proposed by deGroot _et al_ (see scheme A). Clearly, the ability of iron to transfer or accept an electron is of critical importance to lipoxygenase activity. More careful measurement of the products formed will clarify this aspect.

A. deGroot et al (1975).

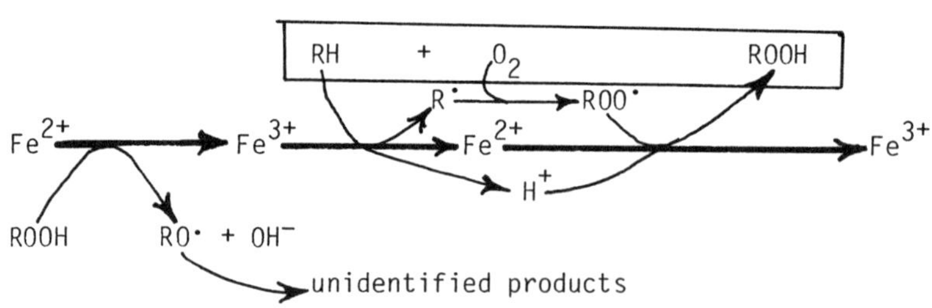

B. Hemler and Lands (1980).

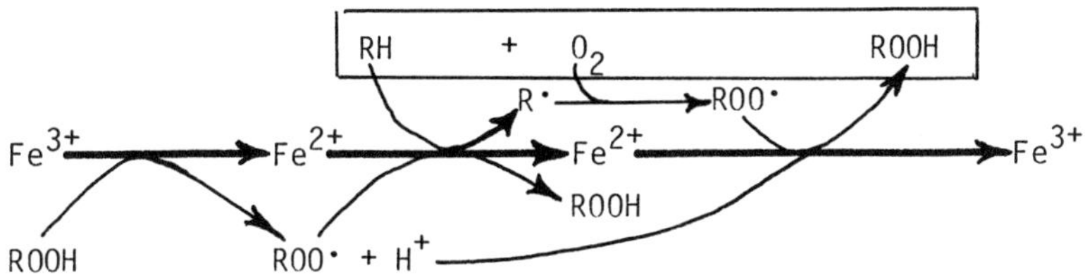

2.4 Self-Catalyzed Inactivation

Smith and Lands (14,40) reported a self-catalyzed loss of soybean lipoxygenase activity which was not attributable to a lack of substrate, to impure substrate, or to product inhibition. They found that the reaction velocity of the lipoxygenase decreased to zero before oxidation of all substrate fatty acid and that fresh enzyme was needed to restore activity. Gibian and Galway (54) proposed that the loss of lipoxygenase activity upon incubation was due to wall absorption. However, if wall absorption caused the loss of activity, we would expect a loss of enzyme during periods when no peroxide activator was present. Studies with glutathione peroxidase present show that this loss did not occur since subsequent addition of N-ethylmaleimide permitted the resumption of lipoxygenase action. Other investigators proposed that the self-limited behavior of the enzyme was attributable to product inhibition (43,54). This possibility is not supported by the experimental data which indicate that after the reaction velocity had dropped to zero, addition of more enzyme caused a resumption to the original maximum velocity for the reaction. It would appear unlikely that, in the presence of the presumed inhibitory reaction product, the fresh enzyme would attain the same maximum velocity. The kinetics of the self-inactivation were first order with respect to enzyme concentration. Smith and Lands (14,40) defined a self-inactivation rate constant (k_2) which was characteristic for each specific polyunsaturated fatty acid in the series of eighteen and twenty carbon acids they studied. A wide variety of polyunsaturated fatty acids exhibited a ten-fold variation in their abilities to catalyze the self-inactivation. The n-3 fatty acids exhibited higher values than the n-6 fatty acids. The dienoic fatty acids, 18:2n-6 and 20:2n-6, were the least effective "suicide substrates", and 20:5n-3 and 22:6n-3 were the most effective (14). Subsequent investigations by other laboratories unfortunately used only linoleic acid as substrate, and linoleate exhibits the lowest tendency to promote the self-catalyzed inactivation reaction. Thus this phenomenon is not clearly indicated by all investigators.

2.5 Calcium Stimulation

In 1969, Koch (55) reported that lipoxygenation of linoleic acid by water extracts of navy bean and soybean was increased 16-fold by the addition of 0.5 mM Ca^{2+} to the reaction mixture and was eliminated by the addition of 0.6 mM EDTA to the reaction mixture. The stimulatory effect of Ca^{2+} was reported to vary for the soybean lipoxygenase isoenzymes. Christopher et al. (56) reported that at pH 7 to 9, lipoxygenase-2 generated 50:50 mixtures of the 13- and 9- hydroperoxide when 0.55 mM Ca^{2+} was present. However, when Ca^{2+} was absent from the lipoxygenase-2 system, the ratio of the 13- to 9-position isomers was 62:38 at pH 7 and 40:60 at pH 9. Soybean lipoxygenase-1 catalyzed lipoxygenation only at the 13-position of linoleic acid in the presence or absence of 0.5 mM Ca^{2+}. A stimulatory effect by calcium ion was reported for the peanut lipoxygenase isoenzymes -1, -2, and -3 (57).

The mechanism whereby Ca^{2+} stimulates lipoxygenases may revolve around an interaction of Ca^{2+} with the fatty acid substrate rather than with the enzyme. Zimmerman and Snyder (58) attempted to resolve whether Ca^{2+} activation of soybean lipoxygenase-2 was due to an action upon the enzyme or the substrate. The activity of soybean lipoxygenase-2 on 1 mM linoleic acid at pH 6.8 was reported as 9.8 μmol O_2/min/ml, but in the presence of equimolar Ca^{2+} the activity increased to 52 μmol O_2/min/ml. The 5-fold increase might reflect some release from substrate inhibition, which can be appreciable at fatty acid concentrations above 0.2 mM (40,41). When 0.02 percent Tween 20, a non-ionic detergent, was added to a system containing enzyme and 1 mM linoleate, the activity increased to 100 μmol O_2/min/ml (58). A Ca^{2+}-linoleate association was suggested as a result of two other effects of Ca^{2+} on linoleate: equimolar Ca^{2+} precipitated the floating fraction of a 265,000 x g sample of 1.4 mM linoleate suspension, and Ca^{2+} failed to activate lipoxygenase-2 when methyl linoleate was the only available substrate (58).

The possibility of a mechanism revolving around the Ca^{2+}:linoleate interaction was extended by Galpin and Allen (59) who used soybean lipoxygenase-1 and horse bean lipoxygenase-2 to conclude that the lipoxygenation rate was independent of micellar linoleate but dependent upon the concentration of non-micellar linoleate. Addition of Ca^{2+}, in a 1:2 ($CaCl_2$:linoleate) molar ratio for systems below 15 μM linoleate caused no effect upon horse bean lipoxygenase-2 activity. However, two-fold and three-fold stimulation of the lipoxygenase was observed with the addition of 75 μM linoleate (1:1, $CaCl_2$:linoleate) and 150 μM linoleate (1:2, $CaCl_2$:linoleate), respectively. Calcium ion was suggested to interfere with micelle formation or turnover rate and thereby increase non-micellar linoleate at concentrations above the CMC (59).

One example of increased mammalian lipoxygenase activity with the addition of Ca^{2+} is the report (60) of a 40-fold increase in the conversion of added free arachidonic acid to 5-HETE and 5,12-diHETE when 20 μM A23187, a calcium ionophore, was added to human polymorphonuclear leukocyte incubations containing 0.87 mM $CaCl_2$. The stimulation by the ionophore could reflect a redistribution of cellular calcium to make it available to the lipoxygenase or substrate. Using the 10,000 x g supernatant of a homogenate of rat basophilic leukemia (RBL-1) cells, Jakschik _et al_. (61) demonstrated an increase in 5-HETE and 5,12-diHETE that was dependent upon calcium concentration (0 to 5 mM Ca^{2+}). These data suggest the possible stimulatory effect of calcium ion upon the 5-lipoxygenase, but similar effects have not been found for the other mammalian lipoxygenases. Because calcium ion is involved in modulating several cellular processes which impinge upon arachidonate metabolism (for example, Ca^{2+} is required for phospholipase A_2 activity which makes substrate arachidonate available for biosynthetic conversion to leukotrienes), direct demonstration of the calcium ion stimulatory effect must await the purification of the mammalian enzymes to homogeneity. For further discussion of the role of Ca^{2+} in leukotriene biosynthesis, the reader is referred to Chapter 4 of this volume.

3. INHIBITORS

Detailed exploration of inhibitor mechanisms may lead ultimately to the effective design of specific anti-lipoxygenase drugs. Tables III and IV briefly summarize inhibitors of representative plant and animal lipoxygenases and their inhibitory concentrations under specific experimental conditions. In evaluating the effectiveness of inhibitors, it is helpful to recall that fatty acid oxygenation reactions exhibit a continuous requirement for a hydroperoxide activator (62), appear to proceed via a free radical chain reaction (48), and undergo self-inactivation during reaction in a manner that could be referred to as having "suicide" substrates.

3.1 Hydroperoxide Abundance and Inhibitor Efficacy

When considering the efficacy of anti-lipoxygenase drugs, it seems helpful to review concepts developed with another fatty acid oxygenase, the cyclooxygenase activity of PGH synthase. For that enzyme, the concentration of lipid hydroperoxides plays a key role in initiating and maintaining the free radical oxygenation reaction. The activation is decreased either by decreasing the amount of hydroperoxide available or by antagonizing the action of lipid hydroperoxide at the enzyme's activator site (62,63). The behavior of the anti-cyclooxygenase drugs discussed below is related to the need for low concentrations (10^{-8} to 10^{-7} M (64) of hydroperoxide to initiate and maintain the free-radical cyclooxygenase reaction. Alterations in this peroxide concentration as a result of activity of intracellular glutathione peroxidase, tissue damage, or drug therapy, might modulate the overall prostaglandin biosythesis. Acetamidophenol, a noncompetitive phenolic inhibitor of cyclooxygenase activity, has been reported to antagonize the activation process as well as block the substrate binding site (62). Phenolic inhibitors seem to lose effectiveness at high levels of peroxide and tend to exert their anti-cyclooxygenase properties only at low levels of peroxide (65).

Lands (66) classified the major nonsteroidal antiinflammatory drugs into three categories related to their effect on cyclooxygenase: rapid reversible competitive inhibitors, time-dependent inactivators, and rapid reversible non-competitive (free radical trapping) inhibitors. Radical trapping agents which behave as reversible non-competitive inhibitors are highly sensitive to ambient hydroperoxide levels in the vicinity of the cyclooxygenase active site (65). Understanding the behavior of such anti-cyclooxygenase drugs with respect to ambient peroxide concentration might assist in the design of effective anti-lipoxygenase compounds. The overall effectiveness of radical trapping or phenolic drugs may need improvement to function at peroxide levels present at a site of tissue injury. Ultimately, comparative I_{50} values for inhibitors of both prostaglandin and leukotriene biosynthesis should be understood in terms of peroxide concentration, and the influence of the ambient hydroperoxide must be accounted for in reporting the effective inhibitory concentrations.

3.2 Antioxidants and Radical Trapping Agents

A large number of recognized lipoxygenase inhibitors may be classified as antioxidants or radical trapping agents (e.g., see Table III and IV). Early evidence for the effectiveness of antioxidants as lipoxygenase inhibitors was provided by Tappel *et al*. (67), using a direct spectrophotometric method to measure initial reaction velocities. Since the reactions catalyzed by the fatty acid oxygenases resemble those in the autoxidation of lipids, the mechanism of inhibition by antioxidants may be similar for both of these processes, i.e. termination of the free radical chain reacation. Removal of the radical trapping antioxidants would permit the reaction to once again continue. Termination of the radical chain reaction antagonizes the hydroperoxide activation of the reaction, and thus the inhibition may be influenced by the local supply of lipid hydroperoxide. For example, the cyclooxygenase activity of PGH synthase was much more readily inhibited by phenolic antioxidant agents when glutathione peroxidase reduced the local availability of hydroperoxides (63,65). It seems reasonable that lipid hydroperoxides would exhibit a similar antagonism of the inhibitory action of antioxidants on lipoxygenase activity (68). Vanderhoek and Lands (68) reported that similar inhibition occurred with soybean lipoxygenase and sheep vesicular gland oxygenase (prostaglandin H synthase) by certain non-competitive inhibitors of vesicular gland dioxygenase: I_{50} values were 3.5 μM for α-naphthol, 56 μM for trimethylhydroquinone, and 190 μM for BHT. For the competitive cyclooxygenase inhibitors, Santoquin (6.5 μM), propylgallate (120 μM), and hydroquinone (180 μM), the relative effectiveness was reversed for soybean lipoxygenase: Santoquin (>200 μM), propylgallate (72 μM), and hydroquinone (22 μM). All of the antioxidants studied exhibited reversible rather than time-dependent, irreversible inhibition.

TABLE III. Soybean Lipoxygenase Inhibitors

Inhibitor	I_{50} (mM)	Reference
α-Naphthol	0.36	(69)
Pyrogallol	8.7	(69)
Phloroglucinol	8.3	(69)
Hydroquinone	5.2	(69)
Propyl gallate	0.32	(69)
NDGA	0.01	(69)
α-tocopherol	0.32	(69)
Quercetin	0.14	(69)
BHA	1.1	(69)
BHT	1.4	(69)
Pyrocatechol	1.0	(69)
Resorcinol	17	(69)
Trimethylhydroquinine	7.2	(68)
Santoquin	0.026	(68)
Saturated Monohydric Alcohols	-	(70)
5,8,11,14-Eicosatetraynoic acid	-	(71)
14,15-Dehydroarachidonic acid	-	(72)

Yoshimoto et al. (73) reported a concentration-dependent decrease in 5-lipoxygenase activity from rat basophilic leukemia cells treated with the flavone derivative, cirsiliol (3',4',5-trihydroxy-6,7-dimethoxyflavone). When the 5-hydroxyl group was modified to a methoxyl group, inhibitory potency decreased approximately 4-fold and converting the 3' and 4' hydroxyl groups to methoxyl groups virtually eliminated all inhibitory potency. Compound A, a derivative of 2-aminophenol, was reported by Miyamoto and Obata (74) to inhibit the 5-lipoxygenase activity from the cytosolic fraction of polymorphonuclear leukocytes with an I_{50} of 0.4 μM. Again, the phenolic structure strongly suggests that these derivatives might function as antioxidants.

Following the report that 4-nitrocatechol can reversibly bind to and inhibit some oxygenases (altering the g=4.3 signal) and inactivate protocatechuate-3,4-dioxygenase by an apparent suicide substrate phenomenon (75), the compound was found to transform the g=6 signal of the presumed Fe^{3+} form of soybean lipoxygenase to a split form of g=4.3. This unexplained reaction of the catechol was regarded to be virtually stoichiometric and apparently irreversible (76). Further study indicated that an initial reversible inhibition of lipoxygenase by this catechol could shift slowly to an irreversible inactivation (76) by interaction with the hydroperoxide-enzyme complex although it did not do so with the native lipoxygenase alone. One clue to the mode of inactivation may be in the reports that catalase could prevent the time-dependent inactivation by NDGA (69) and eicosatetraynoic acid (77). Perhaps some H_2O_2 generated in the incubation mixture caused the inactivation by facilitating a form of reaction inactivation.

3.3 Suicide Substrates

A distinctive property of all fatty acid oxygenases, including lipoxygenases, is their self-catalyzed inactivation by "suicide" substrates (see Section 2.4). The advantage of manipulating such self-inactivation will be reviewed here in light of substrate analogs with potential inhibitory anti-lipoxygenase properties. Inhibition of soybean lipoxygenase by eicosa-5,8,11,14-tetraynoic acid was reported by Downing et al. (71). Preincubation of the enzyme for 15 minutes with 1 μM acetylenic acid completely inactivated enzyme. The acetylenic acid was reported to behave as a slow, irreversible inhibitor of the lipoxygenase. In analogy to the discussion (78) of the mechanism of inhibition of β-hydroxydecanoyl thioester dehydrase by an acetylenic inhibitor, 3-decynoyl-N-acetylcysteamine, Downing et al. (71) proposed that the eicosa-5,8,11,14-tetraynoic acid is converted to an allene. Presumably the allene (or related intermediate) is formed by the enzyme and reacts with the enzyme active site in an irreversible manner.

A key to understanding the ability of a compound to serve as a "suicide" substrate is the peroxide requirement of lipoxygenase. The peroxide requirement is reviewed earlier in Sections 2.2. and 3.1., and the generation and removal of ambient peroxides is considered in Chapters 6 and 8. Vanderhoek and Lands (79) reported that when eicosa-5,8,11,14-tetraynoic acid, was incubated with PGH synthase, peroxide was required for the time-dependent inactivation of the oxygenase activity, and the action seemed analogous to the self-catalyzed inactivation of soybean lipoxygenase caused by various substrates (14). This conclusion, based on the protection of oxygenase activity by glutathione peroxidase, was accompanied by the finding that oxygen was also needed for the inactivation (79). Since the amount of oxygen consumed with the acetylenic analog was below the level of detection, the presumed intermediate seemed very efficient in inactivating the oxygenase.

Further support for an enzyme-catalyzed "suicide" process came from the fact that only the 10- and 13-octadecynoic acid isomers of fourteen isomeric octadecynoic acids tested caused inactivation. The other 12 isomers merely inhibited in a reversible, competitive fashion with a K_I value of about 15 μM, similar to that for oleate (80). Acetylenic acids can inhibit leukotriene formation at steps other than the 5-lipoxygenase. Jakschik (81) reported that all acetylenic acids tested in a series of thirty three isomers inhibited SRS (leukotriene) formation even though some compounds did not inhibit the 5-lipoxygenase. The acids with the acetylenic bond at the 7,8 position tended to be the most effective inhibitors of the 5-lipoxygenase of rat basophilic leukemia cells (RBL-1). Presumably some of the isomers were inactivators of the enzyme whereas others were competitive inhibitors.

An extension of the inhibition with eicosa-5,8,11,14-tetraynoic acid was the development of a series of dehydroarachidonic acid analogs (72). Synthesis of these analogs was undertaken with the concept of developing position-selective inhibitors that might irreversibly block the oxidation of arachidonate by either the 15-, 11- or 5-lipoxygenase enzymes. Aerobic incubation of soy lipoxygenase with 14,15-dehydroarachidonate caused 50% inactivation, whereas no detectable deactivation occurred during anaerobic incubation for several hours (72). The 5,6-, 8,9-, and 11,12-isomers gave no detectable deactivation of the soy enzyme (82). An allenic hydroperoxide was suggested as the possible structure contributing to the enzyme destruction. Corey and Park postulated that scission of its O-O bond might form another intermediate and a hydroxyl radical. Preliminary results indicated that the 5-lipoxygenase was inhibited by another analog, 5,6-DHA (82); presumably by a reactive intermediate attacking the active site (although evidence was not presented). The present data do not yet confirm the concept that position-selective inhibitors that attack only one enzyme can be made. For example, the variable selectivity of the 5-lipoxygenase (2) would make it vulnerable to self-catalyzed inactivation by a variety of substrate analogs. The inactivation of cyclooxygenase by the 10- and 13-isomers of octadecynoic acid suggested that removal of the carbon 12 hydrogen may form the destructive intermediate, whereas removal of hydrogen atoms from carbon 8,11, or 14 and 7,10,13, or 16 was suggested by the inhibition noted with 9,12-octadecadiynoic acid and 5,8,11,

14-eicosatetraynoic acid, respectively (68). Uncertain selectivity in this initial catalytic step of hydrogen atom abstraction makes it difficult to designate which carbon atoms are involved in the formation of the abortive intermediate complex. Hammerstrom (83) reported the inhibition of platelet 12-lipoxygenase by 5,8,11- eicosatrienoic acid (20:3n-9). Fifty percent inhibition of 12-HETE production was reported to occur with the addition of 24 μM 5,8,11-eicosatrienoic acid to a washed human platelet suspension (5 x 10^5/μl) incubated at 37°C for 2 minutes. A different set of data are needed to determine whether the inhibition was an irreversible "suicide" type or merely a competitive substrate phenomenon. It seems important to emphasize that if an inhibitor is to be effective as a "suicide" inactivating agent, the enzyme must have sufficient oxygen and/or hydroperoxide activator to form the destructive intermediate. Otherwise the analog merely binds reversibly to the enzyme site as any other fatty acid would.

3.4 Substrate Analogs

A series of methyl-substituted trienoic acids have been reported to inhibit 5-lipoxygenase. Perchnock et al. (84) synthesized the 7,7- and 10,10-dimethyleicosa-5(Z), 8(Z), 11(Z)-trienoic acids, which gave inhibition of 5-HETE biosynthesis in RBL-1 cells at 100 μM concentration 50 percent and 40 percent, respectively. Also, Ackroyd et al. (85) reported an inhibitory activity of 7,7-dimethyleicosa-5Z,8Z-dienoic acid and its methyl ester. Ackroyd et al. reasoned [as did Perchnock et al.] that abstraction of a hydrogen atom from position 7 of the substrate arachidonate would be prevented by a gem-dimethyl group blocking this site. The dimethyl-dienoic analog at 50 μM inhibited synthesis in RBL-1 cells and human leukocytes 43 percent and 26 percent, respectively. The methyl ester was more potent with inhibition of 74 percent and 50 percent at 50 μM concentration. Both analogs were reported to inhibit the 5-lipoxygenase and not the 12-lipoxygenase from human platelets nor PGH synthase from bovine seminal vesicle microsomes. The original rationale was for an unreactive analog, and the data do not indicate whether the inhibition was due to the formation of an abortive intermediate by a "suicide" substrate or to simple competition of an unreactive analog. If the acid was unreactive, it is difficult to understand the selectivity reported. It seems useful to designate whether the competition observed was more effective than that occurring with endogenous fatty acids such as oleate or palmitate. Another derivative, 5,6-methanoleukotriene A_4 also inhibited 5-lipoxygenase activity and was reported to possess an I_{50} of 3 μM in a guinea pig polymorphonuclear leukocyte system (86). Interestingly, LTA_4, 5-HETE, and 5,12-diHETE were also found to inhibit 5-lipoxygenase activity at IC_{50} = 2 μM. This suggested the possibility of a product-related inhibition that might be reversibly competitive with the activator hydroperoxide. A series of C-7 and C-10 methylated arachidonic acid analogs were reported to inhibit 5-lipoxygenase action (87). The report concluded that the decrease observed in leukotriene biosynthesis may have been due to the inhibition of phospholipase A_2 and not the 5-lipoxygenase. This hypothesis was based upon observations of the effect of the analogs upon a purified phospholipase A_2.

From the variety of substrate analogs discussed above, several patterns in the properties of lipoxygenase inhibitors emerge. The hydroperoxide requirement for efficient self-inactivation by "suicide" substrates means that effective anti-lipoxygenase drugs which behave as "suicide" substrates should have the potential for being metabolized to radical intermediate derivatives. The intermediates may then generate allenic derivatives or other oxidized intermediates that could inactivate the oxygenase *in situ*. The C-6, C-7, and C-10 positions are critical to efficient lipoxygenation and their blockage results in decreased or eliminated oxygenase activity. Blocking the critical carbon atoms to prevent H removal seems likely to give inhibitors that are merely competitive fatty acid analogs. Binding in a covalent manner seems much more likely to have potent biological consequences than would the general competitive hydrophobic binding of a fatty acid to the enzyme.

3.5 Other Compounds

In 1967, Mitsuda *et al*. reported the inhibition of soybean lipoxygenase with monohydric alcohols (70). Reversible inhibition was found to be a function of chain length, and potency of the inhibitor increased with chain length. Mitsuda *et al*. proposed that the compounds occupy the hydrophobic region of the enzyme and antagonize the association of the substrate with the catalytic site. Presumably, the unsaturated moiety of the substrate fatty acid is not allowed access to the active site as a result of such antagonism. Since the alcohols did not prevent inactivation of the lipoxygenase by added hydrogen peroxide Mitsuda et al. (70) proposed that substrate linoleate occupies the catalytic site and does not allow the H_2O_2 access whereas the alcohols might attach to sites other than the catalytic site and allow the active center to be exposed to H_2O_2.

A different type of long-chain alcohol was reported to inhibit the 5-lipoxygenase of guinea pig peritoneal polymorphonuclear leukocytes (88). These compounds contain a methyl substituted quinone ring (and some contained acetylenic groups), and they were effective competitive inhibitors at about 1 μM. Other phenyl substituted heterocyclic compounds, BW 755C (89) and benoxaprofen (90) were reported to block the lipoxygenase of platelets (12-lipoxygenase) and polymorphonuclear leukocytes (5-lipoxygenase), respectively. These inhibitions provide useful new insight into the mechanism of action of antiinflammatory agents that do not cause a direct inhibition of prostaglandin biosynthesis by blocking cyclooxygenase activity.

Dawson et al. (91) reported a 3-fold decrease in EL4 tumor weight in mice treated with benoxaprofen (30 mg/kg/day) for 15 days. In addition benoxaprofen (10^{-4} M) also caused 40-70% inhibition of the growth in vitro of EL4 (mouse T lymphoma), RAJI (human B myeloma) and NB_{100} (human neuroblastoma) cell lines. Adherence of guinea pig polymorphonuclear leukocytes to polycarbonate membranes increased 27% when 10^{-4} M benoxaprofen (an inhibitor of 5-lipoxygenase and PGH synthase activity) was added to assay system, but decreased when BW755c or NDGA (inhibitors of the 5-, 12-, and 15-lipoxygenases) were studied. These data suggest that the products from the 5-lipoxygenase and PGH synthase might cause an increase in cell adherence and their inhibition may impair tumor proliferation.

4. LIPOXYGENASE PRODUCTS AND CARCINOGENESIS

A causal relationship between carcinogenesis and leukotriene biosynthesis has yet to be established. At this time, we must await the designation of a specific role for lipoxygenase products in the biology of cancer, which will occur at some later date when the understanding of the physiological properties of these products has increased. Nevertheless, some experimental evidence can link arachidonate metabolism to tumorigenesis. Levine (99) has reported that agents which inhibit arachidonate metabolism (e.g. dexamethasone) can also inhibit tumor

TABLE IV. Inhibitors of Mammalian Lipoxygenases

Inhibitor	Lipoxygenase Studied	I_{50} (μM)	Ref.
A. Antioxidants and Radical Trapping Agents			
Esculetin(6,7-dihydroxy-coumarin)	12(n-9)	0.65	(92)
Esculin(6-glucoside of esculetin)	12(n-9)	290	(92)
Umbelliferone(7-hydroxy-coumarin)	12(n-9)	500	(92)
4-Hydroxycoumarin	12(n-9)	$>10^4$	(92)
Coumarin	12(n-9)	$>10^4$	(92)
Hydroquinone	12(n-9)	374	(92)
Gallic acid	12(n-9)	393	(92)
Ascorbic acid	12(n-9)	$>10^4$	(92)
Cirsiliol	5(n-16)	0.1	(73)
(3',4',5-trihydroxy-6,7-dimethoxyflavone)	12(n-9)	1	(73)
2-Aminophenol	5(n-16)	15	(74)
Compound A	5(n-16)	0.4	(74)
B. Substrate Analogs			
7,7-dimethyleicosa-5(Z),8(Z),11(Z)-trienoic acid	5(n-16)	100	(84)
10,10-dimethyleicosa-Δ(Z),8(Z),11(Z)-trienoic acid	5(n-16)	100 (I_{40})	(84)
7,7-dimethyleicosa-5(Z),8(Z)-dienoic acid	5(n-16)	50 (I_{26})	(85)
7,7-dimethyleicosa-5(Z),8(Z)-dienoic acid (methyl ester)	5(n-16)	50	(85)
5,6-methanoleukotriene A_4	5(n-16)	3	(86)
7,7-dimehtyl-5(Z),8(Z),11(Z), 14(Z)-eicosatetraenoic acid	5(n-16)	1-3	(87)
5,6-benzoarachidonic acid	5(n-16)	100(I_{94})	(94)
15-Hydroxy-5,8,11,13-eicosatetraenoic acid	5(n-16)	5.7	(95)
15-Hydroxy-5,8,11,13-eicosatetraenoic acid	12(n-9)	0.34	(95)
15-Hydroxy-5,8,11,13-eicosatetraenoic acid	15(n-6)	40	(95)
5,8,11,14-Eicosatetraynoic acid	12(n-9)	4	(83)
5,6-Dehydroarachidonic acid	5(n-16)		(82)

TABLE IV. Inhibitors of Mammalian Lipoxygenases (Continued)

Inhibitor	Lipoxygenase Studied	I_{50} (μM)	Ref.
B. Substrate Analogs (Continued)			
U-60,257 (6,9-Deepoxy-6,9-(Pherylimino)-$\Delta^{6,8}$-Prostaglandin I_1)	5(n-16)	2	(96)
5,8,11-Eicosatriynoic acid	12(n-9)	24	(83)
5,8,11,14,17-Eicosapentaenoic acid	5(n-16)	-	(97)
C. Other Compounds			
Diphenyldisulfide	5(n-16)	1	(98)
Benoxaprofen	5(n-16)	0.4	(90)
BW755C	5(n-16)	8.7-21.8	(93)
BW755C	12(n-9)	7.4	(89)
AA861	5(n-16)	0.8	(88)

growth in cell cultures. Oppositely, agents which can increase arachidonate release (e.g. EGF) also promote tumor growth. The lipoxygenase inhibitors phenidone, NDGA, and BW755C have been found to inhibit the induction of epidermal ornithine decarboxylase (ODC) by the tumor promotor 12-O-tetradecanoylphorbol-13-acetate (TPA) (100). ODC production has been reported to be an essential step in TPA-induced skin tumor production, and data from Nakadate et al (100) strongly suggest that lipoxygenase products are involved in such tumor production.

A possible scheme linking lipoxygenase(s) and lipid hydroperoxides to carcinogenesis is outlined in Figure 3. The 5-lipoxygenase leads to leukotrienes which contribute to the inflammatory response with the aid of prostaglandins. The inflammatory processes lead to the recruitment of more leukotriene-forming cells, which also generate superoxide and hydrogen peroxide. These reduced forms of oxygen can increase the concentration of lipid hydroperoxides, which can be further amplified by the action of the fatty acid oxygenases and then lead to the recruitment of more leukocytes. Thus lipoxygenation can give an amplification of the leukocyte number as well as of the hydroperoxide content.

Guanylate cyclase has been reported to be stimulated by hydroperoxides (101), and more research is needed to determine whether the elevated peroxides from leukocytes can stimulate the cyclase in other

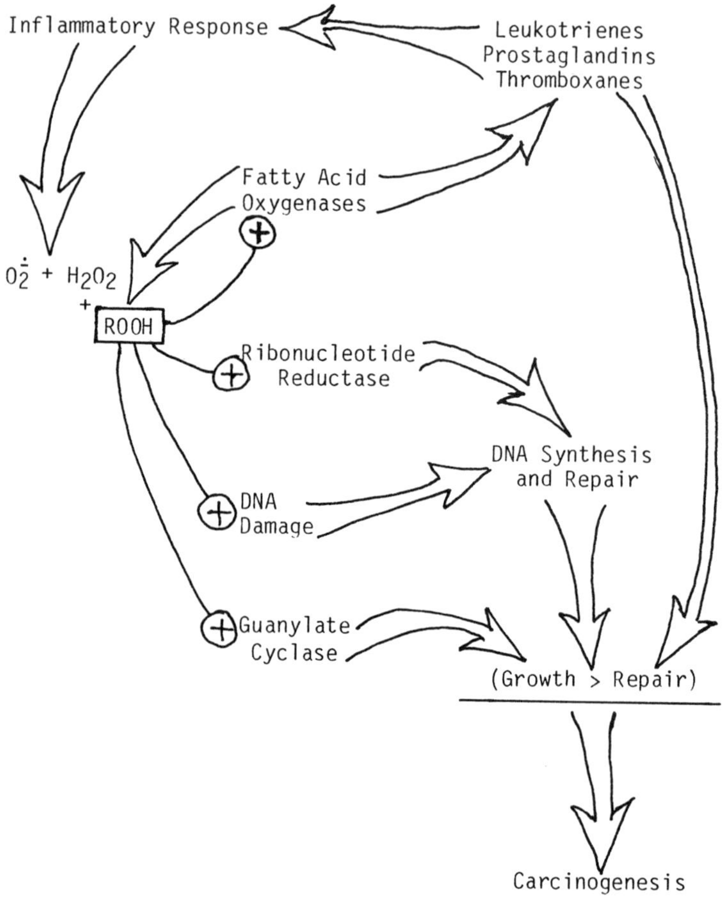

Figure 3. Roles for Radicals in a General Scheme of Carcinogenesis

intact cells. Ribonucleotide reductase catalyzes a radical reaction that requires an oxidative activation (102) to form the deoxyribonucleotides needed for cellular proliferation. Possibly an increase in peroxides can increase the active form of this enzyme.

The degree to which peroxides may stimulate proliferative events will influence the degree to which DNA damage can be repaired before the error is replicated. The model systems that indicate beneficial actions of antioxidants and lipoxygenase inhibitors do not yet permit us to distinguish whether they prevented DNA damage, delayed the replication of possible damage, or blocked eicosanoid formation in a way that facilitated immune surveillance or impaired tissue invasiveness of tumor cells. The many steps that might involve peroxide stimulation make it difficult to assign a specific one to the site of action of the antioxidant inhibitors of lipoxygenase.

Acknowledgement. This review was supported in part by a grant from the United States Public Health Service Grant GM-31494 and a National Institutes of Health Predoctoral Fellowship (AA-7374).

References

1. Borgeat P, Hamberg M, Samuelsson B: Transformation of arachidonic acid and homo-γ-linolenic acid by rabbit polymorphonuclear leukocytes. J. Biol. Chem. (251): 7816-20, 1976.
2. Corey EJ, Albright JO, Barton AE: Chemical and enzymic syntheses of 5-HJPETE, a key biological precursor of slow-reacting substance of anaphylaxis (SRS), and 5-HETE. J. Amer. Chem. Soc. (102): 1435-1436, 1980.
3. Nugtern DH: Arachidonic acid-12-lipoxygenase from bovine platelets. In: Lands, WEM, Smith, WL (eds) Methods Enzymol. Vol. 86. Academic Press, New York, 1982, pp 49-54.
4. Nugtern DH: Arachidonate lipoxygenase in blood platelets. Biochim. Biophys. Acta (380): 299-307, 1975.
5. Maas RL, Turk J, Oates JA, Brash AR: Formation of a novel dihydroxy acid from arachidonic acid by lipoxygenase-catalyzed double oxygenation in rat mononuclear cells and human leukocytes. J. Biol. Chem. (257): 7056-67, 1982.
6. Hamberg M, Samuelsson B: Prostaglandin endoperoxides. Novel transformations of arachidonic acid in human platelets. Proc. Natl. Acad. Sci. USA (71): 3400-04, 1974.
7. Goetzl EJ, Sun FF: Generation of unique mono-hydroxy-eicosatetraenoic acids from arachidonic-acid by human neutrophils. J. Exp. Med. (150): 406-11, 1979.
8. Walker IC, Jones RL, Wilson NH: The identification of an epoxy-hydroxy acid as a product from the incubation of arachidonic acid with washed blood platelets. Prostaglandins (18): 173-8, 1979.

9. Jakschik BA, Kuo CS: Subcellular localization of leukotriene-forming enzymes. In: Samuelsson B, Paoletti R, Ramwell P (eds) Advances in Prostaglandin, Thromboxane, and Leukotriene Research Vol. 11. Raven Press, New York, 1983, pp. 141-145; Jakschik BA, Harper T, Murphy RC, Sun FF: The 5-lipoxygenase and leukotriene forming enzymes. In: Lands WEM, Smith WL (eds) Methods Enzymol. Vol. 86. Academic Press, New York, 1982, pp 30-37.
10. Narumiya S, Salmon JA: Arachidonic acid-15-lipoxygenase from rabbit peritoneal polymorphonuclear leukocytes. In: Lands, WEM, Smith, WL (eds) Methods Enzymol. Vol. 86. Academic Press, New York, 1982, pp 45-47.
11. Parker CW, Aykent S: Calcium stimulation of the 5-lipoxygenase from RBL-1 cells. Biochem. Biophys. Res. Comm. (109): 1011-16, 1982.
12. Ho PPK, Walters CP, Sullivan HR: A particulate arachidonate lipoxygenase in human blood platelets. Biochem. Biophys. Res. Comm. (76): 398-405, 1977.
13. Holman RT, Egwin PO, Christie WW: Substrate specificity of soybean lipoxygenase. J. Biol. Chem. (244): 1149-51, 1969.
14. Smith WL, Lands WEM: The self-catalyzed destruction of lipoxygenase. Biochem. Biophys. Res. Comm. (41): 846-51, 1970.
15. VanOs CPA, Rijke-Schilder GPM, Vliegenthart JFG: 9-L_R-Linoleyl hydroperoxide, a novel product from the oxygenation of linoleic acid by type-2 lipoxygenases from soybean and peas. Biochim. Biophys. Acta (575): 479-84, 1979.
16. Hamberg M, Samuelsson B: On the specificity of the oxygenation of unsaturated fatty acids catalyzed by soybean lipoxidase. J. Biol. Chem. (242): 5329-5335, 1967.
17. J. Sekiya, Aoshima H, Kajiwara T, Togo T, Hatanaka A: Purification and some properties of potato tuber lipoxygenase and detection of linoleic acid radical in the enzyme reaction. Agric. Biol. Chem. (41): 827-832, 1977.
18. Shimizu T, Radmark O, Samuelsson B: Enzyme with dual lipoxygenase activities catalyzes leukotriene A_4 synthesis from arachidonic acid. Proc. Natl. Acad. Sci. USA (81): 689-93, 1984.
19. VanOs CPA, Vente M, Vliegenthart JFG: An nmr shift method for determination of the enantiomeric composition of hydroperoxides formed by lipoxygenase. Biochim. Biophys. Acta (574): 103-111, 1979.
20. Arens D, Seilmeier W, Weber F, Kloos G, Grosch W: Purification and properties of a carotene co-oxidizing lipoxygenase from peas. Biochim. Biophys. Acta (327): 295-305, 1973.
21. Galliard T, Phillips DR: Lipoxygenase from potato tubers. Biochem. J. (124): 431-8, 1971.
22. Zimmerman DC, Vick BA: Specificity of flaxseed lipoxidase. Lipids (5): 392-7, 1970.
23. Matsuda Y, Beppu T, Arima K: Crystallization and positional specificity of hydroperoxidation of fusarium lipoxygenase. Biochim. Biophys. Acta (530): 439-50, 1978.
24. Gardner HW, Weisleder D: Lipoxygenase from zea mays: 9-D-hydroperoxy-trans-10,cis-12-octadecadienoic acid from linoleic acid. Lipids (5): 678-83, 1970.
25. Hamberg M: Steric analysis of hydroperoxides formed by lipoxygenase oxygenation of linoleic acid. Anal. Biochem. (43): 515-26, 1971.
26. Chang CC, Esselman WJ, Clagett CO: The isolation and specificity of alfalfa lipoxygenase. Lipids (6): 100-106, 1971.
27. Pattee HE, Singleton JA, Johns EB: Pentane production by peanut lipoxygenase. Lipids (9): 302-06, 1974.

28. Matthew JA, Chan HW-S, Galliard T: A simple method for the preparation of pure 9-D-hydroperoxide of linoleic acid and methyl linoleate based on the positional specificity of lipoxygenase from tomato fruit. Lipids (12): 324-6, 1977.
29. Grossman S, Trop M, Avtalion R, Pinsky A: Egg plant lipoxygenase: Isolation and partial characterization. Lipids (7): 467-73, 1972.
30. Zimmerman DC, Vick BA: Lipoxygenase in Chlorella pyrenoidosa. Lipids (8): 264-6, 1973.
31. Gardner HW, Christianson DD, Kleiman R: Dimorphotheca sinuata lipoxygenase: Formation of 13-L-hydroperoxy-cis-9,trans-11-octadecadienoicacid from linoleic acid. Lipids (8): 271-6, 1973.
32. Kajiwara T, Nagata N, Hatanaka A, Naoshima Y: Stereoselective oxygenation of linoleic-acid to 13-hydroperoxide in chloroplasts from Thea sinensis. Agric. Biol. Chem. (44): 437-8, 1980.
33. Graveland A: Enzymatic oxidations of linoleic acid and glycerol-1-monolinoleate in doughs and flour-water suspensions. J. Amer. Oil Chem. Soc. (47): 352-61, 1970.
34. Graveland A, Pesman L, vanErde P: Enzymatic oxidation of linoleic acid in barley suspensions. Overdruk uit Tech. Quart. (9): 98-104, 1972.
35. Yabuuchi S, Amaha M: Partial purification and characterization of the lipoxygenase from grains of Hordeum distichum. Phytochem. (14): 2569-72, 1975.
36. Graveland A: Analysis of lipoxygenase nonvolatile reaction products of linoleic acid in aqueous cereal suspensions by urea extraction and gas chromatography. Lipids (8): 599-605, 1973.
37. Spaapen LJM, Verhagen J, Veldink GA, Vliegenthart JFG: The effect of modification of sulfhydryl groups in soybean lipoxygenase-1. Biochim. Biophys. Acta (618): 153-62, 1980.
38. Vliegenthart JFG, Veldink GA: Lipoxygenases. In: Pryor, WA (ed) Free radicals in biology V. Academic Press, New York, 1982, pp 29-64.
39. Lands W, Lee R, Smith W: Factors regulating the biosynthesis of various prostaglandins. Ann. N.Y. Acad. Sci. (180): 107-122, 1971.
40. Smith WL, Lands WEM: Oxygenation of unsaturated fatty acids by soybean lipoxygenase. J. Biol. Chem. (247): 1038-47, 1972.
41. Cook HW, Lands WEM: Further studies of the kinetics of oxygenation of arachidonic acid by soybean lipoxygenase. Can. J. Biochem. (53): 1220-1231, 1975.
42. Haining JL, Axelro B: Induction period in the lipoxidase-catalyzed oxidation of linoleic acid and its abolition by substrate peroxide. J. Biol. Chem. (232): 193-202, 1958.
43. Lagocki JW, Emken EA, Law JH, Kezdy FJ: Kinetic analysis of the action of soybean lipoxygenase on linoleic acid. J. Biol. Chem. (251): 6001-6, 1976.
44. Yokoyama C, Mizuno K, Mitachi H, Yoshimoto T, Yamamoto S, Pace-Asciak CR: Partial purification and characterization of arachidonate 12-lipoxygenase from rat lung. Biochim. Biophys. Acta (750): 237-43, 1983.
45. Egan RW, Tischler AN, Baptista EM, Ham EA, Soderman DD, Gale PH: Specific inhibition and oxidative regulation of 5-lipoxygenase. In: Samuelsson, B, Paoletti, R, Ramwell, P (eds) Advances in Prostaglandin, Thromboxane, and Leukotriene Research Vol. II. Raven Press, New York, 1983, pp 151-157.

46. deGroot JJMC, Veldink GA, Vleigenthart JFG, Boldingh J, Wever R, vanGelder BF: Demonstration by epr spectroscopy of the functional role of iron in soybean lipoxygenase-1. Biochim. Biophys. Acta (377): 71-9, 1975.
47. Lands WEM, Hemler ME: Fatty acid oxygenase kinetics and the role of peroxides 3. Hydroperoxides in inflammation: Cellular and molecular recruitment. In: Caughey, W (ed) Biochemical and clinical aspects of oxygen. Academic Press, New York, 1979, pp 213-226.
48. Hemler ME, Lands WEM: Evidence for peroxide-initiated free radical mechanism of prostaglandin biosynthesis. J. Biol. Chem. (255): 6253-61, 1980.
49. Chan HW-S.: Soya-bean lipoxygenase: An iron-containing dioxygenase. Biochim. Biophys. Acta (327): 32-5, 1973.
50. Roza M, Francke A: Soybean lipoxygenase: An iron-containing enzyme. Biochim. Biophys. Acta (327): 24-31, 1973.
51. Pistorius EK, Axelrod B: Iron, an essential component of lipoxygenase. J. Biol. Chem. (249): 3183-6, 1974.
52. Pistorius EK, Axelrod B, Palmer G: Evidence for participation of iron in lipoxygenase reaction from optical and electron spin resonance studies. J. Biol. Chem. (251): 7144-8, 1976.
53. Slappendel S, Aasa R, Malmstrom BG, Verhagen J, Veldink GA, Vliegenthart JFG: Factors affecting the line shape of the EPR signal of high-spin Fe(III) in soybean lipoxygenase-1. Biochim. Biophys. Acta (708): 259-65, 1982.
54. Gibian MJ, Galaway RA: Steady-state kinetics of lipoxygenase oxygenation of unsaturated fatty acids. Biochemistry (15): 4209-14, 1976.
55. Koch R: Calcium ion activation of lipoxidase. Arch. Biochem. Biophys. (125): 303-7, 1968.
56. Christopher JP, Pistorius EK, Regnier FE, Axelrod B: Factors influencing the positional specificity of soybean lipoxygenase. Biochim. Biophys. Acta (289): 82-7, 1972.
57. Nelson MS, Pattee HE, Singleton JA: Calcium activation of peanut lipoxygenase. Lipids (12): 418-22, 1977.
58. Zimmerman GL, Snyder HE: Role of calcium in activating soybean lipoxygenase 2. J. Agric. Food Chem. (22): 802-5, 1974.
59. Galpin JR, Allen JC: The influence of micelle formation on lipoxygenase kinetics. Biochim. Biophys. Acta (488): 392-401, 1977.
60. Borgeat P, Samuelsson B: Arachidonic acid metabolism in polymorphonuclear leukocytes: Effects of ionophore A23187. Proc. Natl. Acad. Sci. USA (76): 2148-52, 1979.
61. Jakschik BA, Sun FF, Lee L, Steinhoff MN: Calcium stimulation of a novel lipoxygenase. Biochem. Biophys. Res. Comm. (95): 103-110, 1980.
62. Lands WEM, Cook HW, Rome LH: Prostaglandin biosynthesis: Consequences of oxygenase mechanism upon in vitro assays of drug effectiveness. In: Samuelsson, B, Paoletti, R. (eds) Advances in Prostaglandin and Thromboxane Research Vol. 1. Raven Press, New York, 1976, pp 7-17.
63. Lands WEM, Hanel AM: Inhibitors and activators of prostaglandin biosynthesis. In: Pace-Asciak, C, Granstrom, E (eds) Prostaglandins and related substances. Elsevier, Amsterdam, 1983, pp 203-223.
64. Lands WEM, Byrnes MJ: The influence of ambient peroxides on the conversion of 5,8,11,14,17-eicosapentaenoic acid to prostaglandins. Prog. Lipid Res. (20): 287-90, 1981.

65. Hanel AM, Lands WEM: Modification of anti-inflammatory drug effectiveness by ambient lipid peroxides. Biochem. Pharmacol. (31): 3307-11, 1982.
66. Lands WEM: Actions of anti-inflammatory drugs. Trends in Pharmacol. Sci. (March): 78-80, 1981.
67. Tappel AL, Lundberg WO, Boyer PD: The reaction mechanism of soy bean lipoxidase. J. Biol. Chem. (199): 267-81, 1952.
68. Vanderhoek JY, Lands WEM: The inhibition of the fatty acid oxygenase of sheep vesicular gland by antioxidants. Biochim. Biophys. Acta (296): 382-5, 1973.
69. Yasumoto K, Yamamoto A, Mitsuda H: Effect of phenolic antioxidants on lipoxygenase. Agr. Biol. Chem. (34): 1162-8, 1970.
70. Mitsuda H, Yasumoto K, Yamamoto A: Inhibition of lipoxygenase by saturated monohydric alcohols through hydrophobic bondings. Arch. Biochem. Biophys. (118): 664-9, 1967.
71. Downing DT, Ahern DG, Bachta M: Enzyme inhibition by acetylenic compounds. Biochem. Biophys. Res. Comm. (40): 218-23, 1970.
72. Corey EJ, Park H: Irreversible inhibition of the enzymic oxidation of arachidonic acid to 15-(hydroperoxy)-5,8,11(Z),13(E)-eicosatetraenoic acid (15-HPETE) by 14,15-dehydroarachidonic acid. J. Amer. Chem. Soc. (104): 1750-2, 1982.
73. Yoshimoto T, Furukawa M, Yamamoto S, Horie T, Watanabe-Kohno S: Flavonoids: Potent inhibitors of arachidonate 5-lipoxygenase. Biochem. Biophys. Res. Comm. (116): 612-8, 1983.
74. Miyamoto T, Obata T: New Inhibitors of 5-lipoxygenase. In: Shiokawa Y., Katori M, Mizushima Y (eds) Perspectives in prostaglandin research. Excerpta Medica, Amsterdam, 1983. pp. 78-80.
75. Tyson CA: 4-Nitrocatechol as a colorimetric probe for non-heme iron dioxygenases. J. Biol. Chem. (250): 1765-70, 1975.
76. Galpin JR, Teilens AGM, Veldink GA, Vliegenthart JFG, Boldingh J: On the interaction of some catechol derivatives with the iron atom of soybean lipoxygenase. FEBS Lett. (69): 179-82, 1976; Spaapen, LJM, Verhagen J, Veldink GA, Vliegenthart JFG: Properties of a complex of Fe(III)-soybean lipoxygenase-1 and 4-nitrocatechol. Biochim. Biophys. Acta (617): 132-140, 1980.
77. Vanderhoek JY, Lands WEM: Evidence for H_2O_2 mediating the irreversible action of acetylenic inhibitors of prostaglandin biosynthesis. Prostaglandins and Medicine (1): 251-63, 1978.
78. Endo K, Helmkamp GM, Bloch K: Mode of inhibition of β-hydroxydecanoyl thioester dehydrase by 3-decynoyl-N-acetylcysteamine. J. Biol. Chem. (245): 4293-4296, 1970.
79. Vanderhoek JY, Lands WEM: Acetylenic inhibitors of sheep vesicular gland oxygenase. Biochim. Biophys. Acta (296): 374-381, 1973.
80. Lands WEM: Inhibition of prostaglandin biosynthesis. In: Bergstrom S (ed) Advances in Bioscience. Pergamon Press, Oxford, 173, pp. 15-28.
81. Jakschik BA, Disantis DM, Sankarappa SK, Sprecher H: Modulation of leukotriene formation by various acetylenic acids. In: Samuelsson B, Paoletti R (eds) Advances in Prostaglandin, Thromboxane, and Leukotriene Research Vol. 9. Raven Press, New York, 1982, pp 127-35.
82. Corey EJ, Munroe JE: Irreversible inhibition of prostaglandin and leukotriene biosynthesis from arachidonic acid by 11,12-dehydro and 5,6-dehydroarachidonic acids, respectively. J. Amer. Chem. Soc. (104): 1752-4, 1982.

83. Hammarstrom S: Selective inhibition of platelet n-8 lipoxygenase by 5,8,11-eicosatriynoic acid. Biochim. Biophys. Acta (487): 517-19, 1977.
84. Perchonock CD, Finkelstein JA, Uzinskas I, Gleason JG, Sarau HM, Cieslinski LB: Dimethyleicosatrienoic acids: Inhibitors of the 5-lipoxygenase. Tetrahed. Lett. (24): 2457-60, 1983.
85. Ackroyd J, Manro A, Scheinmann F: Synthesis and 5-lipoxygenase inhibitory activity of 7,7-dimethyleicosa-5Z,8Z-dienoic acid. Tetrahed. Lett. (24): 5139-40, 1983.
86. Arai Y, Shimoji K, Konno M, Konishi Y, Okuyama S, Iguchi S, Hayashi M, Miyamoto T, Toda M: Synthesis and 5-lipoxygenase inhibitory activities of eicosanoid compounds. J. Med. Chem. (26): 72-8, 1983.
87. Cohen N, Weber G, Banner BL, Welton AF, Hope WC, Crowley H, Anderson WA, Simko BA, O'Donnell M, Coffey JC, Fiedler-Nagy C, Batula-Bernardo C: Analogs of arachidonic acid methylated at C-7 and C-10 as inhibitors of leukotriene biosynthesis. Prostaglandins (27): 553-62, 1984.
88. Yoshimoto T, Yokoyama C, Ochi K, Yamamoto S, Maki Y, Ashida Y, Terao S, Shiraishi M: 2,3,5-Trimethyl-6-(12-hydroxy-5,10-dodecadiynyl)-1,4-benzoquinone (AA861), a selective inhibitor of the 5-lipoxygenase reaction and the biosynthesis of slow-reacting substance of anaphylaxis. Biochim. Biophys. Acta (713): 470-3, 1982.
89. Higgs GA, Flower RJ, Vane JR: A new approach to anti- inflammatory drugs. Biochem. Pharmacol. (28): 1959-61, 1979.
90. Walker JR, Dawson W: Inhibition of rabbit pmn lipoxygenase activity by benoxaprofen. J. Pharm. Pharmacol. (31): 778-80, 1979.
91. Dawson W., Corvalan JRF, Kitchen EA, Parry MG: 5-Lipoxygenase inhibition in relation to cell movement and cancer. In: Thaler-Dao H, dePaulet AC, Paoletti R (eds) Icosanoids and Cancer. Raven PRess, New York, 1984, pp. 229-234.
92. Sekiya K, Okuda H, Arichi S: Selective inhibition of platelet lipoxygenase by esculetin. Biochim. Biophys. Acta (713): 68-72, 1982.
93. Radmark O, Malmsten C, Samuelsson B: The inhibitory effect of BW755C on arachidonic acid metabolism in human polymorphonuclear leukocytes. FEBS Lett. (110): 213-5, 1980.
94. Pfister JR, Murthy DVK: Synthesis of three potential inhibitors of leukotriene biosynthesis. J. Med. Chem. (26): 1099-1103, 1983.
95. Vanderhoek JY, Bryant RW, Bailey JM: Inhibition of leukotriene biosynthesis by the leukocyte product 15-hydroxy-5,8,11,13-eicosatetraenoic acid. J. Biol. Chem. (255): 10064-66, 1980.
96. Smith RJ, Sun FF, Bowman BJ, Iden SS, Smith HW, McGuire JC: Effect of 6,9-deepoxy-6,9-(phenylimino)-$\Delta^{6,8}$-prostaglandin I_1, (U-60,257), An inhibitor of leukotriene synthesis, On human neutrophil function. Biochem. Biophys. Res. Comm. (109): 943-9, 1982.
97. Prescott SM: The effect of eicosapentaenoic acid on leukotriene B production by human neutrophils. J. Biol. Chem. (259): 7615-21, 1984.
98. Egan RW, Tischler AN, Baptista EM, Soderman DD, Gale PH: Specific inhibition and oxidative regulation of 5-lipoxygenase. In: V Intl. Conf. Prostaglandins. Raven Press, New York, 1982, p 10.

99. Levine L: Stimulation of cellular prostaglandin production by phorbol-esters and growth factors and inhibition by cancer chemo-preventive agents. In: Powles TJ, Bockman RS, Honn KV, Ramwell P (eds) Prostaglandins and cancer: First international conference. Alan R. Liss, New York, 1982.
100. Nakadate T, Yamamoto S, Ishii M, Kato R: Inhibition of 12-O-tetradecanoylphorbol-13-acetate-induced epidermal ornithine decarboxylase activity by lipoxygenase inhibitors: Possible role of product(s) of lipoxygenase pathway. Carcinogenesis (3): 1411-14, 1982.
101. Goldberg ND, Haddox MK: Cyclic GMP metabolism and involvement in biological regulation. Ann. Rev. Biochem. (46): 823-96, 1977.
102. Reichard P, Ehrenberg A: Ribonucleotide reductase-A radical enzyme. Science (221): 514-9, 1983.

3

ENZYMES SYNTHESIZING AND METABOLIZING PROSTANOIDS

SHOZO YAMAMOTO

Most mammalian tissues are capable of synthesizing certain types of oxyeicosanoids (prostaglandin, PG; thromboxane, TX; leukotriene; LT). They are synthesized by a series of enzymes; phospholipases, fatty acid cyclooxygenase (Fig. 1 [1]), PG hydroperoxidase ([2]) and various endoperoxide-metabolizing enzymes. The first three enzymes are considered to be present in all the prostanoid-synthesizing tissues, and PGH_2 with a 9,11-endoperoxide is produced from arachidonic acid which is released from phospholipid. The metabolic map (Fig. 1) diverges at the step of PGH_2 transformation. The bond between the two oxygen atoms of the 9,11-endoperoxide is cleaved. Then each of the two oxygen atoms is bound to another carbon atom forming a bridge in the eicosanoid skeleton, picks up a hydrogen atom to form a hydroxyl group or is bound to its adjacent carbon atom to form a keto group. For example, rearrangement of the oxygen bond brings about isomerization of PGH_2(9,11-endoperoxy) to PGD_2(9-hydroxy-11-keto), PGE_2(9-keto-11-hydroxy), PGI_2(6,9-epoxy) and TXA_2(9,11-epoxy in oxetane-oxane structure). Each stereospecific reaction is catalyzed by a specific enzyme.

The type of oxyeicosanoid synthesized varies from tissue to tissue or from cell to cell. For example, PGE_2 is a predominant product in vesicular gland and kidney medulla. TXA_2 is a major arachidonate metabolite in platelets while PGI_2 is an essentially sole prostanoid produced in artery. Some tissues such as lung produce several types of oxyeicosanoids; PGE_2, PGI_2, TXA_2 and LT's. Such a tissue specificity of oxyeicosanoid biosynthesis is determined by the occurrence of a specific enzyme (or enzymes) in the tissue which produces oxyeicosanoid with a biological activity to modulate the function of the tissue.

W.E.M. Lands (ed.), BIOCHEMISTRY OF ARACHIDONIC ACID METABOLISM.

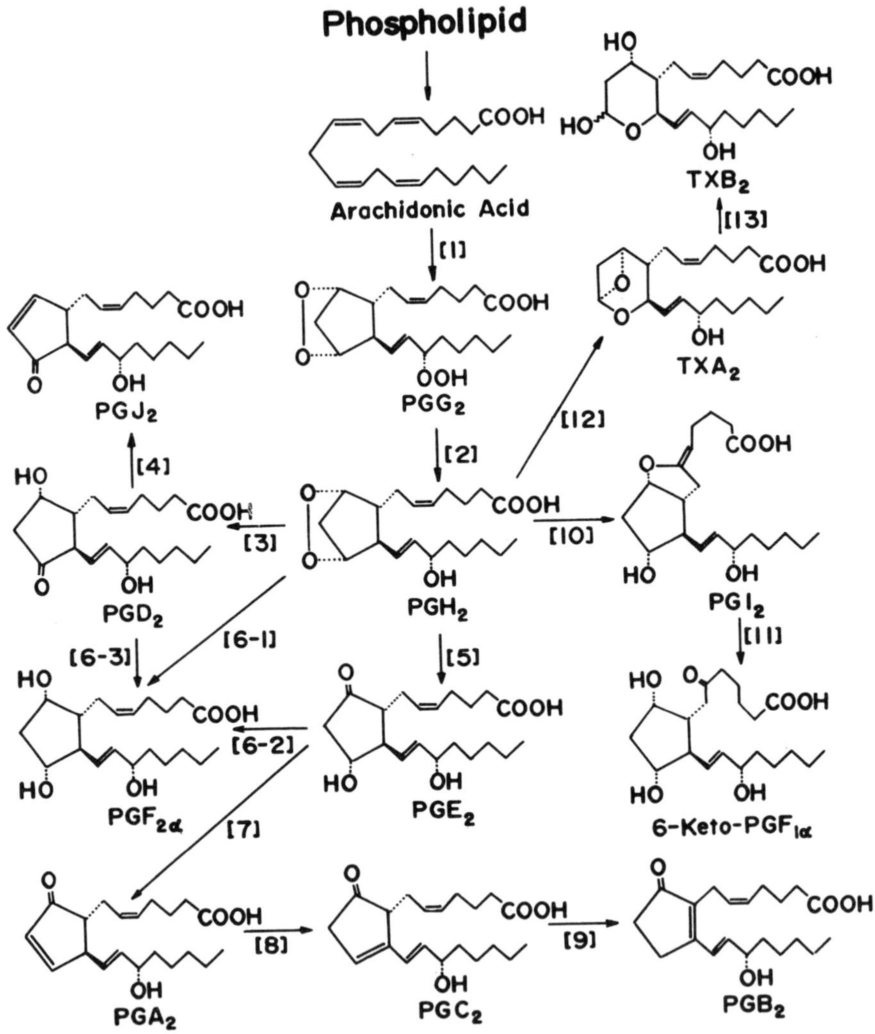

FIGURE 1. A series of enzyme reactions in the arachidonate cascade

The arachidonate cascade is triggered by the reaction of phospholipase, which is activated by a still unknown mechanism closely related to the function of biomembrane. The released arachidonic acid is utilized by cyclooxygenase or lipoxygenase for syntheses of the peroxy precursors of biologically active oxy-eicosanoids. Since cyclooxygenase (1) and 12-lipoxygenase (2) may be inducible enzymes, a regulatory role of these oxygenases should be noted. With crude enzyme preparations in most cases the overall transformation of arachidonic acid to PGD_2, PGE_2 or TXA_2 occurs without a significant accumulation of endoperoxide intermediates, an observation indicating that the produced PG endoperoxide is immediately transformed to individual PG or TX by isomerases abundantly present in various tissues.

References 3-8 are previously published review articles on the enzymes involved in the synthesis and metabolism of prostanoids.

1. PGD SYNTHASE

PGD_2 was earlier considered merely as a degradation product of PG endoperoxide, and its physiological role was almost ignored except for its antiaggregatory activity. The finding that the rat brain contained PGD_2 in a relative abundance (9), stimulated a biochemical study to isolate an enzyme catalyzing PGD_2 synthesis (Fig. 1 [3]) in rat brain. The highly purified PGD synthase from the soluble fraction of rat brain (10) requires no glutathione unlike PGE synthase. A related enzyme that requires glutathione is found in the cytoplasm of rat spleen (11) and mast cells (12). A predominant occurrence of the enzyme in neuroblastoma cells and an insignificant activity in glioma cells suggest a neuromodulator function of PGD_2, and extensive studies have been reported on the role of PGD_2 in the nervous system (13). A number of investigators are now interested in the antineoplastic activity of PGD_2. PGD_2 inhibits the growth or metastasis of various types of tumor cells both in vivo (14,15) and in vitro (16,17). According to a study on the structure-activity relationship, 9-deoxy-Δ^9-PGD_2 is 3 times more potent antineoplastic agent than PGD_2, and a term PGJ_2 was proposed for the compound (18). PGJ_2 is presumed to be synthesized from PGD_2 (Fig. 1 [4]) in a way similar to the conversion of PGE_2 to PGA_2 (see below), but the PGJ synthase has not been well characterized.

2. PGE SYNTHASE

Earlier, when a microsomal fraction was utilized as an enzyme preparation, the enzyme system catalyzing the overall synthesis of PGE_2 from arachidonic acid was referred to as prostaglandin synthetase (EC 1.14.99.1). The term is now out of date since the PG endoperoxide synthase (PGH synthase) converting arachidonic acid to PGH_2 and the PGE synthase transforming PGH_2 to PGE_2 (Fig. 1 [5], are separated by DEAE-cellulose chromatography after solubilization of the whole enzyme system from the microsomes of bovine vesicular gland (19). The solubilized PGE synthase is catalytically active only in the presence of glutathione which is required specifically as a coenzyme for the reaction rather than a stabilizer of the extremely unstable enzyme. However, the glutathione is not oxidized in a stoichiometric amount (20). A hypothesis explains the role of glutathione as a nucleophile in a 1,2-hydride shift during the isomerization of endoperoxide (21). PGE synthase can be also solubilized from the microsomes of sheep vesicular gland, and the enzyme requires glutathione (22).

3. PGF SYNTHASE

$PGF_{2\alpha}$ is one of the oldest members of PG family, and its biological activities have been extensively investigated. However, its origin is not yet definitely known. Three possible pathways have been proposed for the synthesis of $PGF_{2\alpha}$ from PGH_2 (Fig. 1). The endoperoxide can be reduced to give 9- and 11-hydroxyl groups [6-1]. Evidence to support this pathway is the finding that the 9-H of arachidonic acid is retained in $PGF_{2\alpha}$ (23) The microsomes of cow and guinea pig uterus synthesize $PGF_{2\alpha}$ from PGH_2 (24), but the nature of the enzyme and the necessary reductar has not been elucidated. The 9-keto group of PGE_2 can be reduced to a hydroxyl group [6-2], and this conversion of PGE_2 to $PGF_{2\alpha}$ occurs in various tissues. It is of interest that the 9-keto reduction is catalyzed by NADP-linked 15-hydroxy-PG dehydrogenase and both enzyme activities are attributed to the same enzyme (25). $PGF_{2\alpha}$ may be produced also from PGD_2 by the 11-keto reduction of the latter compound [6-3]. PGD_2 11-keto reductase is found in rabbit liver (26,27) and rat lung (28).

4. PGA, PGC AND PGB SYNTHASES

PGA_2 is produced by dehydration of PGE_2 (Fig. 1 [7]), and the enzyme activity responsible for this reaction is found in human serum (29). PGA_2 is transformed to PGC_2 by isomerization of Δ^{10} to Δ^{11} ([8]). The isomerase is present in the serum or plasma of various animals (30). Further isomerization of Δ^{11} to $\Delta^{8(12)}$ transforms PGC_2 to PGB_2 ([9]), and this reaction is catalyzed by an enzyme found in human and rabbit serum (29). It should be noted that PGA_2 and PGB_2 may also be non-enzymatic degradation products formed during careless handling of PGE_2.

5. PGI SYNTHASE (PROSTACYCLIN SYNTHASE)

Isomerization of the 9,11-endoperoxide to a 6,9-epoxide with an 11α-hydroxyl group transforms PGH_2 to PGI_2 (Fig. 1 [10]). PGI_2 is unstable particularly at acidic pH, and it spontaneously converts to a stable compound, 6-keto-$PGF_{1\alpha}$ ([11]). PGI synthase is found in various tissues including blood vessels (Table 2, reference 6), but a paper reports the lack of PGI_2 synthesis in the aorta of chicken (31). The enzyme is localized in the microsomes of artery (32) and other tissues, and can be solubilized with a non-ionic detergent from the aorta of rabbit (33) and pig (34). The solubilized enzyme transforms PGH_1 to 12-hydroxy-8,10-heptadecadienoic acid rather than PGI_1 (33). The enzyme is inhibited by various hydroperoxides of unsaturated fatty acids (32). A spectral change of aortic microsomes caused by interaction with PGH_2 or tranylcypromine (PGI synthase inhibitor) suggested the P-450 nature of PGI synthase (35). A monoclonal antibody raised against PGI synthase is utilized to purify the enzyme to an apparent homogeneity, and the purified enzyme exhibits a major heme peak at 418-420 nm. The enzyme treated with sodium dithionite and carbon monoxide shows a difference spectrum around 440 nm (36).

6. TXA SYNTHASE

TXA_2 is extremely unstable in water and rapidly degrades to TXB_2 (Fig. 1 [3]). As a unique property of TXA synthase, a cleavage of the 9,11-endoperoxide produces 12-hydroxy-5,8,10-heptadecatrienoic acid and malondialdehyde concomitant with the formation of TXA_2. The former two compounds are believed to derive fro PGH_2 rather than TXA_2 and to be produced by the catalysis of TXA synthase itself (37). A theory explains that TXA synthase is not an endoperoxide isomerase but a dismutase-type enzyme (38). PGH_1 is converted predominantly to 12-hydroxy-8,10-heptadecadienoic acid and to TXB_1 only in a small amount (39). Biosynthesis of TXA_2 is found in various tissues (Table 2, reference 6), and local ized in a microsomal fraction of several tissues tested. The enzyme is solubilized and partially purified with the aid of a non-ionic detergent from the microsomes of bovine (40) and human (41) platelets and bovine (42) and porcine (43) lung. A possible involvement of cytochrome P-450 was also presumed in the TXA_2 formation (35). Various compounds including imidazole and its derivatives have been developed as specific inhibitors of the enzyme (7).

7. PROSTANOID METABOLIZING ENZYMES

Administration of radiolabeled prostanoids and analysis of radioactive metabolites in plasma and urine have outlined a genera

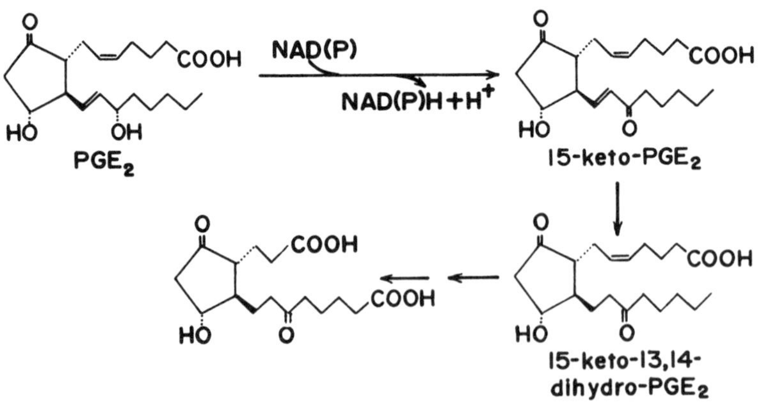

FIGURE 2. Catabolic pathway of prostanoids

metabolic pathway of most prostanoids as shown in Fig. 2. Enzymes responsible for the individual steps were reviewed in ref. 5 and 7. The first and physiologically important step in terms of the biological inactivation of prostanoid is catalyzed by 15-hydroxy-PG dehydrogenase. The enzyme is an NAD- or NADP-linked dehydrogenase specific for the 15-hydroxyl group of prostanoids rather than general alcohols. According to the substrate specificity studies (see Table 2, ref. 7), the NAD-linked dehydrogenases from various sources are more active with PGE, PGF, PGA and PGI, but much less active with PGB and TXB. A dehydrogenase specific for PGA, PGD and PGI has been found, respectively (7). Enzymology of these 15-hydroxy-PG dehydrogenases was reviewed in ref. 7 and 44.

REFERENCES

1. Koshihara Y, Senshu T, Kawamura M, Murota S: Sodium *n*-butyrate induces prostaglandin synthetase activity in mastocytoma P-815 cells. Biochim Biophys Acta (617): 536-539, 1980.
2. Chang WC, Nakao J, Orimo H, Tai HH, Murota S: Stimulation of 12-lipoxygenase activity in rat platelets by 17β-estradiol. Biochem Pharmacol (31): 2633-2638, 1982.
3. Samuelsson B, Granström E, Green K, Hamberg M, Hammarström S: Prostaglandins. Ann Rev Biochem (44): 669-695, 1975.
4. Samuelsson B, Goldyne M, Granström E, Hamberg M, Hammarström S, Malmsten C: Prostaglandins and thromboxanes. Ann Rev Biochem (47): 997-1029, 1978.
5. Lands WEM: The biosynthesis and metabolism of prostaglandins. Ann Rev Physiol (41): 633-652, 1979.
6. Moncada S, Vane JR: Pharmacology and endogenous roles of prostaglandin endoperoxides, thromboxane A_2, and prostacyclin. Pharmacol Rev (30): 293-331, 1979.
7. Yamamoto S: Enzymes in the arachidonic acid cascade. *In*: Pace-Asciak C, Granström E (eds) Prostaglandins and related substances. Elsevier Science Publishers, Amsterdam, 1983, pp 171-202.
8. Maclouf J, Sors H, Rigaud M: Recent aspects of prostaglandin biosynthesis. Biomedicine (26): 362-375, 1977.
9. Abdel-Halim MS, Lundén I, Cseh G, Änggård E: Prostaglandin profiles in nervous tissue and blood vessels of the brain of various animals. Prostaglandins (19): 249-258, 1980.
10. Shimizu T, Yamamoto S, Hayaishi O: Purification and properties of prostaglandin D synthetase from rat brain. J Biol Chem (254): 5222-5228, 1979.
11. Christ-Hazelhof E, Nugteren DH: Purification and characterisation of prostaglandin endoperoxide D-isomerase, a cytoplasmic, glutathione-requiring enzyme. Biochim Biophys Acta (572): 43-51, 1979.
12. Steinhoff MM, Lee LH, Jakschik BA: Enzymatic formation of

prostaglandin D_2 by rat basophilic leukemia cells and normal rat mast cells. Biochim Biophys Acta (618): 28-34, 1980.
13. Hayaishi O: Prostaglandin D_2: A neuromodulator. Advances in Prostaglandin, Thromboxane, and Leukotriene Research (12): 333-337, 1983.
14. Fitzpatrick FA, Stringfellow DA: Prostaglandin D_2 formation by malignant melanoma cells correlates inversely with cellular metastatic potential. Proc Natl Acad Sci USA (76): 1765-1769, 1979.
15. Higashida H, Kano-Tanaka K, Natsuume-Sakai S, Sudo K, Fukami H, Nakagawa Y, Miki N: Cytotoxic action of prostaglandin D_2 on mouse neuroblastoma cells. Int J Cancer (31): 797-802, 1983.
16. Kawamura M, Koshihara Y: Prostaglandin D_2 strongly inhibits growth of murine mastocytoma cells. Prostaglandins Leukotrienes and Medicine (12): 85-93, 1983.
17. Fukushima M, Kato T, Ueda R, Ota K, Narumiya S, Hayaishi O: Prostaglandin D_2, a potential antineoplastic agent. Biochem Biophys Res Communs (105): 956-964, 1982.
18. Fukushima M, Kato T, Ota K, Arai Y, Narumiya S, Hayaishi O: 9-Deoxy-Δ^9-prostaglandin D_2, a prostaglandin D_2 derivative with potent antineoplastic and weak smooth muscle-contracting activities. Biochem Biophys Res Communs (109): 626-633, 1982.
19. Miyamoto T, Yamamoto S, Hayaishi O: Prostaglandin synthetase system — Resolution into oxygenase and isomerase components. Proc Nat Acad Sci USA (71): 3645-3648, 1974.
20. Ogino N, Miyamoto T, Yamamoto S, Hayaishi O: Prostaglandin endoperoxide E isomerase from bovine vesicular gland microsomes, a glutathione-requiring enzyme. J Biol Chem (252): 890-895, 1977.
21. Lands W, Lee R, Smith W: Factors regulating the biosynthesis of various prostaglandins. Ann N Y Acad Sci (180): 107-122, 1971.
22. Nugteren DH, Christ-Hazelhof E: Chemical and enzymic conversions of the prostaglandin endoperoxide PGH_2. Advances in Prostaglandin and Thromboxane Research (6): 129-137, 1980.
23. Hamberg M, Samuelsson B: On the mechanism of the biosynthesis of prostaglandins E_1 and $F_{1\alpha}$. J Biol Chem (242): 5336-5343, 1967.
24. Wlodawer P, Kindahl H, Hamberg M: Biosynthesis of prostaglandin $F_{2\alpha}$ from arachidonic acid and prostaglandin endoperoxides in the uterus. Biochim Biophys Acta (431): 603-614, 1976.
25. Chang DGB, Sun M, Tai HH: Prostaglandin 9-ketoreductase and type II 15-hydroxyprostaglandin dehydrogenase from swine kidney are alternate activities of a single enzyme protein. Biochem Biophys Res Communs (99): 745-751, 1981.
26. Reingold DF, Kawasaki A, Needleman P: A novel prostaglandin 11-keto reductase found in rabbit liver. Biochim Biophys Acta (659): 179-188, 1981.
27. Wong PYK: Purification and partial characterization of

prostaglandin D_2 11-keto reductase in rabbit liver. Biochim Biophys Acta (659): 169-178, 1981.
28. Watanabe K, Shimizu T, Hayaishi O: Enzymatic conversion of prostaglandin D_2 to $F_{2\alpha}$ in the rat lung. Biochemistry International (2): 603-610, 1981.
29. Polet H, Levine L: Metabolism of prostaglandins E, A, and C in serum. J Biol Chem (250): 351-357, 1975.
30. Jones RL, Cammock S, Horton EW: Partial purification and properties of cat plasma prostaglandin A isomerase. Biochim Biophys Acta (280): 588-601, 1972.
31. Claeys M, Wechsung E, Herman AG, Nugteren DH: Lack of prostacyclin biosynthesis by aortic tissue of the chicken. Prostaglandins (21): 739-749, 1981.
32. Salmon JA, Smith DR, Flower RJ, Moncada S, Vane JR: Further studies on the enzymatic conversion of prostaglandin endoperoxide into prostacyclin by porcine aorta microsomes. Biochim Biophys Acta (523): 250-262, 1978.
33. Watanabe K, Yamamoto S, Hayaishi O: Reactions of prostaglandin endoperoxides with prostaglandin I synthetase solubilized from rabbit aorta microsomes. Biochem Biophys Res Communs (87): 192-199, 1979.
34. Wlodawer P, Hammarström S: Some properties of prostacyclin synthase from pig aorta. FEBS Lett (97): 32-36, 1979.
35. Ullrich V, Castle L, Weber P: Spectral evidence for the cytochrome P450 nature of prostacyclin synthetase. Biochem Pharmacol (30): 2033-2036, 1981.
36. DeWitt DL, Smith WL: Purification of prostacyclin synthase from bovine aorta by immunoaffinity chromatography. Evidence that the enzyme is a hemoprotein. J Biol Chem (258): 3285-3293, 1983.
37. Diczfalusy U, Falardeau P, Hammarström S: Conversion of prostaglandin endoperoxides to C_{17}-hydroxy acids catalyzed by human platelet thromboxane synthase. FEBS Lett (84): 271-274, 1977.
38. Anderson MW, Crutchley DJ, Tainer BE, Eling TE: Kinetic studies on the conversion of prostaglandin endoperoxide PGH_2 by thromboxane synthase. Prostaglandins (16): 563-570, 1978.
39. Diczfalusy U, Hammarström S: A structural requirement for the conversion of prostaglandin endoperoxides to thromboxanes. FEBS Lett (105): 291-295, 1979.
40. Yoshimoto T, Yamamoto S, Okuma M, Hayaishi O: Solubilization and resolution of thromboxane synthesizing system from microsomes of bovine blood platelets. J Biol Chem (252): 5871-5874, 1977.
41. Hammarström S, Falardeau P: Resolution of prostaglandin endoperoxide synthase and thromboxane synthase of human platelets. Proc Natl Acad Sci USA (74): 3691-3695, 1977.
42. Wlodawer P, Hammarström S: Thromboxane synthase from bovine lung - Solubilization and partial purification. Biochem Biophys Res Communs (80): 525-532, 1978.
43. Hall ER, Tai HH: Purification of thromboxane synthetase and evidence of two distinct mechanisms for the formation of 12-L-hydroxy-5,8,10-heptadecatrienoic acid by porcine lung microsomes. Biochim Biophys Acta (665): 498-503, 1981.

44. Hansen HS: 15-Hydroxyprostaglandin dehydrogenase. Prostaglandins (12): 647-679, 1976.

4

ENZYMATIC FORMATION OF LEUKOTRIENES

BARBARA A. JAKSCHIK, CHRISTINE G. KUO AND YUE FANG WEI

1. INTRODUCTION

Release of slow reacting substance (SRS) from guinea pig lung treated with cobra venom was described in 1938 (1) and release due to immunological challenge in 1940 (2). Characterization of SRS showed that it was a polar lipid with absorbance in the UV region and that it might contain sulfur (3-6). Work with labeled arachidonic acid indicated that this fatty acid is a metabolic precursor of SRS (7,8).

It was also observed that the major arachidonate metabolites of polymorphonuclear leukocytes are 5S-hydroxy-6,8,11,14-eicosatetraenoic acid (5-HETE) and 5S,12R-dihydroxy-6,8,10,14-eicosatetraenoic acid (leukotriene B_4) (9-11). Additionally two isomeric (5S)-5,12-dihydroxy-6,8,10,14-eicosatetraenoic acids [($E_1E_1E_1Z$)-C-12 epimeric (5,12-diHETE)] and two isomeric 5,6-dihydroxy-7,9,11,14-eicosatetraenoic acids (5,6-diHETE) were formed. Studies with $^{18}O_2$ and $H_2{}^{18}O$ demonstrated that the oxygen of the hydroxy group at C5 originated from molecular oxygen. The oxygen of the hydroxy group at C12 was derived from water (12). The hydroxy acids, epimeric and isomeric to leukotriene (LT) B_4, suggested a labile intermediate in the formation of LTB_4. The labile intermediate could be hydrolyzed non-enzymatically to the dihydroxy acids. Trapping experiments with methanol confirmed the formation of a labile intermediate (12). The labile intermediate was named LTA_4 and the structure of (5S)-5,6-epoxy-7,9,11,14-eicosatetraenoic acid was proposed for it. This structure was subsequently

W.E.M. Lands (ed.), BIOCHEMISTRY OF ARACHIDONIC ACID METABOLISM.

confirmed by chemical synthesis (13). Therefore, in this biosynthetic pathway arachidonic acid is oxygenated to 5-hydroperoxy-6,8,11,14-eicosatetraenoic acid (5-HPETE) by a lipoxygenase. Subsequently the epoxide (LTA_4) is formed by abstraction of the pro-R hydrogen atom at C10 and elimination of a hydroxide ion from the hydroperoxy group (14). The action of a hydrolase on LTA_4 is responsible for LTB_4 formation (12). The major structural difference between LTB_4 and the diHETEs obtained by nonenzymatic hydrolysis of LTA_4 is the double bond configuration at C-6. In LTB_4 this double bond is cis, and with the non-enzymatic products it is trans.

The stimulation of the synthesis of both sets of arachidonate metabolites (SRS and dihydroxy acids) by the calcium ionophore A23187 (7,11,15,16) and similar UV absorbance data suggested a common biosynthetic pathway. The structure of one component of SRS was characterized as (5S,6R)-6-S-glutathionyl-5-hydroxy-(7E,9E,11E,14Z)-eicosatetraenoic acid (LTC_4) (17-19). The structure of the other components of SRS was determined to be (5S,6R)-6-S-cysteinylglycine-5-hydroxy-(7E,9E,11Z,14Z)-eicosatetraenoic acid (LTD_4) (19-21) and (5S,6R)-(7E,9E,11Z,14Z)-6-S-cysteinyl-5-hydroxy-(7E,9E,11Z,14Z)-eicosatetraenoic acid (LTE_4) (21-23). Formation of SRS by a common pathway with LTB_4 via LTA_4 was confirmed by isolating LTA_4 as an intermediate of LTC_4 and LTD_4 formation (24) and converting LTA_4 to LTC_4 biosynthetically (25). Therefore LTA_4 can be converted to LTC_4 by the addition of glutathione by a glutathione-S-transferase. A γ-glutamyl transpeptidase is responsible for the formation of LTD_4 by removing glutamic acid from LTC_4. LTD_4 is further converted to LTE_4 by a dipeptidase which is responsible for the removal of glycine. The formation of leukotrienes via the 5-lipoxygenase is outlined in Fig. 1. The subscripts used in the nomenclature of leukotrienes denote the number of double bonds present. Therefore, leukotrienes derived from arachidonic acid

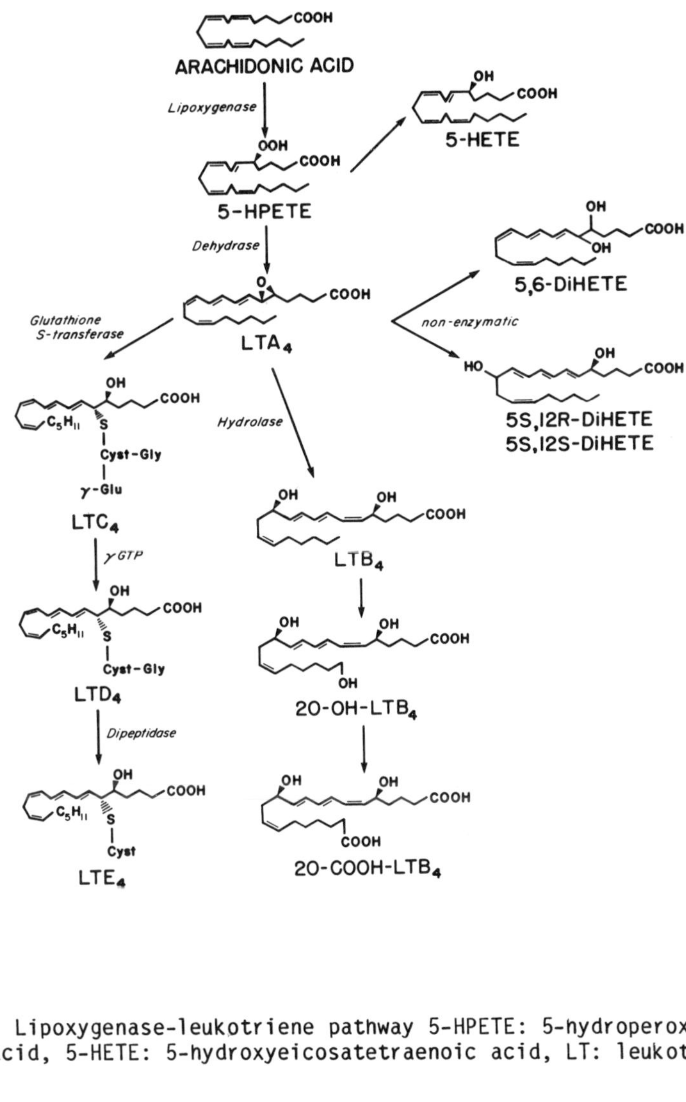

Figure 1. Lipoxygenase-leukotriene pathway 5-HPETE: 5-hydroperoxyeicosatetraenoic acid, 5-HETE: 5-hydroxyeicosatetraenoic acid, LT: leukotriene.

belong to the 4-series, those from eicosapentaenoic acid to the 5-series, and those from eicosatrienoic acid to the 3-series.

We have studied the enzymes of the 5-lipoxygenase-leukotriene pathway utilizing a cell-free system obtained from homogenates of rat basophilic leukemia (RBL-1) cells. We used this enzyme system to study the calcium requirement of the 5-lipoxygenase, the subcellular localization of the different enzymes of the pathway, and their substrate requirements.

2. CALCIUM STIMULATION OF THE 5-LIPOXYGENASE

Earlier experiments with whole RBL-1 cells showed that SRS synthesis from arachidonic acid was markedly potentiated by the calcium ionophore A23187 (7). Similar results were obtained for 5,12-diHETE and 5-HETE formation by RBL-1 cells (26) and by polymorphonuclear leukocytes (27). When RBL-1 cells were homogenized in the presence of 1 mM EDTA, and the 10,000 x g supernatant incubated with arachidonic acid, only insignificant amounts of 5-HETE and 5,12-diHETE (mixture of LTB_4 and 6-trans isomers) were produced as determined by thin layer chromatography. Addition of calcium markedly stimulated the formation of hydroxy acids (28). Similar results were obtained when SRS activity was monitored on the guinea pig ileum (Fig. 3) (29). This enhancement by calcium was dose dependent (Fig. 2). Maximum stimulation was achieved by the readdition of as little as 0.5 mM calcium (28). This suggested that only micromolar concentration of calcium were needed to activate the 5-lipoxygenase. In this preparation it was difficult to determine the actual amount of calcium needed to stimulate the enzyme since it is likely that some calcium is chelated to EDTA during homogenization. Ochi et al (30) found in a partially purified enzyme preparation that maximum stimulation was obtained with 100 μM calcium. Activation of the 5-lipoxygenase by calcium was also observed by Parker and Aykent (31). This stimulation was specific for calcium. Other divalent cations such as magnesium, copper, cobalt, barium, zinc, and iron were ineffective (28,30). The 5-lipoxygenase seems to be the only enzyme in this pathway that is stimulated by calcium. Parker et al., (32) have found that the calcium ionophore A23187 was not required for SRS synthesis when 5-HPETE

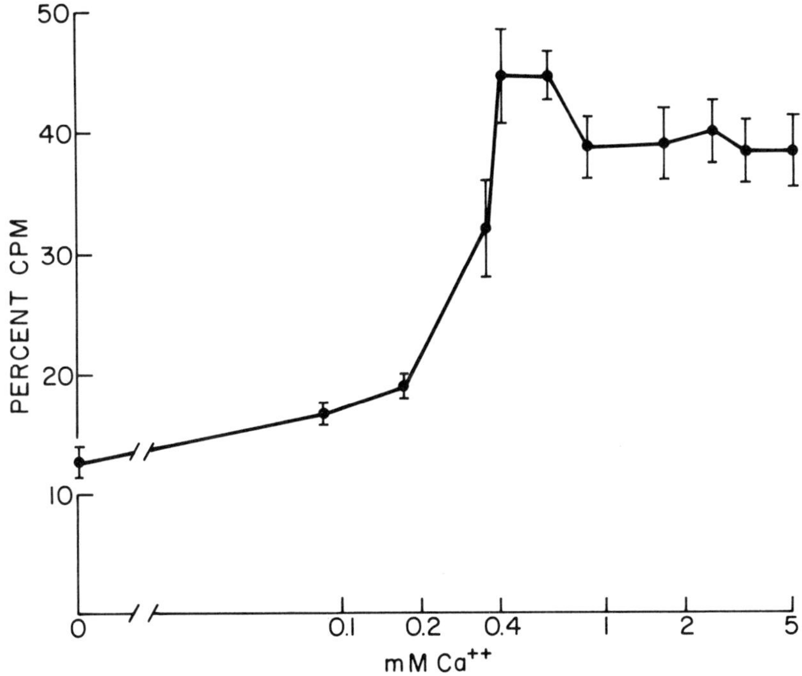

Figure 2. Stimulation of the 5-lipoxygenase by calcium. Supernatant (10,000 x g) of RBL-1 cell homogenates was incubated with 3.3 μM [^{14}C]arachidonic acid with calcium at the concentrations indicated. The reaction was stopped by the addition of acetone to precipitate the protein. The mixture was extracted twice with chloroform at pH 3.2 to 3.5. Thin layer chromatography was performed in the organic phase of ethyl acetate:2,2,4-trimethylpentane:acetic acid:water (110:50:20:100). Radioactive bands were localized by autoradiography, scraped and quantitated by liquid scintillation counting. The results are expressed as percent of total counts in 5-lipoxygenase products, vertical bars, SEM, n=5-11.

was used as the substrate. The conversion of LTA_4 to LTC_4 or LTB_4 also does not require calcium (33-35). Not only is the 5-lipoxygenase the only enzyme in this pathway that is activated by calcium, but also the only lipoxygenase known at the present time that has this property. Another observation which

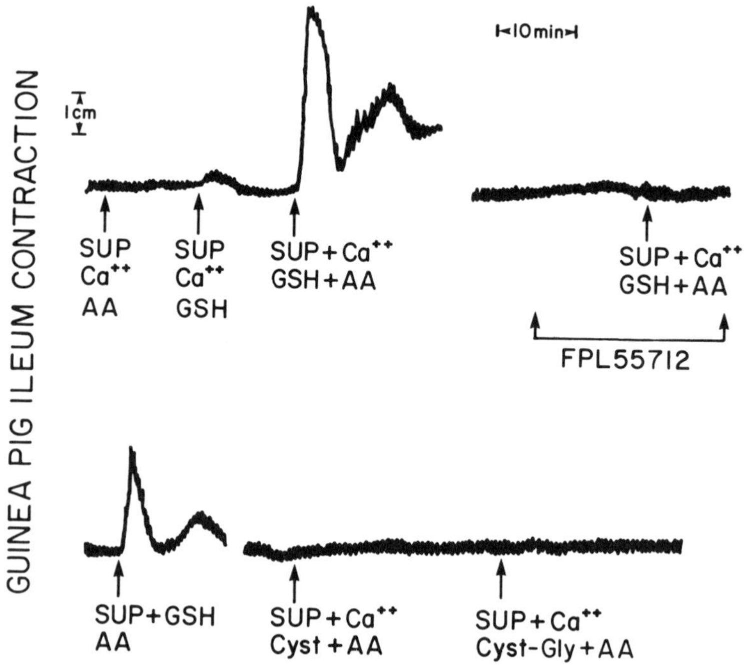

Figure 3. Factors affecting SRS synthesis. Supernatant (10,000 x g) of RBL-1 cell homogenates was incubated for 15 min at 37°C, with the agents indicated at the following concentrations: reduced glutathione (GSH) 1 mM, calcium (Ca^{++}) 2 mM, arachidonic acid (AA) 16 µM, cysteinyl-glycine (cyst-gly) 1 mM, cysteine (cyst) 1 mM, and directly applied to the superfused guinea pig ileum. FPL55712 was applied to the tissue at a concentration of 2 µM.

is peculiar for the 5-lipoxygenase is its stimulation by ATP. This action is shared by other nucleotides but to a lesser degree (30).

3. SUBCELLULAR LOCALIZATION OF THE ENZYMES

The subcellular localization of the various enzymes in the pathway was studied by fractionating the homogenate by differential centrifugation. The whole homogenate of RBL-1 cells produced 5-HETE, LTB_4, 6-trans-isomers of LTB_4, LTC_4, LTD_4 and LTE_4. After centrifugation at 10,000 x g and 100,000 x g, the 5-lipoxygenase, dehydrase (5-HPETE → LTA_4) and LTA-hydrolase (LTA_4 → LTB_4), remained in the supernatant. The conversion of arachidonic acid to 5-HETE, LTB_4 and its 6-trans-isomers was observed predominantly by the 10,000 x g and 100,000 x g supernant (Fig. 4) and not

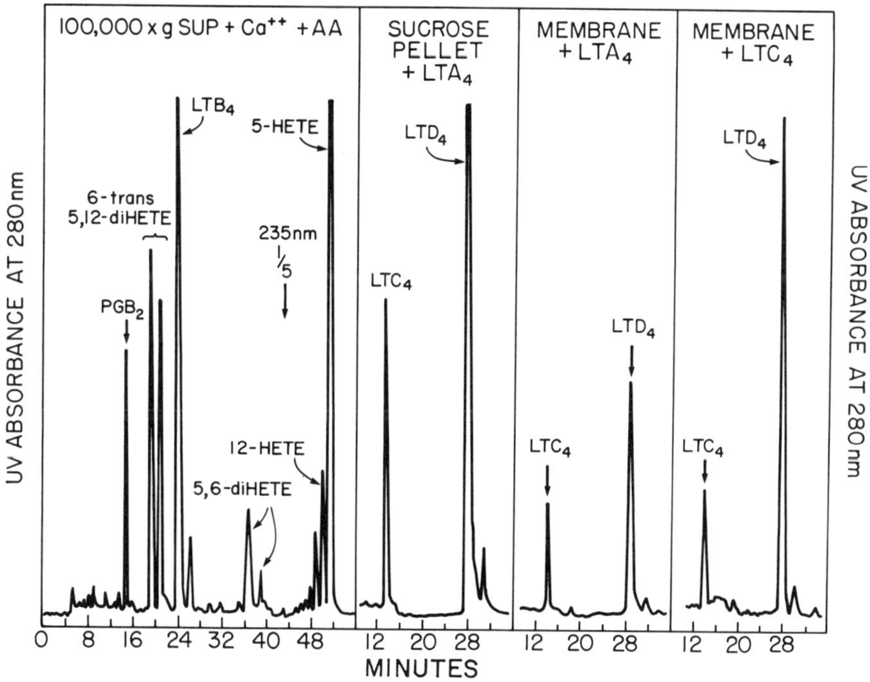

Figure 4. Subcellular localization of leukotriene-forming enzymes. Homogenate of RBL-1 cells was fractionated by differential centrifugation at 10,000 x g and 100,000 x g. The 10,000 x g pellet was further fractionated by sucrose gradient to obtain a crude membrane layer and a pellet (sucrose pellet). Individual subcellular fractions were incubated with the agents indicated (AA:arachidonic acid). The samples were precipitated with ethanol and concentrated on ODS-extraction columns (Baker). If the samples contained peptidoleukotrienes as well as dihydroxy acids, the samples were differentially extracted with methylene chloride. The samples were analyzed on reverse phase HPLC (Ultrasphere C18 column, 5 μm particle size, Beckman Instruments, Inc., flow 1 ml/min). Mobile phase for peptidoleukotrienes, methanol:water:acetic acid (68:32:0.08) pH 6.2; for dihydroxy and monohydroxy acids, methanol:water:acetic acid (72:28:0.05) pH 4.7 (isoratic up to 20 min; then the strength of the mobile phase was increased by a methanol gradient, from 0 to 25% between 20 and 35 min and from 25 to 35% between 40 and 65 min).

by the corresponding pellets. Upon addition of LTA_4 to the 10,000 x g or 100,000 x g pellets or supernatants, only the supernatants generated LTB_4. Therefore the 5-lipoxygenase, dehydrase and hydrolase appear to be cytosolic enzymes (34). The presence of the 5-lipoxygenase in the cytosolic fraction was confirmed by Parker et al (31) and by Ochi et al (30). Egan et al (36) also observed 5-lipoxygenase activity and Maycock et al (37) LTA_4-hydrolase activity in the cytosolic fraction of RBL-1 cells. However, no careful analysis of all the subcellular fractions was reported.

The enzymes responsible for the formation of LTC_4, LTD_4 and LTE_4 sedimented at 10,000 x g centrifugation. The 10,000 x g supernatant made small amounts of LTC_4 and LTD_4 as compared to the whole homogenate. Incubation of the 10,000 x g pellet with LTA_4 and GSH produced all the peptidoleukotrienes (LTC_4, LTD_4, LTE_4). Equivalent results were obtained by bioassay and HPLC analysis (35). Therefore, the LTC_4-forming glutathione-S-transferase, γ-glutamyl transpeptidase ($LTC_4 \rightarrow LTD_4$) and dipeptidase ($LTD_4 \rightarrow LTE_4$) are particulate. The 10,000 x g pellet was further fractionated by sucrose density gradient centrifugation to obtain a crude membrane fraction. This fraction converted LTC_4 to LTD_4 (Fig. 4) and the latter to LTE_4. However, only little conversion of LTA_4 to LTC_4 was observed (Fig. 4). The γ-glutamyl transpeptidase activity in the membrane fraction was time and concentration dependent. The time course reached a plateau at approximately 45 min. The activity was confirmed by serine borate inhibition (38). This compound is known to inhibit γ-glutamyl transpeptidase activity in general (39) and the conversion of LTC_4 to LTD_4 by RBL-1 cells (40). The peptidase activity in the membrane preparation was confirmed by L-cysteine inhibition (38). Sok et al (41) have shown earlier that L-cysteine blocks the conversion of LTD_4 to LTE_4. The experiments with the plasma membrane preparation show that the γ-glutamyl transpeptidase and the dipeptidase are present in this subcellular fraction. Therefore the LTC_4 formed inside the cell may be converted to LTD_4 by the removal of glutamic acid while LTC_4 passes through the membranes to the outside of the cell. This suggests that LTC_4 may have an intracellular role while the function of LTD_4 is mainly extracellular. The localization of these two enzymes in plasma membranes agrees with the localization of these types of enzymes in kidney, lung and liver (42). The localization of the glutathione-S-transferase in the particulate fraction was surprising, because this enzyme is usually cytosolic (43). Recently there have been reports of a microsomal glutathione S-transferase (44). After sucrose gradient centrifugation the major portion of the enzyme activity was found in the pellet. Incubation of this fraction with LTA_4 yielded LTC_4 and some LTD_4. The membrane layer converted LTA_4 also to LTC_4 and LTD_4 but to a far lesser degree (Fig. 4). These results are consistent with the electron microscopic examination of these fractions. The membrane layer contained mainly smooth membrane vesicles. It was contaminated with some granules, a few broken mitochondria, and a small amount of endoplasmic reliculum. The sucrose pellet contained mainly granules, some membrane vescicles, some chromatin and very few mitochondria (38). Since the major component in the sucrose pellet is the granules, the glutathione-S-transferase may be localized in this organelle. The contamination of the membrane fraction by granules and vice versa would explain the presence of some gluthione-S-transferase activity in membrane fraction, and γ-glutamyl transpeptidase activity in the sucrose pellet. The prospects that the LTC_4 forming gluthione-S-transferase is in the granules raises the question of the importance of LTC_4 synthesis for granule release reaction. The soluble fraction of RBL-1 cells has conventional glutathione-S-transferase activity when analyzed spectrophotometrically with adducts of halogenated nitrobenzene, but this enzyme does not seem responsible for LTC_4 formation. Bach et al (45) also have found the LTC_4 forming glutathione-S-transferase to be particulate and that the soluble glutathione-S-transferase in the cells does not convert LTA_4 to LTC_4. These data point out that the LTC_4 forming enzyme is distinct from the conventional gluthione-S-transferases.

4. KINETIC STUDIES OF 5-HETE, LTA_4 AND LTB_4 SYNTHESIS

Utilizing the 10,000 x g supernatant from RBL-1 cell homogenates we studied the time course of formation of 5-HETE, LTA_4 and LTB_4 synthesis and the effect of varying arachidonic acid concentration on the formation of these products. No glutathione was added to the incubation mixture, therefore no peptidoleukotrienes were formed. LTA_4 and LTB_4 produced were analyzed by HPLC and 5-HETE by thin layer chromatography. [^{14}C]arachidonic acid was used as substrate and the amount of products formed calculated from the specific activity. LTA_4 is converted enzymatically to LTB_4, and non-enzymatically to 6-trans-isomers of LTB_4. Therefore, the sum of these products was used to calculate LTA_4 formation. RBL-1 cells also have a 12-lipoxygenase and can produce the double lipoxygenation product 5S,12S-dihydroxyeicosa-6,10-trans-8,14-cis-tetraenoic acid (5S,12S-diHETE). This compound comigrates on reverse phase HPLC with LTB_4. Therefore, the LTB_4 peak was methylated and rechromatographed on normal phase HPLC to separate LTB_4 from 5S,12S-diHETE. Fig. 5 shows the effect of increasing arachidonic acid concentration on product formation. As the concentration of arachidonic acid increased, the ratio of the different products formed changed markedly. At lower concentrations of arachidonic acid a greater percentage of LTA_4 was converted to LTB_4 than at higher concentrations of arachidonic acid. The formation of LTB_4 leveled off at 30 μM arachidonate, whereas LTA_4 synthesis leveled off only at 100 μM, and 5-HETE had not reached a plateau

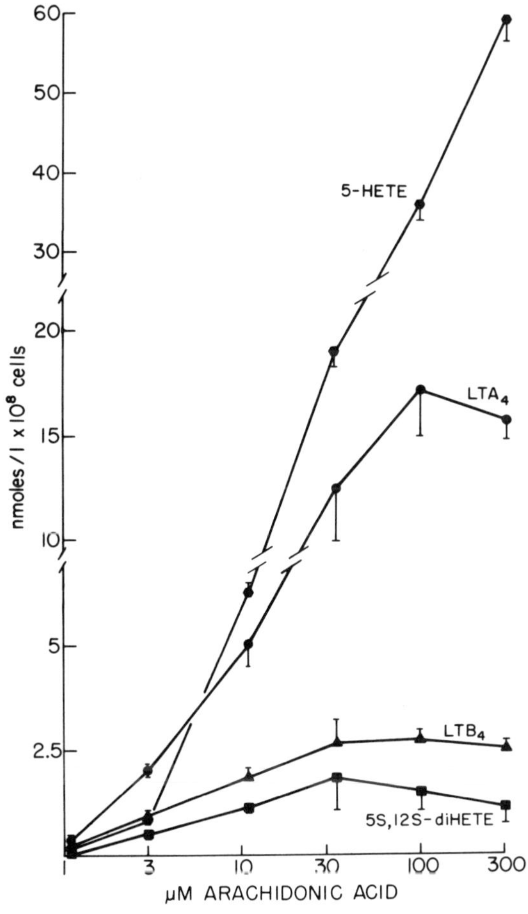

Figure 5. Effect of the variation of arachidonic concentration on 5-lipoxygenase products. Supernatant (10,000 x g) of RBL-1 cell homogenates was incubated with 1.5 mM calcium, [^{14}C]arachidonic acid at concentrations indicated, at 37°, for 15 min. The samples for LTA_4 and LTB_4 were processed as described in the legend of Fig. 4. The LTB_4 peak was methylated and rechromatographed to separate LTB_4 from 5S,12S-diHETE. The samples for 5-HETE were evaluated by thin layer chromatography (see legend of Fig. 2). The products were quantitated from the number of counts in each peak or band, and the specific activity of the substrate. Vertical bars, SEM (n=4). [Reproduced from (46) with permission of Academic Press].

at 300 μM arachidonate (Fig. 5) (34). These experiments suggest that this enzyme system has a much larger capacity to synthesize 5-HETE and LTA_4 than LTB_4. Although large amounts of LTA_4 were formed at higher concentrations of arachidonic acid, the amount of LTB_4 synthesized was limited. Accumulation of nonenzymatic hydrolysis products of LTA_4 were also observed under conditions where the peptidoleukotrienes were formed (whole homogenate with glutathione and arachidonic acid) (35). These findings are in contrast to the arachidonate cyclooxygenase pathway where the endoperoxide synthetase is the rate limiting step. The poor conversion of LTA_4 to LTB_4 raised the question of cofactor requirement. A number of agents which are known to enhance hydrolase activities were tested, but none increased LTB_4 synthesis (34).

The time course of formation of the different products measured was very similar (Fig. 6). The plateau was reached by approximately 10 min. In order to test whether the enzymes were still active when the time course reached its plateau, more arachidonic acid was added after 15 min of incubation. No additional LTA_4 or LTB_4 synthesis occurred. This suggested that the 5-lipoxygenase and/or LTA-synthetase were inactivated by arachidonic acid or its products. Further investigation of this phenomenon showed that preincubation of the 10,000 x g supernatant with arachidonic acid caused a loss of enzyme activity in a time dependent fashion (Table I). No such inactivation was observed by incubation without arachidonic acid present. This inactivation could be caused also by 8,11,14-eicosatrienoic acid but not by the saturated acid palmitic (Table I). RBL-1 cells convert 8,11,14-eicosatrienoic acid to hydroxy acids (47) but do not oxygenate palmitic acid. These data suggested that a product of a polyunsaturated acid can inactivate the 5-lipoxygenase.

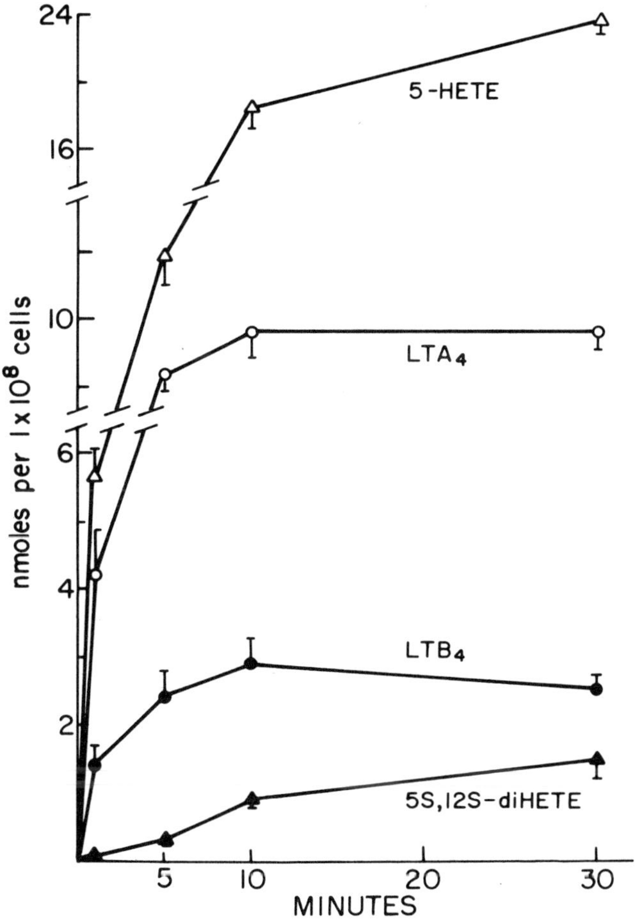

Figure 6. Time course of the biosynthesis of 5-HETE, LTA_4, LTB_4 and 5S,12S-diHETE. Supernatant (10,000 x g) of RBL-1 cell homogenates was incubated with 1.5 mM calcium and 30 μM [^{14}C]arachidonic acid for the time indicated. Samples were processed and products quantitated as described in the legend of Fig. 5. Vertical bars, SEM (n=3-5). [Reproduced from (46) with permission of Academic Press].

TABLE I

INACTIVATION OF ENZYMES

% OF CONTROL*

	Arachidonic†		8,11,14-Eicosatrienoic†		Palmitic†	
	5 min.	10 min.	5 min.	10 min.	5 min.	10 min.
5-HETE	30 ± 8	8 ± 2	49 ± 2	10 ± 1.1	170 ± 5	176 ± 11
diHETE¶	16 ± 4	4 ± 1	36 ± 4	8 ± 0.8	206 ± 18	203 ± 15

	15-HPETE**		
	1 μM	4 μM	10 μM
5-HETE	111 ± 3	74 ± 6	43 ± 13
diHETE¶	53 ± 4	24 ± 2	12 ± 4

* The data are expressed as products formed as percent of control (preincubation of 10,000 x g supernatant of RBL-1 cell homogenate without exogenous fatty acid added, followed by incubation with 17 μM arachidonic acid for 15 min. The data were analyzed by thin layer chromatography.

† The 10,000 x g supernatant was preincubated with 10 μM fatty acid for the time indicated.

** The 100,000 x g supernatant was preincubated with 15-HETE for 15 min at the concentration indicated.

¶ The amount of diHETE is approximately equivalent to LTA_4 formation.

Preincubation of the 10,000 x g supernatant with 15-hydroperoxyeicosatetraenoic acid inactivated the 5-lipoxygenase in a concentration dependent manner (Table I). The enzyme inactivation observed did not seem to confine itself to the 5-lipoxygenase but also seemed to include the dehydrase which is responsible for LTA_4 synthesis. As demonstrated in Fig. 5, the ratio of LTA_4 to 5-HETE increased as the substrate concentration decreased. Therefore if only the 5-lipoxygenase was inactivated and not the dehydrase, more LTA_4 then 5-HETE should have been formed during the time course of inactivation. However, the reverse was observed. The loss of dehydrase activity was further demonstrated with the preincubations with 15-HPETE. Again a greater decrease in LTA_4 than in 5-HETE levels was observed with all concentration of 15-HPETE tested (Table I) (34). The findings indicate that hydroperoxides formed from arachidonic acid are probably the inactivating species. Inactivation of the 5-lipoxygenase by peroxides has also been reported by Egan et al (36). Preincubation of the 10,000 x g supernatant with aracnidonic acid followed by incubation with LTA_4 demonstrated no change in LTB_4 formation. This indicates that the LTA_4 hydrolase is not inactivated (34).

The kinetics for the synthesis of the double lipoxygenase product 5S,12S-diHETE were somewhat different than for LTB_4. At low concentrations of arachidonic acid only a relatively small amount of this product was formed, and early in the time course (1 min) barely detectable levels of 5S,12S-diHETE were present. This suggests that high concentrations of substrate may be required for the formation of this compound.

5. SUBSTRATE SPECIFICITY

To determine the substrate specificity for the 5-lipoxygenase various ^{14}C-labelled polyenoic fatty acids were incubated with 10,000 x g supernatant in the presence of calcium and the products were analyzed by thin layer chromatography. Fatty acids with double bonds at 5,8,11 were good substrates for the 5-lipoxygenase and dehydrase (diHETE formation) (Fig. 7). These fatty acids were also readily converted to bioassayable peptidoleukotrienes. There was a striking conversion of 20:3(5,8,11), 20:4(5,8,11,14), 20:5(5,8,11,14,17), 19:4(5,8,11,14) and 18:4(5,8,11,14) fatty acids by the 5-lipoxygenase and dehydrase in contrast to fatty acids with their first double bond at Δ^4, Δ^6, Δ^7, or Δ^8. The latter fatty acids were oxygenated to some extent to monohydroxy acids, but only insignificant amounts of diHETE or SRS activity were produced. Chain length did not seem

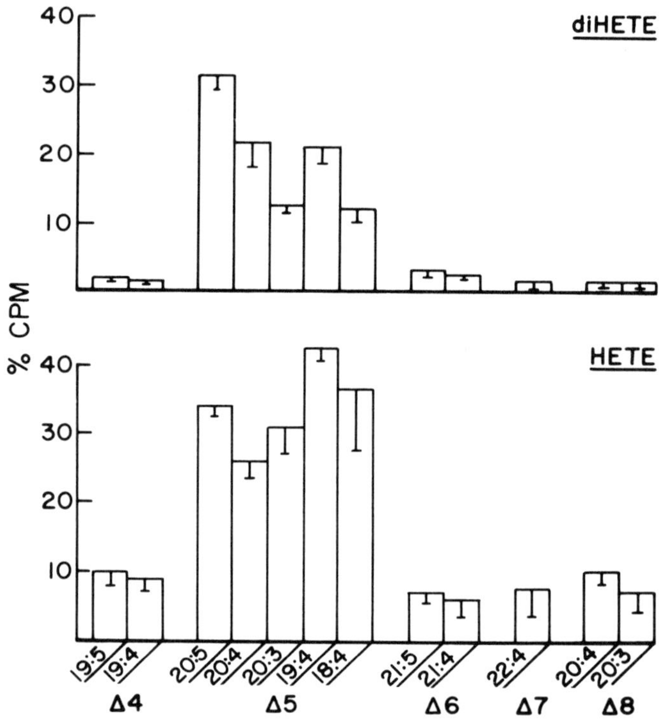

Figure 7. Fatty acid structural requirement for leukotriene biosynthesis. Supernatant (10,000 x g) of RBL-1 cell homogenates was incubated with the various fatty acids (^{14}C-labeled). Products were determined as described in the legend of Fig. 2. Bars denote SEM (n=4) [Reproduced from (46) with permission of Academic Press].

to have an effect on 5-lipoxygenase activity. The dehydrase activity was somewhat deminished when the chain length was reduced to 18 carbons. A similar reduction in dehydrase activity was observed when the number of double bonds was reduced to three (47). The unsaturation requirement at C5,8,11 is preculiar for the 5-lipoxygenase and has not been observed for the cyclooxygenase or 12-lipoxygenase in platelets (48).

The products obtained from the various polyenoic fatty acids were further analyzed by reverse phase HPLC. Fatty acids with the double bond at Δ^4, Δ^6, Δ^7, and Δ^8 showed no detectable conversion to LTA_4, LTB_4 or peptidoleukotrienes (Fig. 8 and 9). HPLC analysis of the Δ^5 polyenoic fatty acids addressed the question of the number of double bonds needed for LTB_4 formation. Incubation of 5,8,11-eicosatrienoic acid with 10,000 x g supernatant of RBL-1 homogenates showed that this fatty acid can be readily converted to LTA_3, but only insignificant amounts of LTB_3 were formed (Fig. 8, Table II). Normal phase HPLC analysis of the methylated LTB_3 peak from reverse phase HPLC indicated that the major portion of the latter was the double lipoxygenase product 5S,12S-diHETE (90%). This indicates that double bonds at C5,8,11 are sufficient for the dehydrase (LTA synthesis), but the double bonds at C14 is necessary for efficient LTB production (49). An additional double bond at C17 (5,8,11,14,17-eicosapentaenoic acid) does not influence any of the enzyme activities leading to LTB_5 formation (Fig. 7,8). The observations regarding the double bond requirement for the hydrolase is of biological interest since 5,8,11-eicosatrienoic acid accumulates in essential fatty acid deficiency. This condition is known to be associated with a decreased inflammatory response including reduced edema formation and leukocyte infiltration. The lack of formation of LTB, a potent chemotactic factor, could be responsible for the reduced leukocyte infiltration. LTB_4 is known to cause plasma exudation when PGE_2 is present. 5,8,11-Eicosatrienoic acid can be converted to peptidoleukotrienes (Fig. 9). LTC_4 and LTD_4 contract bronchial and arterolar smooth muscle and cause plasma exudation. LTC_3 and LTD_3 can contract smooth muscle. It is not known whether they can cause an increase in vascular permeability. The conversion of eicosapentaenoic acid (50,51) and of 5,8,11-eicosatrienoic acid (52) has been reported by other investigators by stimulating mouse

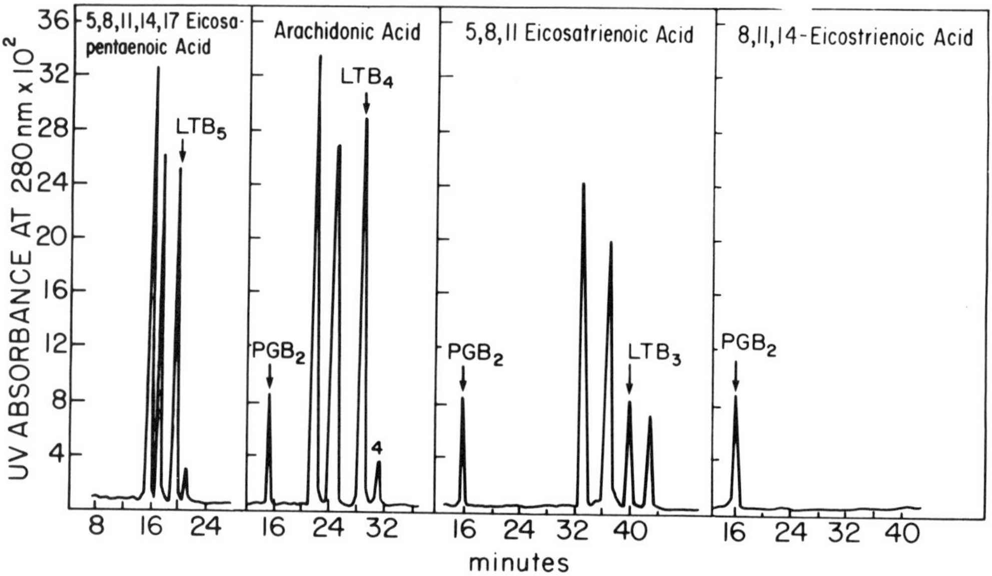

Figure 8. Effect of number of double bonds on LTB biosynthesis. Supernatant (10,000 x g) of RBL-1 cell homogenates was incubated with 1.5 mM calcium and 30 μM fatty acid as indicated. Samples were processed and analyzed as per legend of Fig. 4. The figure demonstrates the reverse phase HPLC analysis of the products formed.

mastocytoma cells with the calcium ionophore A23187. Hammarström (53) has also demonstrated leukotriene formation from 8,11,14-eicosatrienoic acid by the mouse mastocytoma. However the relative amounts formed as compared with arachidonic acid conversion were not indicated. It is also not clear whether this leukotriene synthesis was via the 5-lipoxygenase or another lipoxygenase.

In order to study the requirement of the double bond at C8 and C11 we incubated 5,8,14- and 5,11,14-eicosatrienoic acid with 10,000 x g supernatant of RBL-1 cell homogenate. 5,11,14-Eicosatrienoic acid was not

TABLE II

DOUBLE BOND REQUIRMENT FOR THE LIPOXYGENASE-LEUKOTRIENE PATHWAY

nmol/1 x 10^8 CELLS *

Fatty Acid	5-HETE	LTA	LTB
5,8,11,14-eicosatetraenoic	17.6 ± 1.8	10.2 ± 1.2	2.1 ± 0.2
5,8,11-eicosatrienoic	____**	3.1 ± 0.06	0.07 ± 0.02
5,8,14-eicosatrienoic	18.9 ± 1.3	nd†	nd
5,11,14-eicosatrienoic	nd	nd	nd

* 10,000 x g supernatant of RBL-1 cell homogenate was incubated with each one of the fatty acids (30 μM) in the presence of 1 mM Ca^{++}.

** 5-HETE formation was not determined for 5,8,11-eicosatrienoic acid at 30 μM concentration. At 7 μM substrate concentration the amount of 5-HETE was comparable to that obtained from arachidonic acid.

† nd: not detected.

oxygenated by the 5-lipoxygenase and 5,8,14-eicosatrienoic acid was converted to 5-HPETE but not to LTA (54). Therefore double bonds at C5 and 8 are necessary for the 5-lipoxygenase. Data illustrating the double bond requirements for the different enzymes in the pathway leading to LTB_4 formation are summarized in Table II. These observations indicate that the number of double bonds necessary at each step of the 5-lipoxygenase-leukotriene pathway varies and increases with each step. The double bonds at C5 and C8 are needed for the 5-lipoxygenase, C5,8 and 11 for the dehydrase and C5,8,11 and 14 for the LTA-hydrolase.

In contrast to the stringent requirements for a polyenoic fatty acid to be converted to LTB, the requirements with regard to the fatty acid portion for the glutathione-S-transferase are not as strict. It seems that any

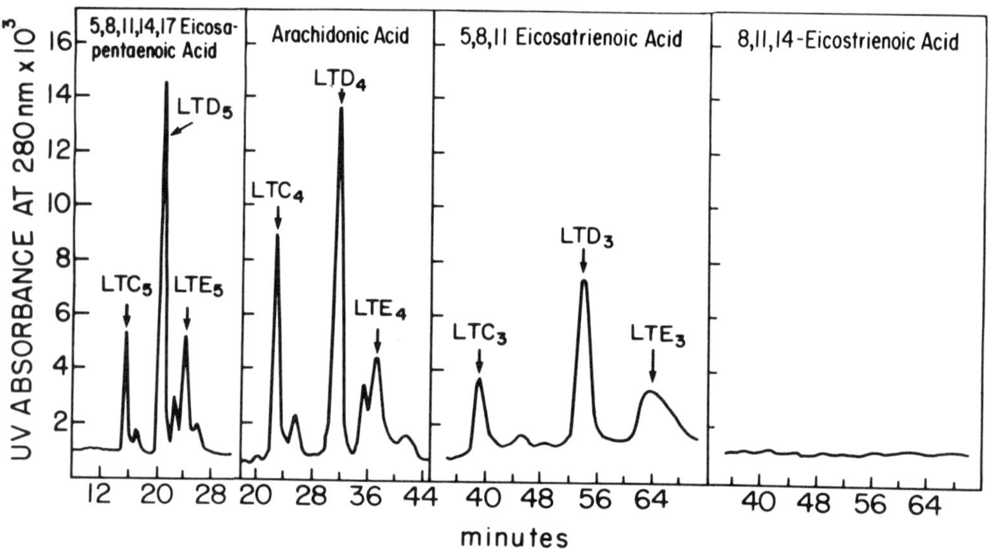

Figure 9. HPLC analysis of peptidoleukotrienes formed from 5,8,11,14,17-eicosapentaenoic, arachidonic, 5,8,11-eicosatrienoic and 8,11,14-eicosatrienoic acid. Methods as per legend of Fig. 4.

polyenoic fatty acid that is converted to LTA will be converted to LTC. 5,8,11-Eicosatrienoic acid was readily converted to LTC_3 and LTD_3 (Fig. 9). This low specificity of the glutathione-S-transferase was further illustrated in a study testing a series of acetylenic fatty acid varying in chain length, number and position of double bonds (55) as inhibitors of leukotriene formation. Any of the 33 acetylenic acids investigated inhibited LTC_4 formation, while only some were 5-lipoxygenase inhibitors.

With regard to the peptide portion, the glutathione-S-transferase appears to have more stringent requirements. Only glutathione seems to serve as a substrate, cysteine or cysteinylglycine will not substitute (Fig. 3) (29). In spite of the apparent lack of specificity for the fatty acid substrate, the glutathione-S-transferase involved in LTC formation is quite different from conventional glutathione-S-transferases. As mentioned earlier, the cytosolic glutathione-S-transferase in RBL-1 cells does not convert LTA_4 to LTC_4 (38,45). The work by Bach et al (45) also showed very clearly that adducts of halogenated nitrobezene which are routinely used to measure glutathione-S-transferase activity were not a substrate for the LTC forming enzyme. Moreover, most of the inhibitors of conventional glutathione-S-transferases did not inhibit LTC_4 synthesis. Therefore, the enzyme involved in LTC_4 synthesis is a specific glutathione-S-transferase distinct in substrate requirement and subcellular localization.

6. SUMMARY

The study of the enzymes of the 5-lipoxygenase-leukotriene pathway by a cell-free enzyme system has demonstrated a number of specific properties for this pathway. Calcium markedly stimulates the 5-lipoxygenase but not other lipoxygenases. The structural requirements of polyunsaturated fatty acids as substrate for the different enzymes in the pathway are quite stringent. Only fatty acids with the first double bond at Δ^5 are acted upon. For the 5-lipoxygenase double bonds at C5 and C8 are necessary, for the dehydrase (LTA synthesis) double bonds at C5,8 and 11, and for the LTA-hydrolase double bonds at C5,8,11 and 14. The glutathione-S-transferase seems to convert any LTA formed, but is quite specific for glutathione. Cysteine or cysteinylglycine does not serve as substrate. The glutathione-S-transferase involved in LTC_4 synthesis is specific for this pathway and does not act on adducts of halogenated nitrobenzenes conventionally used in glutathione-S-transferase assays. Furthermore, most conventional glutathione-S-transferases do not convert LTA_4 to LTC_4. The subcellular localization of the different enzymes of the pathway is in various parts of the cell. The 5-lipoxygenase, dehydrase and LTA-hydrolase are cytosolic. The glutathione-S-transferase, γ-glutamyl-transpeptidase and dipeptidase are particulate enzymes, sedimenting at 10,000 x g. The γ-glutamyl transpeptidase and dipeptidase are bound to plasma membranes. The localization of the glutathione-S-transferase is not certain, circumstantial evidence suggests the granules as the site. These findings characterize the leukotriene forming enzymes as a unique group of arachidonic acid metabolizing enzymes.

Acknowledgements

This investigation was supported by NIH grant R01-HL 21874 and P30-CA16217-03.

References

1. Feldberg W, Kellaway CH: Liberation of histamine and formation of lysolecithin-like substances by cobra venom. J Physiol (Lond) (94): 187-226, 1938.
2. Kellaway CH, Trethewie ER: Liberation of a slow reacting smooth muscle stimulating substance in anaphylaxis. J Exp Physiol (30): 121-145, 1940.
3. Strandberg K, Uvnäs B.: Purification and properties of the slow reacting substance formed in the cat paw with compound 48/80. Acta Physiol Scand (82): 358-374, 1971.
4. Orange RP, Murphy RC, Karnovsky ML, Austen KF: The physiochemical characteristics and purification of slow reacting substance of anaphylaxis. J Immunol (110): 760-770, 1973.
5. Morris HR, Taylor GW, Piper PJ, Sirois O, Tippins JK: Slow-reacting substance of anaphylaxis. Purification and characteristics. FEBS Letters (87): 203-206, 1978.
6. Parker CW, Jakschik BA, Huber MM, Falkenhein SF: Characterization of slow reacting substance as a family of thiolipids derived from arachidonic acid. Biochem Biophys Res Comm (89): 1186-1192, 1979.
7. Jakschik BA, Falkenhein S, Parker CW: Precursor role of arachidonic acid in release of slow reacting substance from rat basophilic leukemia cells. Proc Natl Acad Sci USA (74): 4577-4581, 1977.
8. Bach MK, Brashler JK, Gorman RR: On the structure of slow reacting substance of anaphylaxis: evidence for synthesis from arachidonic acid. Prostaglandins (14): 21-38, 1977.
9. Borgeat P, Hamberg M, Samuelsson B: Transformation of arachidonic acid and homo-γ-linolenic acid by rabbit polymorphonuclear leukocytes. J Biol Chem (251): 7816-7820, 1976.
10. Borgeat P, Samuelsson B: Transformation of arachidonic acid by rabbit polymorphonuclear leukocytes. Formation of a novel dihydroxy-eicosatetraenoic acid. J Biol Chem (254): 2643-2646, 1979.

11. Borgeat P, Samuelsson B: Metabolism of arachidonic acid in polymorphonuclear leukocytes. Structural analysis of novel hydroxylated compounds. J Biol Chem (254): 7865-7869, 1979.
12. Borgeat P, Samuelsson B: Arachidonic acid metabolism in polymorphonuclear leukocytes: Unstable intermediate in the formation of dihydroxy acids. Proc Natl Acad Sci USA (76): 3213-3217, 1979.
13. Radmark O, Malmsten C, Samuelsson B, Clark DA, Giichi G, Marfat A, Corey EJ: Leukotriene A: Streochemistry and enzymatic conversion to leukotriene B. Biochem Biophys Res Comm (92): 954-961, 1980.
14. Panossian A, Hamberg M, Samuelsson B: On the Mechanism of biosynthesis of leukotrienes and related compounds. FEBS Letters (150): 511-513, 1982.
15. Bach MK, Brashler JR: In vivo and in vitro production of a slow reacting substance in the rat upon treatment with calcium ionophore. J Immunol (113): 2040-2044, 1974.
16. Jakschik BA, Kulczycki A Jr, MacDonald HH, Parker CW: Release of slow reacting substance (SRS) from rat basophilic leukemia (RBL-1) cells. J Immunol (119): 618-623, 1977.
17. Murphy RC, Hammarström S, Samuelsson B: Leukotriene C: A slow reacting substance from murine mastocytoma cells. Proc Natl Acad Sci USA (76): 4275-4279, 1979.
18. Hammarström S, Murphy RC, Samuelsson B, Clark DA, Mioskowski C, Corey EJ: Structure of leukotriene C: Identification of amino acid part. Biochem Biophys Res Comm (91): 1266-1272, 1979.
19. Parker CW, Huber MM, Hoffman MK, Falkenhein SF: Characterization of two major species of slow reacting substance from rat basophilic leukemia cells as glutathionyl thioethers of eicosatetraenoic acid oxygenated at the 5-position. Evidence that peroxy groups are present and important for spasmogenic activity. Prostaglandins (18): 673-686, 1979.
20. Morris HR, Taylor GW, Piper PJ, Tippens JR: Structure of slow reacting substance of anaphylaxis from guinea pig lung. Nature (285): 104-106, 1980.
21. Parker CW, Falkenhein SF, Huber MM: Sequential conversion of the glutathionyl side chain of slow reacting substance (SRS) to cysteinyl-glycine and cysteine in rat basophilic leukemia cells stimulated with A23187. Prostaglandins (20): 863-887, 1980.
22. Sok DE, Pai JK, Atrache V, Kang YC, Sih CJ: Enzymatic inactivation of SRS-cys-gly (leukotriene D). Biochem Biophys Res Comm (101): 222-229.
23. Lewis RA, Drazen JM, Austen KF, Clark DA, Corey EJ: Identification of the C(6)-S-conjugate of leukotriene A with cysteine as a naturally occurring slow reacting substance of anaphylaxis (SRS-A). Importance of the 11-cis geometry for biologic activity. Biochem. Biophys Res Comm (96): 271-277, 1980.
24. Hammarström S, Samuelsson B: Detection of leukotriene A_4 as an intermediate in the biosynthesis of leukotriene C_4 and D_4. FEBS Letters (122): 83-85, 1980.
25. Radmark O, Malmsten C, Samuelsson B: Leukotriene A_4: Enzymatic conversion to leukotriene C_4. Biochem Biophys Res Comm (96): 1679-1687, 1980.
26. Jakschik BA, Lee LH, Shuffer G, Parker CW: Arachidonic acid metabolism in rat basophilic leukemia (RBL-1) cells. Prostaglandins (16): 733-748, 1978.
27. Borgeat P, Samuelsson B: Arachidonic acid metabolism in polymorphonuclear leukocytes: Effects of ionophore A23187. Proc Natl Acad Sci USA (76): 2148-2152, 1979.

28. Jakschik BA, Sun FF, Lee LH, Steinhoff MM: Calcium stimulation of a novel lipoxygenase. Biochem Biophys Res Comm (95): 103-110, 1980.

29. Jakschik BA, Lee LH: Enzymatic assembly of slow reacting substance. Nature (287): 51-52, 1980.
30. Ochi K, Yoshimoto T, Yamamato S: Arachidonate 5-lipoxygenase of guinea pig peritoneal polymorphonuclear leukocytes. J Biol Chem (258): 5754-5758, 1983.
31. Parker CW, Aykent S: Calcium stimulation of the 5-lipoxygenase from RBL-1 cells. Biochem Biophys Res Comm (109): 1011-1016, 1982.
32. Parker CW, Koch D, Huber MM, Falkenhein SF: Incorporation of radiolabel from [1-^{14}C]5-hydroperoxyeicosatetraenoic acid into slow reacting substance. Biochem Biophys Res Comm (94): 1037-1043, 1980.
33. Jakschik BA, Kuo CG: Subcellular localization of leukotriene-forming enzymes. Adv Prostaglandin Thromboxane Leukotriene Res (11): 141-145, 1983.
34. Jakschik BA, Kuo CG: Characterization of Leukotrienes A_4 and B_4 biosynthesis. Prostaglandins (25): 767-782, 1983.
35. Jakschik BA, Harper T, Murphy RC: Leukotriene C_4 and D_4 formation by particulate enzymes. J Biol Chem (257): 5346-5349, 1982.
36. Egan RW, Tischler AN, Baptista EM, Ham EA, Soderman DD, Gale PH: Specific inhibition and oxidative regulation of 5-lipoxygenase. Adv Prostaglandin Thromboxane Leukotriene Res (11): 151-157, 1983.
37. Maycock AL, Anderson MS, DeSousa DM, Kuehl FA Jr: Leukotriene A_4: Preparation and enzymatic conversion in a cell-free system to leukotriene B_4. J Biol Chem (257): 13911-13914, 1982.
38. Kuo CG, Lewis MT, Jakschik BA: Leukotriene D_4 and E_4 formation by plasma membrane bound enzymes (submitted).
39. Tate SS, Meister A: Serine-borate complex as a transition state inhibitor of γ-glutamyl transpeptidase. Proc Natl Acad Sci USA (75): 4806-4809, 1978.
40. Örning L, Hammarström S: Inhibition of leukotriene C and leukotriene D biosynthesis. J Biol Chem (255): 8023-8026, 1980.
41. Sok DE, Pai JK, Atrache V, Sih CJ: Characterization of slow reacting substance (SRSs) of rat basophilic leukemia (RBL-1) cells: Effect of cysteine on SRS profile. Proc Natl Acad Sci USA (77): 6481-6485, 1980.
42. Tate SS, Thompson GA, Meister A: Recent studies on γ-glutamyl transpeptidase. In: Arias TM, Jakoby, WB (eds). Glutathione Metabolism and Function, Raven Press, New York, 1976, pp. 45-55.
43. Wolkoff AW, Weisiger RA, Jakoby WB: The multiple roles of glutathione transferases (ligandins) In: Popper H, Schaffner F (eds) Prog Liver Dis (6): 214-224, 1979.
44. Morgenstern R, Meyer J, DePierre JW, Ernster L: Characterization of rat-liver microsomal glutathione-S-transferase activity. Europ J Biochem (104): 167-174, 1980.
45. Bach MK, Brashler JR, Morton DR: Solubilization and characterization of the leukotriene C_4 synthetase of rat basophilic leukemia cells: A novel particulate glutathione-S-transferase. Arch Biochem Biophys (230): in press, 1984.
46. Jakschik, BA, Kuo CG: Characterization of leukotriene formation. In: Charkrin LW, Bailey DM: The Leukotrienes, Academic Press, Inc. New York, 1984, pp. 141-164.
47. Jakschik BA, Sams AR, Sprecher H, Needleman P: Fatty acid structural requirements for leukotriene biosynthesis. Prostaglandins (20): 401-410, 1980.

48. Needleman P, Wyche AA, LeDue LE, Jakschik BA, Sankarapa SK, Sprecher H: Fatty acids as sources of potential "magic bullets" for the modification of platelet and vascular function. Prog Lipid Res (20): 415-42, 1981.
49. Jakschik BA, Morrison AR, Sprecher H: Products derived from 5,8,11-eicosatrienoic acid by the 5-lipoxygenase-leukotriene pathway. J Biol Chem (258): 12797-12000, 1983.
50. Hammarstrom S: Leukotriene C_5: a slow reacting substance derived from eicosapentaenoic acid. J Biol Chem (255): 7093-7094, 1980.
51. Murphy RC, Pickett WC, Culp BR, Lands WEM: Tetraene and pentaene leukotrienes: Selective production from murine mastocytoma cells after dietary manipulation. Prostaglandins (22): 613-622, 1981.
52. Hammarström S: Conversion of 5,8,11-eicosatrienoic acid to leukotrienes C_4 and D_4. J Biol Chem (256): 2275-2279, 1981.
53. Hammarström S: Conversion of dihomo-γ-linolenic acid to an isomer of leukotriene C_3, oxygenated at C-8. J Biol Chem (256): 7712-7714, 1981.
54. Wei YF, Evans RW, Morrison AR, Sprecher H, Jakschik BA: Double bond requirement for the 5-lipoxygenase pathway (submitted).
55. Jakschik BA, DiSantis DM, Sankarappa SK, Sprecher H: Modulation of leukotriene formation by various acetylenic acids. Adv Prostaglandin Thromboxane Leukotriene Res (9): 127-135, 1982.

5

CELLULAR AND SUBCELLULAR COMPARTMENTATION OF PROSTAGLANDIN AND THROMBOXANE SYNTHESIS

WILLIAM L. SMITH

1. INTRODUCTION

A current general perception of prostaglandin synthesis and function is as follows. Prostaglandins are formed in short-lived bursts by virtually all cells in response to specific cell surface stimuli; prostaglandins then exit the cells in which they are formed and act via receptors on the surface of parent or neighboring cells to elicit responses, usually activation of adenylate cyclase. Thus, prostaglandins are viewed as local hormones or autocoids. The latter concept derives from three general observations: (a) biologically active prostaglandins are rapidly catabolized to less active 15-keto derivatives during passage through the circulation, most notably by lung and kidney (1-5); (b) the serum concentrations of prostaglandin derivatives considered to be biologically active are below those necessary to elicit myotropic responses (i.e. $\leq 10^{-10}$ M (6-9)) and (c) prostaglandins are formed by all organs and not by a central exocrine gland (10-12).

Prostaglandins can have systemic effects when produced in large quantities in pathological states such as in human mastocytosis (13) or in the case of the VX carcinoma in rabbits (14). However, there are few examples of systemic effects of prostaglandins in healthy animals. One is that of PGI_2 produced by the uterus acting as a circulating vasodepressor in pregnant dogs (15).

Because prostaglandins are considered to act in close proximity to their sites of synthesis, the issue of compartmentation of synthesis is of special importance. In this chapter we will address two questions: (a) which cells synthesize and respond to prostaglandins? and (b) where within prostaglandin-forming and prostaglandin-responsive cells do the prostaglandin biosynthetic enzymes and prostaglandin receptors reside?

Defining sites of prostaglandin biosynthesis both at the cellular and subcellular levels suggests potential sites of action. In fact, showing that a cell forms prostaglandins essentially establishes a cellular site of action since most, if not all, cells which synthesize prostaglandins can respond in some way to their own prostaglandin products (16-18). Localizing prostaglandin biosynthetic enzymes and prostaglandin receptors at the subcellular level provides insights into how the metabolism of the intermediates is channeled topologically and how the synthesis of prostaglandins is coordinated with receptor binding. Information about the subcellular location of prostaglandin biosynthetic enzymes, along with data gained from studies on the mechanism and turnover of prostaglandin-forming enzymes in vitro, is also important in developing a composite scheme of how prostaglandin synthesis is regulated in intact cells.

2. CELLULAR LOCALIZATION OF PROSTAGLANDIN AND THROMBOXANE FORMATION

Studies conducted by Van Dorp, Nugteren and coworkers during the late 1960s indicated that homogenates prepared from each of approximately 20 different rat organs and tissues form cyclooxygenase products (10,11). These and related studies led to the perception that all cells are capable of forming prostaglandins albeit in greatly different amounts. It is now clear, however, from work with isolated cell populations that this generalization is incorrect. Perhaps only 10-20% of all cells form prostaglandins. Examples of formed elements in blood which do not synthesize cyclooxygenase products include erythrocytes and T lymphocytes (12,20). Immunohistochemical studies on kidney sections (21,22) and studies with dissected segments of the renal tubule (23) have indicated that while collecting tubules and thin limb epithelia form prostaglandins there is no cyclooxygenase activity in other parts of the tubule including proximal tubules, distal convoluted tubules or the ascending thick limb of Henle's loop (21-23). Among cultured cell lines which are apparently incapable of prostaglandin synthesis are LLC-PK_1 cells originating from pig kidney (24), some rat glioma cells (25) and HeLa cells of human origin (25).

In general, the formation of cyclooxygenase products by organ homogenates results from the biosynthetic activity of only a few cell types within the organ. Endothelial cells from large arteries and veins form PGI_2 and perhaps PGE_2 (26-29). Microvessels also produce prostaglandins (30-32). Moreover, all smooth muscle cells, both vascular and nonvascular, synthesize PGI_2 (33). Synthesis of prostaglandins by smooth muscle and endothelial cells accounts for most or all of the prostaglandin formation observed in homogenates of some organs. For example, heart homogenates form prostaglandins despite the fact that cardiac muscle, which accounts for most of the tissue mass, has little or no capacity for prostaglandin production (34,35).

Other than in the cases of endothelial and smooth muscle cells, it is usually not possible to predict which cells within an organ will synthesize prostaglandins. For example, there are no apparent correlates between the embryonic origins or functional capacities of cells and their ability to form prostaglandins (12). The only way to verify that a given cell type forms cyclooxygenase products is empirically. It does appear, however, that once the sites of synthesis are determined in an organ from one species, the data can be extrapolated to other species. For example, there are seven cell types in rabbit kidney which contain PGH synthase (21-23). These include renal medullary interstitial cells, vascular smooth muscle and endothelial cells, glomerular epithelial and mesangial cells, and epithelial cells comprising the thin limb and collecting tubules. With a few exceptions this same pattern of synthesis is seen in kidneys from other species including sheep, cow, dog, rat and guinea pig (22).

3. SUBCELLULAR LOCATION OF PROSTAGLANDIN-FORMING ENZYMES AND PROSTAGLANDIN RECEPTORS

The location of prostaglandin-forming enzymes is of interest for two reasons. First, because of the paucity of data on receptors, the location of synthesis may suggest sites of action of prostaglandins. Second, by knowing the location of these enzymes, it is possible to deduce which putative regulators of enzyme activity are important in whole cells in controlling prostaglandin formation. For example, proteins which are found in the circulation and which inhibit PGH synthase activity _in vitro_ are probably not important regulators of this enzyme, which is located primarily within cells.

The biosynthesis of prostaglandins and thromboxanes can be considered to proceed in three distinct phases (36). First is the mobilization of arachidonate from membrane phosphoglycerides occurring in response to interaction of a protease or hormone with the cell surface; second is the oxygenation of free arachidonate to yield the prostaglandin endoperoxides PGG_2 and PGH_2, and third is the metabolism of newly formed PGH_2 to one of the biologically active forms of prostaglandins (i.e. PGD_2, PGE_2, $PGF_{2\alpha}$, PGI_2 and TxA_2). We will discuss the information currently available on the location of prostaglandin receptors and of enzymes involved in the conversion of arachidonate to active prostaglandin derivatives.

3.1 Prostaglandin Receptors. For a prostaglandin to act on a cell, the appropriate receptor-response machinery must exist, and the prostaglandin must be present at a high enough concentration to occupy a sufficient number of receptor sites to elicit the response. The concentration of biologically active prostaglandins in the circulation is about 10^{-10} M (6-9). The minimum concentration of prostaglandins necessary to elicit most biological responses, and in particular, cAMP synthesis is 10^{-8}-10^{-9} M (16-18, 37). If this latter concentration of prostaglandin is taken as that required to elicit a response, prostaglandins are only likely to act prior to dilution in the circulation -- in other words, quite near their sites of synthesis. Prostaglandins are found to elicit (or inhibit) cAMP formation in all cells which synthesize prostaglandins. This finding is consistent with a local action of these hormones since it indicates that a biochemical response apparatus is found at the cellular site of synthesis.

Prostaglandin binding activities are associated with membrane fractions from a variety of different cells and tissues (38-40). In most cases it has not been possible to assert that measurements of prostaglandin binding are identical to measurements of prostaglandin receptors. Attempts to correlate receptor binding with biological responses have been made in the case of $PGF_{2\alpha}$ binding to membranes from the corpus luteum (38) and in the case of PGI_2 binding to platelet membranes (39). In the case of the $PGF_{2\alpha}$ binding protein, there is a good correlation between the abilities of a large series of prostaglandins to bind this protein and to cause luteal regression in animals (38). In the case of the PGI_2 binding activity of platelet membranes there is a correlation between the ability of prostaglandins to bind these membranes, to elicit cAMP synthesis and to inhibit platelet aggregation (39). In the case of the PGI_2 binding activity the K_D value for the low affinity PGI_2 receptor is similar to the EC_{50} value for cAMP formation.

The $PGF_{2\alpha}$ binding protein of corpus luteum is a membrane-bound protein which, based on differential centrifugation techniques, is associated primarily, but not exclusively with the plasma membrane (38). There is also indirect evidence that PGI_2 interacts with a receptor located on the platelet cell surface to cause activation of adenylate cyclase. PGE_1 immobilized by attachment to Sepharose beads inhibits platelet aggregation (41), and PGE_1 and PGI_2 compete with one another in activating platelet adenylate cyclase (39). A third example of a biochemical response involving interaction of a prostaglandin with the cell surface is in the case of the inhibition of ADH-induced cAMP formation in renal collecting tubule cells by PGE_2 (42). These three examples indicate that there are receptors for PGE_2, $PGF_{2\alpha}$ and PGI_2 on the outer surface of the plasma membranes of at least some of the cells that respond to prostaglandins. There is no definitive data on the subcellular location of PGD_2 or TxA_2 receptors.

It is not clear whether there are intracellular prostaglandin receptors. Rao and coworkers have shown that PGE_1 and $PGF_{2\alpha}$ binding activities are associated with several internal membrane fractions prepared by careful fractionation of bovine corpora lutea (43). Membrane fractions exhibiting specific binding included the endoplasmic reticulum, the nuclear membrane, several Golgi fractions and lysosomes. No specific binding to mitochondria was detected. It is not known whether the binding of prostaglandins to these subcellular fractions represents interactions of prostaglandins with functional receptors. However, this is an intriguing possibility based on evidence for multiple intracellular sites of prostaglandin formation (see below).

Prostaglandins do not appear to be taken up to an appreciable extent by most cells (44). Thus, when an intact cell or a tissue containing intact cells is found to be responsive to an exogenous prostaglandin, one can presume that this response results from the binding of the prostaglandin to a cell surface receptor. Unfortunately, the apparent lack of prostaglandin uptake by intact cells makes it difficult to assess the significance of prostaglandin binding to putative intracellular receptors.

3.2 Location of Phospholipid Pools Which are Sources of Arachidonate Mobilized for Prostaglandin Formation. Hormonal and in some cases proteolytic stimuli of prostaglandin production lead to the specific release of arachidonic acid and not other fatty acids commonly found at the sn-2 position of phosphoglycerides (45-47). Arachidonic acid released in response to hormonal stimuli is derived primarily from phosphatidylinositol (PI) and phosphatidylcholine (PC) (48,49). PI and PC are ubiquitous components of all mammalian cells. When phospholipids of tissues or cultured cells are prelabeled with $[^3H]$arachidonic acid and then stimulated to release arachidonate, a larger percentage of radioactivity is released than would be predicted on the basis of the mass of arachidonate that is released (48). Thus, there are two or more metabolic pools of each of these precursors. The physical location of these pools within cells is not known. Moreover, although both PI and PC appear to have an asymmetric distribution within membranes (50), this sidedness is not absolute and both PI and PC molecules are found on both surfaces on all cellular membranes. Thus, arachidonate could conceivably be mobilized from the membranes of any or all cellular organelles.

3.3 Location of Lipases Involved in Arachidonate Mobilization. The release of arachidonate from PI appears to involve the sequential actions of a PI-specific phospholipase C (51) and diglyceride and monoglyceride lipases (52). The PI-specific phospholipase C of platelets is found in the supernatant fraction prepared from sonicates of platelets (51) and presumably is present in the cytoplasm of cells. One would anticipate that activation of PI-specific phospholipase C could lead to the hydrolysis of inositol phosphate from PI molecules on the cytoplasmic surface of virtually all cellular membranes. Thus, the locations of diglyceride lipase and monoglyceride lipase would determine the sites of arachidonate release. In platelets 95% of glyceride lipase activity is membrane-bound, but the subcellular location has not been defined (52).

There is speculation that arachidonate is mobilized from PC through the action of a phospholipase A_2 perhaps acting on a pool(s) of PC species rich in arachidonic acid. Phospholipase A_2 activities have been found in virtually all intracellular membrane fractions (53). Arachidonate-rich pools of PC are more highly concentrated on the outer surface of lipid bilayers (50). Thus arachidonate originating from these pools would be expected to appear on the cytoplasmic surface of membranes such as the endoplasmic reticulum particularly as it curves about the nucleus.

It should be noted that virtually all membrane-bound enzymes associated with phospholipid metabolism are found on the outer or cytoplasmic surface of the endoplasmic reticulum (54). Studies on the location of various phospholipids and the lipases which catalyze the hydrolysis of arachidonate from these lipids is at least consistent with the concept that the arachidonate which is mobilized for prostaglandin formation originates from the cytoplasmic surface of one or more cellular membranes.

3.4 Location of PGH Synthase. The prostaglandin biosynthetic enzyme whose subcellular location has been studied most extensively is PGH synthase. PGH synthase catalyzes the formation of PGH_2 from arachidonic acid. There is indirect evidence for efficient metabolic coupling of arachidonate release and oxygenation. That is, arachidonic acid does not accumulate following treatment of cells with so called "specific" or "physiological" stimuli which cause the selective mobilization of this fatty acid, but arachidonate does accumulate following treatment of cells with "nonspecific" or "pathological" stimuli which cause the release of all fatty acids located at the sn-2 position of phosphoglycerides (45-49). Therefore, as a first approximation, the location of PGH synthase may reflect sites of selective arachidonate mobilization.

The subcellular location of PGH synthase activity has been investigated by differential centrifugation techniques using bovine thyroid glands (55) and rabbit renal medulla (56). The location of the PGH synthase antigen has been examined by immunocytochemistry on cultures of Swiss mouse 3T3 cells (57), bovine endothelial cells (58) and in sections of vascular and nonvascular smooth muscle from several tissues and species (33,59). Data from the differential centrifugation studies have indicated that most of the PGH synthase activity is associated with the nuclear and microsomal fractions. However, in each case the microsomal fractions contained significant plasma membrane contamination.

Localization of the PGH synthase antigen in 3T3 cells by immunoelectron microscopy clearly indicates that PGH synthase antigenic reactivity is located on both the nuclear membrane and the endoplasmic reticulum of these cells (57). PGH synthase-positive staining was not noticeable above control levels on either the plasma membrane or mitochondrial membrane. Cultures of bovine endothelial cells were examined by light microscopy after treatment with the same staining procedures employed to determine the location of PGH synthase in 3T3 cells. The endothelial cells showed light cytoplasmic and dense perinuclear staining -- a pattern identical to that seen at the light microscope level with 3T3 cells (56). Moreover, this same pattern of cytoplasmic and perinuclear staining has been noted by immunocytofluorescence in a number of other cell types including renal collecting tubules (21,22). Thus, in most cells the PGH synthase appears to be present predominantly on the endoplasmic reticulum and nuclear membrane.

Subsequent studies have been conducted with intact microsomes from sheep vesicular glands to determine if the PGH synthase is located on the lumenal or cytoplasmic face of the endoplasmic reticulum (60). When intact microsomal vesicles were treated with a nonpenetrating protease, the PGH synthase was inactivated. Furthermore, when the vesicles were labeled by treatment with [^{3}H]acetyl salicylate, the radioactivity was solubilized from the vesicles by protease treatment. Moreover, a series of monoclonal antibodies directed against PGH synthase caused immmunoprecipitation of the vesicles. These studies have established that the active site and at least three different antigenic determinants of PGH synthase are on the surface of these vesicles. Thus, the PGH synthase is located on the cytoplasmic side of the endoplasmic reticulum. Presumably, the PGH synthase which is located on the nuclear membrane is on the outer membrane of the nuclear envelope. The outer membrane of the envelope forms a continuum with the endoplasmic reticulum and most enzymes located on this outer membrane are also found in the ER (50).

More recent immunocytofluorescence studies on smooth muscle tissue have indicated that the PGH synthase antigen is present on the plasma membrane of at least some cells (33). The location of PGH synthase was examined as a part of a study of the distribution of PGI_2 synthase in a variety of vascular and nonvascular smooth muscles. As will be mentioned again below, all smooth muscle cells stained for the PGI_2 synthase antigen. Smooth muscle from most tissues and organs showed no detectable PGH synthase-positive staining. This is because the level of PGH synthase in most smooth muscle is below the detection limit of the immunocytofluorescence technique (although PGH synthase can be detected in these same tissues using immunoradiometric assays (59)). In those smooth muscles in which the PGH synthase can be detected, it is found on either the plasma membrane or nuclear membrane. Thus, the PGH synthase is located on the plasma membrane of some cells. These results also serve to emphasize that an inability to visualize the PGH synthase antigen on a cellular membrane by immunocytochemical techniques should not be taken as evidence that there is no enzyme associated with the organelle. Similarly, although there is no apparent cytoplasmic staining of smooth muscle cells using anti-PGH synthase antibodies, most of the PGH synthase activity in smooth muscle homogenates is actually associated with the microsomal fraction and probably resides in the endoplasmic reticulum (61). Smooth muscle has a relative paucity of ER in comparison to most cells and the lack of cytoplasmic staining is likely a reflection of this situation. The studies on smooth muscle suggest that the PGH synthase is associated with several different membrane systems -- nuclear, reticular and cell surface. These findings with smooth muscle raise the question of whether PGH synthase has this same distribution in all cells, but that the antigen is simply not present at detectable levels in many cases.

3.5 Location of Enzymes that Metabolize Prostaglandin Endoperoxides. Cells that form prostaglandins can release measureable amounts of PGH_2 to the outside of the cell (62). Moreover, when PGH_2 is added to cultured cells, it can be converted to PGH_2 metabolites (27). Thus, it is not certain whether PGH_2 is metabolized principally within the cells in which it is synthesized or by closely neighboring cells.

Four lines of evidence suggest that in most cases PGH_2 is converted to active prostaglandin products in the same cell in which it is synthesized. First, cells which stain for the PGH synthase antigen also stain positively with antibodies directed against PGH_2-metabolizing enzymes. Endothelial cells stain positively for both PGI_2 synthase and PGH synthase (33,59), platelet cells stain positively for both TxA synthase and PGH synthase, and vesicular gland epithelial cells stain for both PGH-PGE isomerase and PGH synthase (64). Second, in various cells and tissues the activities (measured at saturating substrate levels) of the enzymes which catalyze the isomerization of PGH_2 are usually 5-10 times higher than the PGH synthase activity. This is true for the TxA synthase/PGH synthase pair in platelets (65), for the PGH-PGE isomerase/PGH synthase pair in medullary collecting tubules (37) and for the PGI_2 synthase/PGH synthase pair in aortic smooth muscle (59). The one clear exception is bovine aortic endothelia where there are comparable amounts of PGI_2 synthase and PGH synthase activities (59). Third, significant accumulation of PGH_2 does not occur when microsomes are incubated with arachidonic acid unless these incubations are performed in the presence of inhibitors (e.g. histidine in the case of TxA synthase of platelets (65)) or in the absence of cofactors (e.g. reduced glutathione in the case of sheep vesicular glands (66)). Fourth and finally, there is some evidence that the PGH synthase and PGH_2-metabolizing enzymes are associated with the same membrane systems (33,61,64,67).

Evidence that PGH synthase and PGH_2-metabolizing enzymes are associated with the same membrane systems exists for PGH-PGE isomerase (64), PGI_2 synthase (33) and TxA synthase (66). All of these enzymes are membrane-bound, sedimenting at 6×10^6 g.min. Differential centrifugation techniques indicate that TxA synthase and PGH synthase activities are both present in the dense tubule fraction derived from platelets (67). TxA synthase activity present in platelet membranes is inactivated by treatment of the membranes with Proteinase K suggesting that this enzyme, like PGH synthase, is located on the cytoplasmic surface of these membranes (58). PGI_2 synthase in all smooth muscle cells has been localized on the plasma and nuclear membranes by immunocytofluorescence (33). This pattern of staining is the same as that observed for PGH synthase (when it is detected). Monoclonal antibodies which precipitate glutathione-dependent PGH-PGE isomerase quantitatively coprecipitate PGH synthase and PGH-PGE isomerase activities from intact membrane vesicles prepared from sheep vesicular gland (64). This latter result would be expected only if both proteins were associated with the same membrane systems and, in fact, only if both proteins were present on the cytoplasmic face of the ER. The only situation where the topological relationship between PGH_2 synthesis and PGH_2 isomerization is not clear occurs in the case of PGD synthesis. Enzyme activities catalyzing the PGH-PGD isomerase reaction are soluble proteins apparently located in the cytoplasm of cells (68,69).

3.6 Prostaglandin Release from and Entry into Cells. Early studies on prostaglandin biosynthesis indicated that tissues which were homogenized in organic solvents to effectively block prostaglandin synthesis contained very low levels of these lipids (10). This observation suggests that prostaglandins are not stored by cells. It was later shown that stimulation of cultured cells to synthesize prostaglandins led to the rapid appearance of the products into the culture medium (25). Although prostaglandins exit cells quickly, it is often not appreciated that prostaglandins do not simply diffuse across cell membranes at appreciable rates. For example, exogenous prostaglandins do not enter either red blood cells, which do not synthesize prostaglandins (44), or renal collecting tubule cells, which do synthesize prostaglandins (42). The observations on red blood cells indicate that lipid bilayers are impermeable to prostaglandins. The studies on collecting tubule cells suggest that there is either an antiport or active transport system to extrude prostaglandins synthesized within cells. One point of confusion, is that intact cultured cells readily convert PGH_2 to active prostaglandin derivatives (27). Thus, either PGH_2, unlike other prostaglandins, can enter cells or there is some enzyme on the outer surface of the plasma membrane which will utilize this substrate.

4. SUMMARY

The data currently available on the cellular location of prostaglandin synthetic systems indicate that prostaglandins are formed by some but not all animal cells. These products are rapidly diluted in the circulation indicating that they act at or near their sites of synthesis. Synthesis of prostaglandins appears to take place primarily, but perhaps not exclusively, intracellularly within the confines of the cell membrane. Interestingly, there appear to be multiple sites of prostaglandin synthesis within cells. There are two ways to rationalize multiple sites of synthesis and, in particular, to account for the fact that prostaglandins seem to be produced on both the plasma membrane and the endoplasmic reticulum. The first explanation is that there are multiple sites of action. In support of this possibility are the observations that prostaglandin binding activities are associated with several organelles. The second possibility is that the biosynthetic system present on the plasma membrane uses substrates (i.e. arachidonate and/or PGH_2) derived from neighboring cells while the endoplasmic reticulum system forms prostaglandins only in response to cell surface stimuli. This possibility is consistent with the efficient metabolism of exogenous arachidonic acid and PGH_2 by intact cells.

There are several questions which are of obvious importance with regard to the location of prostaglandin synthesis and function. First, what is the location of prostaglandin receptors. This question can be answered by immunocytochemistry once antibodies to the receptor proteins are prepared. Second, what are the topological relationships between stimuli of arachidonate release, arachidonate release itself and arachidonate oxygenation. One possible way to begin addressing this question would be to fuse two cells which form different prostaglandin products in response to different stimuli and then to determine which product(s) is formed by the hybrid cell in response to different stimuli. A final question is, on what side of the plasma membrane are PGH synthase and/or PGH_2 metabolizing enzymes located? This can be addressed by determining if antibodies to these enzymes bind to intact cells and if metabolism of exogenous PGH_2 is altered by treatment of cells with nonpenetrating proteases.

Acknowledgements: Work performed in the authors laboratory was supported by USPHS Grants HD10013, AM22042 and HL29685. Sherry L. Huey and Arlyn Garcia-Perez provided thoughtful comments on this manuscript.

REFERENCES

1. Ferreira SH, Vane JR: Prostaglandins: Their disappearance from and release into the circulation. Nature (216):868-873, 1967.
2. Anggard E, Samuelsson B: Prostaglandins and related factors: 28. Metabolism of PGE_1 in guinea pig lung: the structures of two metabolites. J Biol Chem (239):4097-4102, 1964.
3. Shimizu T, Watanabe K, Tokumoto H, Hayaishi O: Isolation of $NADP^+$-dependent PGD_2-specific 15-hydroxyprostaglandin dehydrogenase from swine brain. Meth Enz (86):147-152, 1982.
4. Korff JM, Jarabek J: Isolation and properties of an $NADP^+$-dependent PGI_2-specific 15-hydroxyprostaglandin dehydrogenase from rabbit kidney. Meth Enz (86):152-155, 1982.
5. Oates JA, Roberts LJ, Sweetman BJ, Maas RL, Gerkens JF, Taber DF: Metabolism of the prostaglandins and thromboxanes. In Samuelsson B, Ramwell PW, Paoletti R (eds) Advances in prostaglandin and thromboxane research. Vol. 6, Raven Press, New York, 1980, pp 35-41.
6. Granstrom E, Samuelsson B: Radioimmunoassays for prostaglandins. In Frohlich JC (ed) Advances in prostaglandin and thromboxane research. Vol. 5, Raven Press, New York, 1978, pp 1-53.
7. Dunn MJ, Liard JF, Dray F: Basal and stimulated rates of renal secretion and excretion of prostaglandins E_2, $F_{2\alpha}$ and 13,14-dihydro-15-keto F_α in the dog. Kidney Int (13):136-143, 1978.
8. Christ-Hazelhof E, Nugteren DH: Prostacyclin is not a circulating hormone. Prostaglandins (22):739-746, 1981.
9. Fitzgerald GW, Brash AR, Falardeau P, Oates JA: Estimated rate of prostacyclin secretion in the circulation of normal man. J Clin Invest (68):1272-1276, 1981.
10. Jouvenaz GH, Nugteren DH, Beerthuis RK, Van Dorp DA: A sensitive method for the determination of prostaglandins by gas chromatography with electron-capture detection. Biochim Biophys Acta (202):231-234, 1970.
11. Van Dorp DA: Recent developments in the biosynthesis and the analyses of prostaglandins. Ann NY Acad Sci (180):181-199, 1971.
12. Smith WL: Localization of enzymes responsible for prostaglandin formation. In Willis AL, Vickerey BH, Pace-Asciak C (eds) CRC handbook of prostaglandins and related lipids. CRC Press, Boca Raton, Florida, in press.

13. Roberts LJ, Sweetman BJ, Lewis RA, Austen KF, Oates JA: Increased production of prostaglandin D_2 in patients with systemic mastocytosis. N Eng J Med (303):1400-1406, 1980.
14. Hubbard WC, Hough AJ, Johnson RM, Oates JA: The site of VX_2 tumor transplantation affects the development of hypercalcemia in rabbits. Prostaglandins (19):881-889, 1980.
15. Gerber JG, Payne NA, Murphy RC, Nies AS: Prostacyclin produced by the pregnant uterus in the dog may act as a circulating vasodepressor substance. J Clin Invest (67):632-636, 1981.
16. Gorman RR, Fitzpatrick FA, Miller OV: Reciprocal regulation of human platelet cAMP levels by thromboxane A_2 and prostacyclin. In George WJ, Ignarro LJ (eds) Advances in cyclic nucleotide research. Vol. 9, Raven Press, New York, 1978, pp 597-609.
17. Garcia-Perez A, Smith WL: Use of monoclonal antibodies to isolate cortical collecting tubule cells: AVP induces PGE release. Am J Physiol (244):C211-C220, 1983.
18. Torikai S, Kurokawa K: Distribution of prostaglandin E_2-sensitive adenylate cyclase along the rat nephron. Prostaglandins (21):427-438, 1981.
19. Stokes JB, Kokko JP: Inhibition of sodium transport by prostaglandin E_2 across the isolated perfused rabbit collecting tubule. J Clin Invest (59):1099-1104, 1977.
20. Kennedy MS, Stobo JD, Goldyne ME: In vitro synthesis of prostaglandins and related lipids by populations of human peripheral blood mononuclear cells. Prostaglandins (20):135-145, 1980.
21. Smith WL, Wilkin GP: Immunochemistry of prostaglandin endoperoxide-forming cyclooxygenases: The detection of cyclooxygenases in rat, rabbit and guinea pig kidneys by immunofluorescence. Prostaglandins (13):873-892, 1977.
22. Smith WL, Bell TG: Immunohistochemical localization of the prostaglandin-forming cyclooxygenase in renal cortex. Am J Physiol (235):F451-F457, 1978.
23. Currie MG, Needleman P: Distribution of prostaglandin synthesis along the rabbit nephron. Fed Proc (41):1545, 1982.
24. Hassid A: Transport-active renal tubular epithelial cells (MDCK and LLC-PK_1) in culture. Prostaglandin biosynthesis and its regulation by peptide hormones and ionophore. Prostaglandins (21):985-992, 1981.

25. Levine L, Hinkle PM, Voelkel EF, Tashjian AH: Prostaglandin production by mouse fibrosarcoma cells in culture: inhibition by indomethacin and aspirin. Biochem Biophys Res Comm (47):888-896, 1972.

26. Skidgel RA, Printz MP: PGI_2 production by rat blood vessels: diminished prostacyclin formation in veins compared to arteries. Prostaglandins (16):1-16, 1978.

27. Ingerman-Wojenski C, Silver MJ, Smith JB, Macarek E: Bovine endothelial cells in culture produce thromboxane as well as prostacyclin. J Clin Invest (67):1292-1296, 1981.

28. Hong SL: Effect of bradykinin and thrombin on prostacyclin synthesis in endothelial cells from calf and pig aorta and human umbilical vein. Thrombosis Res (18):787-795, 1980.

29. Weksler BB, Marcus AJ, Jaffe EA: Synthesis of prostaglandin I_2 (prostacyclin) by cultured human and bovine endothelial cells. Proc Nat Acad Sci (74):3922-3926, 1977.

30. Gerritsen ME, Printz MP: Prostaglandin D synthase in microvessels from the rat cerebral cortex. Prostaglandins (22):553-566, 1981.

31. Gecse A, Ohlecz A, Mezei Z, Telegdy G, Joo F, Dux E, Karnushina I: Prostacyclin and prostaglandin synthesis in isolated brain capillaries. Prostaglandins (23):287-297, 1982.

32. Birkle DL, Wright KF, Ellis CK, Ellis EF: Prostaglandin levels in isolated brain microvessels and in normal and norepinephrine - stimulated cat brain homogenates. Prostaglandins (21):865-872, 1981.

33. Smith WL, DeWitt DL, Allen ML: Bimodal distribution of the prostaglandin I_2 synthase antigen in smooth muscle cells. J Biol Chem (258):5922-5926, 1983.

34. Ahmada G, Sobel BE, Needleman P: Synthesis of prostaglandins by cultured myocytes and cardiac fibroblasts. J Mol Cell Cardiol (12):685-700, 1980.

35. Needleman P, Smith WL: unpublished observations.

36. Pace-Asiak C, Smith WL: Enzymes in the biosynthesis and catabolism of the eicosanoids: prostaglandins, thromboxanes, leukotrienes and hydroxy fatty acids. In Boyer PD (ed) The enzymes. Vol. 16, Academic Press, New York, 1983, pp 544-603.

37. Grenier FC, Rollins TE, Smith WL: Kinin-induced prostaglandin synthesis by renal papillary collecting tubule cells in culture. Am J Physiol (241):F94-F104, 1981.

38. Siegel AM: Receptors for PGI_2 and PGD_2 on human platelets. Meth Enz (86):179-192, 1982.
39. Hammarstrom S: A receptor for prostaglandin $F_{2\alpha}$ from corpora lutea. Meth Enz (86):202-209, 1982.
40. Smigel M, Fleischer S: Characterization and localization of prostaglandin E_1 receptors in rat liver plasma membranes. Biochim Biophys Acta (332):358-373, 1974.
41. Sinha AK, Colman RW: Persistence of increased platelet cAMP induced by prostaglandin E_1 after removal of the hormone. Proc Nat Acad Sci (77):2946-2950, 1980.
42. Garcia-Perez A, Smith WL: Apical-basolateral membrane asymmetry in canine cortical collecting tubule cells: bradykinin, AVP, PGE_2 interrelationships. J Clin Invest, in press.
43. Rao ChV, Mitra SB: Distribution of PGE and $PGF_{2\alpha}$ receptor proteins in the intracellular organelles of bovine corpora lutea. Meth Enz (86):192-202, 1982.
44. Bito LZ, Baroody RA: Impermeability of rabbit erythrocytes to prostaglandins. Am J Physiol (229):1580-1584, 1974.
45. Bills TK, Smith JB, Silver MJ: Metabolism of [^{14}C]arachidonic acid by human platelets. Biochim Biophys Acta (424):303-314, 1976.
46. Bills TK, Smith JB, Silver MJ: Selective release of arachidonic acid from the phospholipids of human platelets in response to thrombin. J Clin Invest (60):1-6, 1977.
47. Hsueh W, Isakson PC, Needleman P: Hormone selective lipase activation in the isolated rabbit heart. Prostaglandins (13):1073-1091, 1977.
48. Rittenhouse-Simmons S, Deykin D: In Gordon JL (ed) Platelets in biology and pathology. Vol. 2, Elsevier/North-Holland, Amsterdam, 1981, pp 349-371.
49. Neufeld E, Majerus PW: Arachidonate release and phosphatidic acid turnover in stimulated human platelets. J Biol Chem (258):2461-2467, 1983.
50. DePierre JW, Ernster L: Enzyme topology of intracellular membranes. Ann Rev Biochem (46):201-262, 1977.
51. Rittenhouse SE: Preparation of selectively labeled phosphatidylinositol and assay of phosphatidylinositol - specific phospholipase C. Meth Enz (86):3-11, 1982.

52. Majerus PW, Prescott SM: Characterization and assay of diacylglycerol lipase from human platelets. Meth Enz (86):11-17, 1982.

53. Dennis EA: Phospholipases. In Boyer PD (ed) The enzymes. Vol. 16, Academic Press, New York, 1983, pp 308-356.

54. Coleman RA, Bell RM: Topography of membrane-bound enzymes that metabolize complex lipids. In Boyer PD (ed) The enzymes. Vol. 16, Academic Press, New York, 1983, pp 605-625.

55. Freidman Y, Lang M, Burke G: Further characterization of bovine thyroid prostaglandin synthase. Biochim Biophys Acta (397):331-341, 1975.

56. Anggard E, Bohman SO, Griffin JE, Larsson C, Maunsbach AB: Subcellular localization of the prostaglandin system in the rabbit renal papilla. Acta Physiol Scand (84):231-243, 1972.

57. Rollins TE, Smith WL: Subcellular localization of prostaglandin-forming cyclooxygenase in swiss mouse 3T3 cells by electron microscopic immunocytochemistry. J Biol Chem (255):4872-4875, 1980.

58. Smith WL, Rollins TE, DeWitt DL: Subcellular localization of prostaglandin forming enzymes using conventional and monoclonal antibodies. In Holman RT (ed) Progress in lipid research. Vol. 20, Pergamon Press, Oxford, 1982, pp 103-110.

59. DeWitt DL, Day JS, Sonnenburg WK, Smith WL: Concentrations of prostaglandin endoperoxide synthase and prostaglandin I_2 synthase in the endothelium and smooth muscle of bovine aorta. J Clin Invest (72):1882-1888, 1983.

60. DeWitt DL, Rollins TE, Day JS, Gauger JA, Smith WL: Orientation of the active site and antigenic determinants of prostaglandin endoperoxide synthase in the endoplasmic reticulum. J Biol Chem (256):10375-10382, 1981.

61. Sonnenburg WK, Smith WL: unpublished observations.

62. Marcus AJ, Weksler BB, Jaffe EA, Broekman MJ: Synthesis of prostacyclin from platelet-derived endoperoxides by cultured human endothelial cells. J Clin Invest (66):979-986, 1980.

63. Smith WL: unpublished observations.

64. Tanaka Y, Smith WL: unpublished observations.

65. Nugteren DH, Buytenhek M, Christ-Hazelhof E, Moonen P, van der Ouderaa FJ: Enzymes involved in the conversion of endoperoxides. In Holman RT (ed) Progress in lipid research. Vol. 20, Pergaman Press, Oxford, 1982, pp 169-172.

66. Graff G: Preparation of PGG_2 and PGH_2. Meth Enz (86):376-386, 1982.

67. Gerrard JM, White JG: Prostaglandins and thromboxanes: "middlemen" modulating platelet function in hemostasis and thrombosis. In Spaet TH (ed) Progress in hemostasis and thrombosis. Vol. 4, Greene and Stratton, New York, 1978, pp 87-125.

68. Shimizu T, Yamamoto S, Hayaishi O: Purification of PGH-PGD isomerase from rat brain. Meth Enz (86):73-77, 1982.

69. Christ-Hazelhof E, Nugteren DH: Isolation of PGH-PGD isomerase from rat spleen. Meth Enz (86):77-84, 1982.

6

ROLE OF ACTIVE OXYGEN IN THE FORMATION OF LIPID HYDROPEROXIDES

PAUL J. MARSHALL

1. INTRODUCTION

Interest in the role of dietary fat on tumorgenesis has been stimulated by the epidemiological data showing a high degree of positive correlation between the incidence of breast and colon carcinomas and the per capita consumption of fat (1,2). There appears to be some distinction between the effects of saturated and unsaturated fats. Carroll and his group demonstrated that polyunsaturated fats were more effective than saturated fats in promoting the ability of 7,12-dimethylbenz(a)anthracene to produce mammary tumors (3,4). Other groups have found that polyunsaturated fats promoted growth and proliferation of transplantable tumors (5,6). McCay and his group have demonstrated that diets containing significant levels (20% caloric) of polyunsaturated fats facilitated tumor formation (7). Although polyunsaturated fats appear to participate in the promotional stage of carcinogenesis as evidenced by their effectiveness when administered several weeks after 7,12-dimethylbenz(a)anthracene (7), the precise mechanism by which polyunsaturated fats enhance tumor growth is unknown.

Polyunsaturated fats serve as substrates for both enzymatic and nonenzymatic peroxidative attack, and the resulting lipid peroxides have been implicated in the pathophysiology of a number of disease processes including carcinogenesis. The mechanism by which lipid peroxides participate in disease is unclear. Some investigators have suggested that the pathological consequences of lipid peroxidation may be reflected in the loss of membrane integrity and function (8). For example, loss of lysosomal membrane would result in the liberation of hydrolytic enzymes (9) which, in turn, would lead to digestion and further cellular damage. Others have pointed to the reactivity of the hydroperoxides at initiating destructive free radical-mediated reactions. With respect to the development of carcinomas, Marnett and his coworkers have suggested that

W.E.M. Lands (ed.), BIOCHEMISTRY OF ARACHIDONIC ACID METABOLISM.

lipid hydroperoxides could indirectly be involved in tumor initiation by the activation of procarcinogens (10,11). They have shown that benz(a)anthracene, benzo(a)pyrene, and benzidine can be oxygenated in a hydroperoxy fatty acid-dependent cooxidation reaction. In addition, decomposition of lipid peroxides can produce a conjugated dialdehyde, malondialdehyde, that has the potential to crosslink proteins and nucleic acids. Shambeiger et al. (12) reported that malondialdehyde was carcinogenic in mice. Marnett and Tuttles (13) also suggested that the malondialdehyde, which was generated as a result of prostaglandin synthesis, directly contributed to the initiation of carcinogenesis.

Peroxidized lipids serve as an intercellular signal (14) and enhance the synthesis of other mediators such as prostaglandins (15,16). Prostaglandins, in particular prostaglandin E, are elevated in both experimental and human tumors (17-19), and may act in promoting tumorgenesis directly by serving as growth factors for cell proliferation or indirectly by suppressing immunosurveillance and cytotoxicity (19,20). Lipid hydroperoxides serve as activators for PGH synthase and increase the rate at which prostaglandins are synthesized (15,16). Thus, the presence of lipid peroxides could be a cause for the increased quantities of prostaglandins made by tumors.

Due to the correlation of lipid peroxidation with a number of disease states (21,22), interest in the mechanism of lipid peroxidation has heightened in the past ten years (23,24). Lipid hydroperoxides can be stimulated by both nonenzymatic mechanisms as well as enzymatic events. With respect to the latter, lipid hydroperoxides are direct products of cyclooxygenase (see Chapter 1) and lipoxygenase (see Chapter 2) and have been implicated as products of microsomal cytochrome P450 (25,26). A number of indirect enzymatic events are known to promote lipid peroxidation by the generation of reactive intermediates. Active oxygen metabolites are produced by xanthine oxidase (27,28), an enzyme associated with cell injury (29), and by NADPH oxidase and myeloperoxidase, enzymes associated with activated phagocytes (30). Other indirect enzymatic events include the cyclic reduction and oxidation of certain xenobiotics (31) and anthraquinone antibiotics (32,33) by microsomal NADPH-cytochrome P450 reductase. Nonenzymatic lipid peroxidation can occur as a result of the administration of toxic agents that diminish the level of cellular glutathione (34,35), or in the presence of peroxides such as hydrogen peroxide and cumene peroxide (36). These various processes that give rise to hydroperoxides share common features in their mechanism of peroxidation. The focus of this review will be on two such features: the role of activated oxygen and iron in the initiation of lipid peroxidation.

2. DETECTING HYDROPEROXIDES AND INTERMEDIATES

2.1 Lipid hydroperoxide assays

As with all biochemical investigations, careful consideration of the analytical methods is important in determining the value of the results. An overview of the methods used to detect lipid peroxides will be offered before a consideration of the mechanisms of lipid peroxidation is presented. Many techniques exist for measuring lipid peroxidation and lipid peroxides, and a summary is presented in Table 1. Most of the present techniques have important limitations that restrict the usefulness of the resulting data and conclusions. Some investigators have recognized these limitations and have confirmed their results by employing several different techniques.

2.1.1 Indirect assays

The thiobarbituric acid assay (37,38) is the most widely employed technique to measure lipid peroxidation because of its facility. This assay has been modified several times in recent years (39,40) and has been the subject for two reviews (41,42). The assay measures the reaction product of thiobarbituric acid with a conjugated dialdehyde such as malondialdehyde. The malondialdehyde-like substance is a secondary product that is formed upon the decomposition of peroxidized fatty acids, and is generated from peroxidized lipid with a 1-10% efficiency (43). Because of this recovery problem, many investigators report only the absorbance of the chromophore product and fail to determine the recovery efficiency for their system. Care must be exercised during the processing of the sample especially when iron is present in order to prevent inadvertent peroxidation (40,44). Furthermore, a number of compounds (e.g. sugars, acetaldehydes, and alcohols) can react to give a chromophore. For example, in studies with xanthine oxidase and acetaldehyde, acetaldehyde was found to interfere with the thiobarbituric acid assay (41). Because the conversion efficiency of lipid

Table 1: Techniques for Detecting Hydroperoxides

Technique	Sensitivity	Limitations	Reference
Indirect Assays			
Thiobarbituric Acid Assay	10 nmol	Poor and variable recoveries; reacts with sugars, aldehydes and alcohols.	37,38
Diene conjugation	1 µM	Detects hydroxy lipids and other conjugated compounds.	41
Volatile hydrocarbons	pmole	Cannot measure steady state level.	45-46
Direct Assays			
Iodometric	1 µM		41,49
Radiotracer	<pmol	Labelled fatty acid must distribute evenly in lipid pool.	
Hydroperoxide oxidation of dichlorofluorescin	25 pmol	Must be performed anaerobically; antioxidants and hydroxyl radical scavengers interfere.	50
Activation of Cyclooxygenase	5 pmol	Protein (>1 mg/ml) interferes.	51

hydroperoxide is poor and variable, it is difficult to compare the extent of lipid peroxidation occuring between tissues. Recovery efficiencies must be evaluated for each sample, and the rates of malondialdehyde formation should be determined to obtain an assessment of peroxidation. Furthermore, α-tocopherol should be used to inhibit unwanted peroxidation (43). Addition of α-tocopherol after the incubation will help to distinguish whether the malondialdehyde-like substance was produced during the incubation or arose as a result of sample handling.

In addition to malondialdehyde formation, other assays have made use of other secondary products formed or secondary processes occurring as a result of lipid peroxidation. Measurement of ethane and pentane (45,46) has been employed to assess lipid peroxidation. These gases are a sensitive index of lipid hydroperoxide metabolism. Unfortunately, measurment of gases gives only a general index of the rate of peroxidation during an incubation period; it cannot measure the amount of peroxides present at a particular time in a sample. Chemiluminescence (47) and fluorescence (48) are exceptionally sensitive techniques, but they measure secondary reaction processes and, like the measurement of gases, cannot be used for steady state measurements of peroxide content. Because both the measurements of gases and light emission are restricted in the sample handling and processing, they are not amenable to many experimental conditions.

The measurement of absorbance at 232 nm detects the cis, trans conjugated diene resulting from peroxide formation (41). The technique (with an extinction coefficient of 22,000) can measure steady state levels of lipid dienes in the range of approximately 10^{-6} M. This type of assay, however, cannot differentiate between a peroxy and hydroxy conjugated lipid and can lead to an overestimation of the amount of peroxide present. Absorbance measurement also suffers from interferences from a number of conjugated and aromatic compounds (e.g. xanthine, urate, and nucleotides).

2.1.2 Direct assays

Assays for the direct measurement of lipid hydroperoxides are available but have drawbacks limiting their utility. Iodometric measurements provide a direct assay for lipid peroxides (49), but this method lacks sensitivity, being able to measure concentrations of hydroperoxide above approximately 1 μM (41). Radiochemical techniques combined with thin layer chromatography provide a way to measure peroxidation of radiolabeled precursors, but requires a preincubation period to incorporate the radiolabel into the lipid pool.

Two recently-developed assays are promising in that they avoid many of the pitfalls of the other assays and offer a high degree of sensitivity and specificity in measuring lipid hydroperoxides. Cathcart et al. (50) developed a fluorescent assay for lipid hydroperoxides based on the stoichiometric oxidation of dichlorofluorescin with hematin as catalyst. Under optimal conditions, the assay was reported to detect as little as 25 pmoles of hydroperoxide. Although an attractive alternative to other assays, there are important limitations to the method. To diminish the possibility of peroxidation during the 50 min incubation, the reaction with dichlorofluorescin is performed anaerobically.

Antioxidants present in micromolar quantities, as well as many alcohols, can quench the reaction by scavenging the reaction intermediates (hydroxyl radical and iron-oxo complexes) involved in the oxidation of the fluorescent agent. In spite of these problems, this method may prove promising especially as a post-column derivatization technique for high pressure liquid chromatography.

Our group has developed a direct enzymic assay to quantitate lipid hydroperoxides (51). The assay makes use of the activation of cyclooxygenase activity by hydroperoxides, and can detect as little as 5 pmoles of lipid hydroperoxides in a three milliliter assay volume. Biological samples can be assayed directly and rapidily. The only limitation found so far has been some interference by high concentrations of proteins that bind fatty acids. Bovine serum albumin, for example, was found to inhibit the assay when present at a concentration of 1 mg/ml.

2.2 Radical intermediates

The use of Electron Paramagnetic Resonance (EPR) in conjunction with "spin trap" agents has been used to identify intermediate radicals involved in lipid peroxidation. This is an extremely sensitive technique and has been useful in demonstrating the presence of transitory radical intermediates such as hydroxyl radical (52,53). However, the technique depends upon the assumption that the interaction between the spin trap agent and the radical is specific. Spin trap agents can enter into several side reactions in biological systems or in other complex reaction mixtures (23,54). Also, although signifying the presence of the intermediate, the detection of an EPR signal does not necessarily mean that the identified intermediate participates in lipid peroxidation. The intermediate may be generated coincidently with the occurence of lipid peroxidation. The signal generated from the putative spin trap product of hydroxyl radical with 5,5-dimethyl-1-pyrroline-1-oxide (DMPO) could have arisen from the oxidation of the spin trap by ferric ion, the breakdown of the DMPO-superoxide spin adduct, or the lipid hydroperoxy radical adduct in the presence of ethyldiaminetetraacetate (EDTA)-chelated ferrous ion (54).

2.3 Scavengers of reactive intermediates

The use of inhibitors has been an important tool in identifying intermediates and the mechanism involved in lipid peroxidation. However, caution must be exercised in extrapolating results obtained with the use of inhibitors from well-defined systems to more physiological systems. Inhibitors of lipid peroxidation can be divided into two categories: compounds which react with certain intermediates (e.g. mannitol reacts with hydroxyl radicals or diphenylfuran reacts with singlet oxygen) and enzymes that reduce "active oxygen" to more stable species (e.g. superoxide dismutase which reduces superoxide to hydrogen peroxide or catalase which reduces hydrogen peroxide to water).

As with spin trap agents, the specificity of inhibitors (especially with respect to scavenging compounds) is merely assumed in many reports. Most scavenger compounds, however, can react with more than one radical species. Thus, inhibition of a free radical-induced effect does not prove the involvement of only one species in the process. Conversely, the lack of protective effect by scavenger does not necessarily prove that the radical of interest was not involved in peroxidation. A secondary radical may result from the reaction between the scavenger compound and the primary radical. This secondary radical may then initiate peroxidation (55).

Concern should also be exercised over the interpretation of results when using scavenger enzymes. For example, the inhibition of peroxidation upon the addition of catalase may be due to the antioxidant thymol which is added as a stabilizing agent in commercial catalase preparations (23). Also, the effectiveness of superoxide dismutase in inhibiting peroxidation may not necessarily be due to the direct dismutation of superoxide. Superoxide dismutase may recognize an active oxygen complex such as ADP-Fe^{+3}-O_2^- and subsequently reduce the oxygen in this complex. In addition, if the proximate radical complex involved in initiating lipid peroxidation were to dissociate or break down before reacting with lipid, the resulting products might be scavenged by added agents, and the active radical complex would be prevented from reforming. Since the end result would be the inhibition of peroxidation, it would appear that the scavenged product, rather than the complex, was directly involved in peroxidation.

Another problem with the interpretation of results from experiments using inhibitors is the assumption that the inhibitor has access to the site of radical generation and/or lipid peroxidation. This assumption may not be warranted especially in the case where enzymes are used as scavengers These enzymes are often hydrophilic and thus restricted to the cytosolic or aqueous phase of the incubation. Therefore, peroxidation occurring in a hydrophobic environment, such as in membranes or micelles, would be inaccessible to these enzymes (56). Soluble enzymes would only be effective when either the generation of the radical intermediates or the peroxidation were occurring in hydrophilic compartment.

The mechanism of peroxidation can be affected by the buffer in which the peroxidation is performed. Tris buffers can have a complex action on peroxidation. On the one hand, Tris buffers being polyhydroxylated buffer can scavenge hydroxy radicals and thereby diminish the likelihood that peroxidation will be initiated by hydroxyl radicals (44). Tris buffers can also effect the chelation of iron complexes and thus their redox potentials (57). Phosphate buffers may also alter the chelation state as well as the redox potential of iron. Some investigators have avoided these pitfalls by either performing their experiments or confirming their results in isotonic saline.

In addition to buffers, the presence of other additives can influence the mechanism of peroxidation. The initiation by both hydroxyl radicals and singlet oxygen is especially vulnerable to quenching by organic compounds and, in particular, solvents used as carriers or as lipid-dispersing agents. These initiating radicals are highly reactive with rates of reactions approaching those for diffusion limited reactions. For singlet oxygen to initiate peroxidation, it must be generated in close proximity to the lipid; otherwise, it will react readily with whatever it collides. The use of xanthine, the substrate for xanthine oxidase as well as its product from the reacation, urate, can scavenge hydroxyl radical, and thereby block the initiation process by hydroxyl radicals (44). Other additives, such as dimethoxyethane, ethanol or dimethylsulfoxide, can reduce the contribution of hydroxyl radicals in initiating lipid peroxidation.

3. MECHANISM OF PEROXIDATION

Hydroperoxides can be formed directly from polyunsaturated fatty acids by an enzymatic oxygenation process or nonenzymatically as a consequence of oxidative free radical attack. The latter event may be important in initiating both events since all fatty acid oxygenases studied in detail require a lipid peroxide activator (see Chapter 1 for cyclooxygenase and Chapter 2 for lipoxygenase). These enzymes can then amplify catalytically a small initial amount of activating hydroperoxide by forming larger amounts of specific hydroperoxide products.

The nonenzymatic peroxidation of lipid is generally regarded to be initiated by attack of a free radical (X$^\bullet$) on polyunsaturated fatty acid (Figure 1). This attack results in the abstraction of the allylic hydrogen. The fatty acid radical formed subsequently rearranges to a cis,trans-diene concomitant with the incorporation of oxygen to yield a hydroperoxy radical (23,58). The abstraction of an allylic hydrogen is perferred over other methylene hydrogens due to a lower bond energy (86 vs. 96 kcal; respectively) and the resonance stabilization of the radical intermediate. The newly formed hydroperoxy radical can itself abstract hydrogen from another fatty acid propagating the reaction. It has been estimated that such a reaction sequence goes through 8 to 14 propagation cycles before being terminated (59).

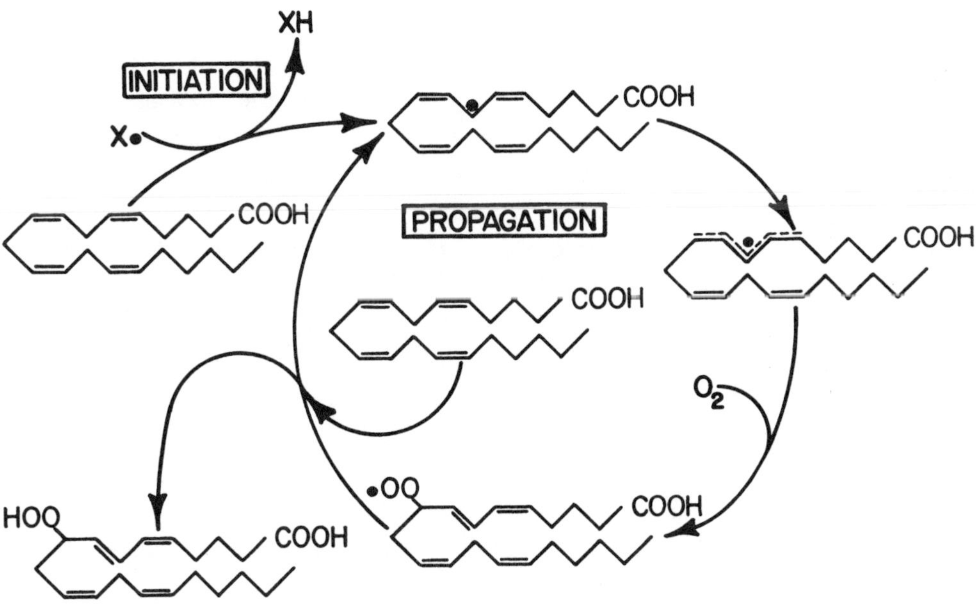

Figure 1. Nonenzymatic formation of lipid hydroperoxides.

Even though the allylic hydrogens of the polyunsaturated fatty acid can be considered to be prone to reaction, a strong oxidizing agent ($X^{\cdot}$) is still required for their abstraction. Such agents can arise by several mechanisms (e.g. radiation, drug metabolism); but, in most biological situations, the production of free radicals or factors possessing properties of free radicals occurs during one electron transfers involving the reduction of oxygen (Table 2).

Table 2: Proposed Factors in the Generation of the Initiating Factors ($X\cdot$) for Lipid Peroxidation

Initiation	Source
$OH\cdot$	Fe^{+2}, H_2O_2
$^1O_2$	$O_2^{\dot{-}}$, ROOH, EDTA-Fe
Fe_2-O (ferryl ion)	ADP-Fe^{+2}-O_2, ADP-Fe^{+3}
$RO\cdot$	ROOH, chelated-Fe^{+2}

There is considerable controversy as to the mechanism of initiation of lipid peroxidation by partially reduced or "activated" oxygen. The three initiation processes most often discussed are: (1) the Haber-Weiss, Fenton-catalysed reaction giving rise to the singlet oxygen or hydroxyl radical; (2) an active oxygen-iron complex (perferryl or ferryl iron), and (3) iron-mediated homolytic cleavage of the hydroperoxide O-O bond. All of these mechanisms share the requirement for a transition metal capable of undergoing a univalent reduction.

The Haber-Weiss reaction is often invoked as the mechanism whereby the generation of hydroxyl radical or singlet oxygen that can subsequently initiate peroxidation.

$$O_2^{\bar{}} + H_2O_2 \rightarrow OH^- + {}^1O_2 + OH^{\cdot} \qquad \text{Equation 1.}$$

There are several reasons for implicating this reaction. First, the lipid peroxidation has been associated with the appearance of superoxide ($O_2^{\bar{}}$) and/or hydrogen peroxide (H_2O_2) (22,27). Secondly, scavenger enzymes that catabolize these reduced oxygen species (such as superoxide dismutase and catalase) prevent lipid peroxidation in some cases. Superoxide results from the one electron reduction of oxygen. This reduction can be catalysed by a number of flavin-containing enzymes such as NADPH oxidase and xanthine oxidase (27) as well as by the mitochrondrial electron transport chain (22). Hydrogen peroxide can result from the spontaneous dismutation of superoxide, which occurs at the rate of 10^7 M/sec at acidic pH, or it can be generated enzymatically from superoxide dismutase, xanthine oxidase, glucose oxidase as well as enzymes associated with peroxisomes. Superoxide and hydrogen peroxide are relatively stable reduced oxygen species. Lipid membranes are readily permeable to hydrogen peroxide (60) and superoxide (as its conjugate acid, $HO_2^{\cdot}$) (27). Neither superoxide nor hydrogen peroxide is thought to directly cause lipid peroxidation, this requiring a more reactive oxygen species such as hydroxyl radical($OH^{\cdot}$) or singlet oxygen ($^1O_2$) resulting from the Haber-Weiss reaction.

Although the Haber-Weiss reaction is thermodynamically exothermic, a transition metal (Fenton reagent) is needed for catalysis (59); otherwise, the reaction is too slow to account for the formation of the oxidizing agents (61). Iron is most often considered in this regard because it is widely available in biological system. The general scheme for the Fenton reagent production of a hydroxyl radical is shown below:

$$O_2^- + M^n \longrightarrow O_2 + M^{(n-1)} \qquad \text{Equation 2.}$$

$$H_2O_2 + M^{(n-1)^2} \longrightarrow OH^{\cdot} + OH^- + M^n \qquad \text{Equation 3.}$$

It should be noted that the need for superoxide as a reductant in Equation 2 can be circumvented by direct reduction of the metal.

3.1 Singlet oxygen

Kellogg and Fridovich investigated the nature of the initiating factors in lipid peroxidation using acetaldehyde and xanthine oxidase to generate both superoxide and hydrogen peroxide (44). The peroxidation of linolenate was measured by the increased absorbance at 233 nm (due to diene conjugation) and by the analysis of lipid peroxides with thin layer chromatography. Acetaldehyde was used as the substrate for xanthine oxidase because xanthine and its product urate absorb in the UV and can act as scavengers of free radicals. Superoxide dismutase or catalase inhibited lipid peroxidation in this system as did the scavengers of singlet oxygen, 2,5-dimethylfuran (1 mM) and β-carotene (10 μM). However, hydroxyl radical scavengers such as mannitol

(0.1 M) or t-butylalcohol (0.1 M) did not prevent lipid peroxidation. Furthermore, when 2,5-dimethylfuran was exposed to a known source of singlet oxygen, it produced a characteristic product. This product was also observed when 2,5-dimethylfuran was exposed to the xanthine oxidase system, and its accumulation was prevented by superoxide dismutase or catalase but not by scavengers of hydroxyl radicals. The lack of effect of hydroxyl radical scavengers in these experiments may have been due to the presence of 10% dimethoxyethane which itself may scavenge hydroxyl radicals.

Kellogg and Fridovich extended their earlier studies by using liposomes and washed human erythrocytes as targets for peroxidation (62). In these systems, xanthine oxidase caused peroxidation and hemolysis. The presence of superoxide dismutase or catalase in the incubation buffer protected the liposomes from peroxidation and the erythrocytes from lysis. Scavengers of singlet oxygen such as histidine and 2,5-dimethylfuran also protected. In contrast with the result with linolenate, mannitol (10 mM) did provide protection for both the liposomes and erythrocytes against radical attack. Kellogg and Fridovich proposed that superoxide and hydrogen peroxide cooperate in producing both singlet oxygen and hydroxyl radical which were the proximate causes for lipid peroxidation and hemolysis.

Dixit et al. (63) also suggested that singlet oxygen was responsible for the NADPH-driven lipid peroxidation in epidermal microsomes. Scavengers of singlet oxygen such as histidine (5 mM), 2,5-dimethylfuran (5 mM), and β-carotene (10 μM) decreased the amount of malondialdehyde formed by 60% whereas scavengers of hydroxyl radicals such as mannitol (5 mM), benzoate (5 mM), and ethanol (5 mM) were ineffective. Catalase decreased the formation of malondialdehyde, but superoxide dismutase did not. In this experiment the endogenous transition metal was most likely reduced by NADPH, circumventing the need for superoxide. Thus, it would appear that only hydrogen peroxide was needed to initiate peroxidation. However, the mechanism for the formation of the proposed singlet oxygen remains obscure. As an alternative explanation of these results, the observed microsomal peroxidation could have been due to initiation by preexisting lipid hydroperoxides. Other reports have suggested that singlet oxygen is involved in the hydroperoxide-dependent peroxidation catalysed by reduced iron (see section 3.4.).

King et al. (64) provide evidence for the formation of singlet oxygen during the NADPH-driven peroxidation of liver microsomes. Their evidence was based on the enzyme-dependent formation of dibenzoylethylene from diphenylfuran, a reaction which is regarded to be specific for singlet oxygen. Both lipid peroxidation and the formation of singlet oxygen were related to the amount of ferric iron present. Singlet oxygen may be responsible for part of the NADPH-driven lipid peroxidation because diphenylfuran (1 mM) inhibited membrane peroxidation by 50%. However, xanthine oxidase could not be associated with singlet oxygen generation since the product of diphenylfuran with singlet oxygen was not detected. Furthermore, these authors acknowledged that, although singlet oxygen can initiate lipid peroxidation, the formation of singlet oxygen could have occurred as a result of the degradation of preexisting lipid hydroperoxides and thus be involved in the propagation of peroxidation.

Svingen et al. (65) found that diphenylfuran (0.2 mM) did not prevent lipid peroxidation induced by xanthine oxidase and ADP-ferric iron (17 mM; 0.1 mM) when measured iodometrically or by malondialdehyde formation. This lack of inhibition was also witnessed when NADPH-cytochrome P450 reductase was used in place of xanthine oxidase. Thus, singlet oxygen did not appear to be involved in the initiation of peroxidation. However, in both cases superoxide dismutase inhibited lipid peroxidation, thus suggesting the peroxidation occurred through a superoxide dismutase-sensitive reaction.

3.2 Hydroxyl radical

As indicated above, the hydroxyl radical is a strong oxidizing agent capable of initiating lipid peroxidation. Evidence that the hydroxyl radical is the proximate cause in initiating lipid peroxidation comes from several groups. Fong et al. (56,66) investigated the possible role of hydroxyl radical in the peroxidation of lysosomal membranes. Lysis occurred in the presence of ADP-ferric iron from the actions of either xanthine and xanthine oxidase or partially-purified NADPH-cytochrome P450 reductase. During membrane lysis, peroxidation of the lipid was observed as assayed by the formation of malondialdehyde. Both the peroxidation and the lysis could be inhibited by the addition of either Mn^{+2} (a superoxide scavenger), catalase, or hydroxyl radical traps (ethanol, benzoate, or mannitol) (66). Paradoxically, exogenous superoxide dismutase increased the peroxidation and lysis initiated by xanthine oxidase (by 200%) or microsomal P450 reductase (by 20%) but decreased peroxidation when purfied P450 reductase was used (56). These two distinct actions of superoxide dismutase can be explained by the possibility that, in the case where superoxide dismutase inhibited peroxidation, the accumulation of superoxide was rate limiting and the reduction of iron by superoxide was slow. Therefore, superoxide dismutase acted as an effective scavenger. In the situation where superoxide dismutase enhanced peroxidation, either the flux of superoxide was sufficient to reduce ferric to ferrous iron, or the reduction of iron occurred by a different mechanism. In this system the action of superoxide dismutase may serve to enhance the formation of hydrogen peroxide that can then react with the ferrous iron to yield hydroxyl radical.

McCord and Day (67) have examined the generation of the hydroxyl radical by measuring the absorbance change of tryptophan following the attack by the hydroxyl radical. Using xanthine and xanthine oxidase, they demonstrated that tryptophan was oxidized only in the presence of EDTA-ferrous iron (10-30 μM). A lag of several minutes that occurred before oxidation commenced could be eliminated by the addition of hydrogen peroxide (10 μM). The oxidation of tryptophan was inhibited by catalase or by the hydroxyl radical scavengers mannitol and ethanol.

Koster and Slee (69) suggested that hydroxyl radical was responsible for the oxidation of various cellular fractions. By adding NADPH (0.04-4 mM) and ADP-ferric iron (2 mM; 0.12 mM, respectively) to rat liver microsomes, these investigators found a time-dependent increase in malondialdehyde, chromolipid and protein polymerization. The inclusion of catalase (0.6 mg/ml) or thiourea (60 mM) was effective in preventing these increases.

The demonstration that hydroxyl radical can occur in biological solutions containing ferrous iron and hydrogen peroxide has been demonstrated using EPR techniques. Piette and his coworkers (68) have also explored the role of hydroxyl radical in peroxidation by using the hydroxyl radical spin trap 5,5-dimethyl-1-pyrroline-1-oxide (DMPO). They demonstrated that the addition of DMPO inhibited peroxidation, as measured by malondialdehyde formation, that the strength of the resulting EPR signal parrelleled the progress of peroxidation, that the superoxide dismutase enhanced the EPR signal and that the addition of the hydroxyl radical thiourea inhibited the EPR signal. In this system the generation of the spin adduct signal required EDTA-ferrous iron. Floyd (53) studied the effect of nucleotide complexes of ferrous iron on hydroxyl radical formation using the spin trap agent DPMO which reacts very rapidly with hydroxyl radical (3.2 10^9/M/sec). He found that the triphosphate nucleotides were 20% more effective in catalysing the formation of hydroxyl radical than the dinucleotides. The amount of hydroxyl radical trapped as a function of the nucleotide level increased in a sigmoidal fashion beginning at a nucleotide/ferrous iron ratio of two and rose to a ratio of five. Monophosphate nucleotides did not increase hydroxyl radical formation, and the ferric iron whether chelated by tri- or diphosphate nucleotides was ineffective in hydrogen peroxide catalysis to yield hydroxyl radical. Floyd tested the possible interference of divalent cations, Mg^{+2} and Ca^{+2}, on the ability of the nucleotide-ferrous iron complex to catalyze the formation of hydroxyl radical because these divalent cations are associated with nucleotides in the cell. At concentrations of 2 mM, Mg^{+2} and Ca^{+2} decreased the spin trap signal by 8% and 27%, respectively. Floyd concluded that the hydroxyl radical could be formed as a reaction between hydrogen peroxide and the chelated ferrous iron in biological solutions. These results are in contrast with those of Tien et al. (28) who could not observe the formation of the hydroxyl radical in the presence of ADP and found that the resulting peroxidation was not sensitive to scavengers of hydroxyl radicals.

Aust and his coworkers have investigated the role of the Haber-Weiss reaction in the initiation of lipid peroxidation (70). When $FeCl_2$ (0.2 mM) and hydrogen peroxide (0.1 mM) were added to phospholipid liposomes, there was an increase in the peroxidation as determined by both malondialdehyde formation and iodometric measurements of lipid peroxides. Catalase, mannitol (10 mM), benzoate (10 mM), and TrisCl (10 mM) prevented this increase thus implicating the involvement of hydroxyl radical. Peroxidation that depended upon the enzymatic generation of superoxide by xanthine oxidase was also studied both with phospholipid liposomes and detergent-dispersed linoleate. With either target peroxidation was not observed in the absence of iron. The addition of EDTA-chelated (0.11 mM) or ADP-chelated (0.5 mM) ferric iron

(0.1 mM) to either target system resulted in peroxidation. The combination of these two forms of chelated iron increased the extent of peroxidation by 100-fold over the control levels. Only in the system with detergent-dispersed linoleate and EDTA-ferric iron acting as the Fenton reagent was the involvement of hydroxyl radical in the initiation of peroxidation noted. When ADP-ferric iron or the combination of EDTA-/ADP-ferric iron was used as the Fenton reagent, catalase or hydroxyl radical scavengers were ineffective in preventing peroxidation, whereas superoxide dismutase was able to prevent the xanthine oxidase-driven peroxidation in each of the chelated iron systems. The involvement of the hydroxyl radical and the mechanism of initiation of peroxidation appeared to be different. The difference between the two systems suggested a correlation between the proximity of hydroxyl radical generation and its ability to initiate peroxidation. In the system with liposomes, hydroxyl radical generation occurs in the aqueous phase. It is unlikely that the highly reactive hydroxyl radical would diffuse away from the site of formation to the liposomes to initiate peroxidation. However, when the lipid is dispersed by detergent it may be more accessible to the hydroxyl radical and thus more likely to be attacked.

3.3 Perferryl- and ferryl-ion oxygenation

The involvement of the perferryl ion in intiating peroxidation was first proposed by both Ernster and his coworkers (71) and Heaton and Uri (72) over twenty years ago. More recently, Aust and his coworkers as well as several other groups have expressed renewed interest in this mechanism as a means to resolve the myraid of conflicting results over the initiation of peroxidation. The perferryl ion complex can be represented as the binding of oxygen to reduced iron:

$$Fe^{+2}O_2 \leftrightarrow Fe^{+3}O_2^{\overline{\cdot}} \qquad \text{Equation 4.}$$

The metal involved in the binding can be any transition metal capable of undergoing one electron redox reaction. The complex of oxygen and the reduced metal imparts a free radical characteristic to the complex, and being highly electronegative, this complex has been proported to be a strong oxidant. Since the perferryl ion has a free radical characteristic, the spin restriction between the ground state of oxygen and the unsaturated fatty acid are circumvented, and the peroxidation reaction can occur.

Aust and his coworkers have proposed the perferryl ion complex as the initiating species of lipid peroxidation when their results indicated that products of the Haber-Weiss reaction could account for peroxidation only in highly defined conditions. Using EPR techniques, they confirmed that the hydroxyl radical could be generated in the presence of hydrogen peroxide (0.1 mM) and $FeCl_2$ (0.15 mM) and that the resulting lipid peroxidation as measured by malondialdehyde could be prevented by mannitol (10 mM) (28,70). However, when the nucleotide chelator ADP (1.7 mM) was added, the EPR signal was drastically reduced, and the resulting lipid peroxidation was no longer sensitive to mannitol inhibition. Moreover, no EPR signal was observed when ADP-ferric iron was reduced by NADPH and NADPH-cytochrome P450 reductase (28). They concluded that the hydroxyl radical can promote lipid peroxidation, but, when iron chelators (which are prevelant in biological tissues) are added, the mechanism of lipid peroxidation may not involve hydroxyl radicals.

Pederson and Aust (73,74) presented evidence to suggest that the perferryl ion initiated the NADPH-driven peroxidation. Using either microsomal lipids or liposomes with exogenous NADPH-cytochrome P450 reductase, they found that the addition of both EDTA-ferric iron and ADP-ferric iron was necessary to promote peroxidation (73). The inclusion of catalase or mannitol (0.25 mM), sodium formate (0.5 M), or 2-methyl-2-propanol (0.1 M) did not prevent peroxidation, thus demonstrating that the participation of hydrogen peroxide or hydroxyl radical was not necessary for peroxidation in that system (74). Furthermore, the NADPH-driven peroxidation in both microsomes and liposomes did not depend on the production of superoxide or singlet oxygen as witnessed by the inability of superoxide dismutase or the singlet oxygen scavenger 1,3-diphenylisobenzofuran to prevent peroxidation. They concluded that the requirement for ADP-ferric iron and NADPH, coupled with the known ability of NADPH-cytochrome P450 reductase to reduce ADP-chelated iron (75,76), indicated that the ADP-ferrous iron was probably directly involved in initiation (74).

Svingen et al (65,77) have proposed that ADP-perferryl ion is responsible for the initiation of both the NADPH-driven and the superoxide-driven lipid peroxidation. These investigators have studied the initiation in hydroperoxide-free liposomes. Peroxidation was initiated by using ADP-ferrous iron, or ADP-ferric iron along with NADPH plus cytochrome P450 reductase, or ADP-ferric iron in the presence of xanthine oxidase. Lipid peroxidation was not promoted by ferrous, ferric or ADP-ferric iron. Peroxidation was inhibited by superoxide dismutase but not by catalase or hydroxyl radical scavengers. Since superoxide by itself was unable to initiate peroxidation, it was concluded that initiation ocurred by the perferryl ion. The perferryl ion could have been formed by the reduction of the ADP-ferric iron followed by the binding of oxygen or by the reduction of ADP-ferric iron by superoxide.

Nakano and his coworkers (32,57,78) examined the involvement of the hydroxyl radical and perferryl ion in adriamycin-promoted lipid peroxidation. Adriamycin as well as other anthracycline antibiotics undergo a one electron reduction by microsomal NADPH cytochrome P450 reductase in a redox cycling pathway (32,78,79). Using a system that contained NADPH, purified cytochrome P450 and ADP-ferric iron-adriamycin complex in TrisCl buffer, no hydroxyl radical was detected by EPR spin trapping techniques; but peroxidation still occurred (57). When a phosphate buffer replaced the Tris buffer, the ADP-ferric iron and adriamycin complex dissociated, no lipid peroxidation occurred, but a hydroxyl radical was detected by EPR. In this system, the hydroxyl radical may have reacted with water-soluble compounds before attacking the phospholipid micelles. It seems most likely that the NADPH-driven peroxidation is initiated by a perferryl ion complex rather than hydroxyl radical.

Recently direct initiation of lipid peroxidation by the perferryl ion complex has been questioned on thermodynamic grounds. W.H. Koppenol pointed out that the perferryl ion complex is not a strong enough oxidizing agent and could only serve as a generator of hydrogen peroxide and hydroxyl radicals (80). However, the ferryl ion complex is thermodynamically a strong oxidizing agent. The formation of the ferryl ion has been proposed to occur by the complexing of ferrous iron to the perferryl ion complex:

$$Fe_2^{+2}O_2 + Fe^{+2} \longrightarrow Fe^{+2}O_2Fe^{+2} \qquad \text{Equation 5.}$$

$$Fe^{+2}O_2Fe^{+2} \longrightarrow 2Fe^{+2}O \qquad \text{Equation 6.}$$

It should be noted that the ferryl ion can be formed by either NADPH-driven or superoxide-driven reactions (65,81).

Aust and his coworkers (82,83) have examined the initiation of peroxidation by the ferryl ion complex. When ADP-ferrous iron (1.7 mM;

0.1 mM) was incubated with liposomes in 30 mM NaCl (pH 7.0), a seven minute lag time was seen before the generation of malondialdehyde was observed. Addition of hydrogen peroxide (0.1 mM) or increasing the pH decreased the lag time for the ADP-ferrous iron-induced peroxidation, and the addition of hydroxyl radical scavengers did not prevent peroxidation. The addition of ADP-ferric iron also reduced the lag time as well as increased the rate of production of malondialdehyde in a dose-response fashion with complete elimination of the lag time observed with a ratio of ADP-ferric iron to ADP-ferrous iron of two. These results suggested that an intermediate product had to accumulate before peroxidation could proceed. The products of ADP-ferrous iron autooxidation include superoxide, hydrogen peroxide, hydroxyl radical, and ADP-ferric iron. In order to determine which autooxidation product was necessary, AMP was used as a chelator to obtain reproducible lag times. The addition of AMP-ferric iron decreased the lag time for AMP-ferrous iron-induced peroxidation. The inclusion of mannitol (0.5M) or the scavenger enzymes catalase and superoxide dismutase had no effect on the formation of malondialdehyde by either AMP-ferrous iron alone or the combination of AMP-ferrous iron and AMP-ferric iron. To examine the possibility that the iron-dependent rates represent the breakdown of pre-existing hydroperoxides rather than initiation, they examined initiation using hydroperoxide-free arachidonic acid micelles (82). The results confirmed that AMP-ferric iron directly stimulated the rate of malondialdehyde production by AMP-ferrous iron. Since the effect of the ferric-chelate cannot be explained through an initiation reaction dependent upon preexisting hydroperoxides or by enhancing oxy radical production, they concluded that the ferric-chelate participated in the formation of an initiator with the ferrous-chelate and oxygen perhaps through a ferryl ion complex.

3.4 Hydroperoxide-dependent oxygenation

The initiation and propagation of lipid peroxidation can occur by the homolytic cleavage of existing hydroperoxides; and, once lipid hydroperoxides are present in liposomes, the predominant mechanism of initiation is their breakdown to form free radicals. Aust et al. (23) have argued that the most likely mechanism by which the free radicals are formed is through a ferrous iron-promoted decomposition of the lipid hydroperoxide:

$$ROOH + chelated\text{-}Fe^{+2} \rightarrow RO^{\cdot} + OH^{-} + chelated\text{-}Fe^{+3} \quad \text{Equation 7.}$$

Although ferric-iron can promote the reductive activation of the hydroperoxide, the rate of this reaction is slow because the two reactants have opposing pH maxima for activity. The hydroperoxide requires ionization (pK = 10.8) to be reactive, but the ferric iron is insoluble above pH 4.

In certain conditions Svingen et al. (77) demonstrated that the hydroperoxide-driven mechanism of peroxidation accounted for greater than 90% of the lipid hydroperoxides formed. In the presence of ADP-ferric iron (1.7 mM; 0.1 mM), the rate of malondialdehyde formation and lipid hydroperoxide formation in the NADPH-driven peroxidation was 0.3 nmol/min/ml and 1.8 nmol/min/ml, respectively. When EDTA-ferric iron, DTPA-ferric iron, or ferric cytochrome P450, was included in the reaction, the rates were increased over ten-fold. These enhanced rates were not inhibited by superoxide dismutase, by a scavenger of singlet oxygen diphenylfuran (0.2 mM), or by a scavenger of hydroxyl radical benzoate (40 mM). Inhibition of peroxidation was achieved by butylatedhydroxytoluene (1 mM) indicating the free radical nature of the propagation.

In the propagation cycle the catalyst must be able to efficiently catalyze the breakdown of the hydroperoxide to reactive intermediates. Using cumene hydroperoxide or partially peroxidized liposomes as a source of hydroperoxides to assay the catalytic activity of various ADP-chelated or EDTA-chelated complexes, only EDTA-ferrous iron could catalyze the efficient breakdown of the hydroperoxides (84). Thus, although EDTA-ferrous iron is limited in the initiation of lipid peroxidation (70,84), it can catalyze the propagation cycle (Figure 1).

Thomas et al. (85) demonstrated that the rate of xanthine oxidase catalyzed peroxidation of linoleic acid depended directly upon the initial hydroperoxide concentration. Linoleic acid pretreated with $NaBH_4$ to reduce pre-existing hydroperoxides resulted in a much slower rate of peroxidation than untreated linoleic acid. The addition of superoxide dismutase prevented peroxidation. These authors suggested that the superoxide generated by xanthine oxidase reacted with the preexisting lipid hydroperoxides to produce an alkoxy radical ($RO^{\cdot}$) and that this alkoxy radical propagated the peroxidation. In these experiments DTPA (0.1 mM) was included to chelate metals present in the phosphate buffer. The iron contamination was reported to be less than 1 μM, and, since peroxidation in the presence of DTPA was not significantly stimulated by the addition of $FeCl_3$ (2.8 μM), iron-mediated peroxidation was not considered to be significant under these experimental conditions. These results suggest that hydroperoxides react directly with superoxide, circumventing the need for a transition metal to catalyze the reaction. This proposed reaction mechanism is similar to the Haber-Weiss reaction so that the finding of peroxidation occurring in the absence of a transitional metal needs further elucidation.

4. ACTION OF CHELATORS

A common property for the mechanisms of peroxidation is the catalytic need for iron. Iron is required as the Fenton reagent in the Haber-Weiss reaction in order to generate hydroxyl radicals at a significant rate (61). Iron is needed to complex with oxygen to form the perferryl-ion or the highly reactive ferryl-iron complex. Also, ferrous iron is needed to catalyze the homolytic decomposition of lipid hydroperoxides to alkoxy radicals which has been suggested as the radical intermediate in the hydroperoxide-driven peroxidation (23). Copper can replace iron as a catalysis in peroxidation (86). In the presence of superoxide and hydrogen peroxide, copper acts like a Fenton reagent. However, copper can also scavenge electrons, and its ability to catalyze peroxidation is compromised by EDTA-chelation (86). Other transition metals that undergo univalent redox reactions, such as cobalt and manganese, can also participate in the initiation of peroxidation. Despite the adequancy of other transition metals most of the attention has been focused on iron because it has been found to be an active promotor of peroxidation, and its concentration _in vivo_ is higher than the other transition metals.

The requirement for iron was first demonstrated by Ernster and his colleagues (71) who found that the addition of iron chelated by ADP was required for NADPH-dependent lipid peroxidation in microsomes. Subsequently, many researchers investigating the NADPH-dependent peroxidation (76,87,88) and other systems promoting peroxidation (28,89) have found that the addition of iron was necessary for peroxidation. Several reports have argued against the requirement for a transition metal (44,62,85). As indicated in Section 3.4., Thomas et al. (85) reported that the superoxide-driven, hydroperoxide-dependent peroxidation was not significantly enhanced by the addition of $FeCl_3$ (2.8 μM) chelated by EDTA (0.1 mM). Since the initial contamination of iron in their buffer was less than 1 μM, the addition of this iron chelate represented in sizable increase in the iron concentration and, therefore, might have been expected to provoke a faster rate of peroxidation. Therefore, their results suggest that iron may not mediate the peroxidative attack of active oxygen in all cases. Alternatively, the lack of effect on peroxidation upon iron addition may have been due to the 30-fold excess of chelators present (EDTA, 0.1 mM; DTPA, 0.1 mM). Large ratios of chelator to iron have been found to prevent both the hydroxyl radical-mediated and the perferryl-ion mediated peroxidation (84,90).

Most mechanisms proposed for the initiation of peroxidation require iron to be reduced to the ferrous form as an initial step. The addition of either ferric iron or chelated-ferric iron alone is insufficient to promote peroxidation (77). Whereas, the rate of lipid peroxidation is first order with respect to the addition of ADP-ferrous iron. When a reductant such as NADPH is added with ADP-ferric irons, the resulting rate of peroxidation is comparable to that of ADP-ferrous iron.

The reduction of ferric iron appears to occur through a number of pathways. With xanthine oxidase or with NADPH oxidase, ferric iron is reduced by superoxide as evidenced by the ability for superoxide dismutase to prevent reduction and peroxidation. With the NADPH-driven peroxidation, it has been suggested that superoxide is generated in the incubation containing either the NADPH cytochrome P450 reductase or liver microsomes and then reduces the iron (66,69,91). However, these reports are based on indirect evidence, and other investigators have reported that superoxide dismutase did not inhibit the NADPH-dependent peroxidation promoted by ADP-ferric ion (65,92,93). Therefore, it would appear that EDTA- or ADP-chelated-ferric iron is reduced directly by the reductase. Furthermore, direct deduction of the iron may also be the mechanism through which ascorbate (94) and glutathione (95) initiate lipid peroxidation.

Chelators have been widely employed to prevent ferric iron from precipitating at physiologic pH. _In vivo_ iron is most often found in a complex. Extracellular iron is typically associated with heme-proteins or with the iron transport protein, transferritin (96). The largest intracellular forms of iron are found to be associated with ferritin and proteins (97,98). A variety of small biological compounds can complex iron such as heme, citrate, succinate and nucleotides. Primosigh and Thomas (99) have demonstrated the presence of a soluble intracellular compounds containing iron that appear to serve as a proximate source of iron for heme proteins and ferritin. Wills demonstrated that this pool of soluble iron could have a role in lipid peroxidation and that inhibition of peroxidation could be achieved by the chelator desferrioxamine (100).

Chelation has been found to enhance, suppresss or not affect the rate of metal-promoted peroxidation depending on the nature of the chelate, the metal, the redox state of the metal, and the chelate-metal complex formed. Tappel et al. (101) demonstrated that iron chelated into heme promoted lipid peroxidation both in vitro and in vivo. Pederson et al. (74) found that either EDTA-chelated or ADP-chelated ferric iron as opposed to non-chelated ferric iron increased the rate of the NADPH-dependent peroxidation in intact microsomes or in liposomes. Superoxide-dependent lipid peroxidation has also been shown to be promoted by iron complexes with EDTA and/or ADP (28,65,102). Conversely, Samuni et al. (86) found that EDTA blocked the actions of copper as a Fenton reagent. Tien et al (28,70) demonstrated that ADP inhibited the formation of the hydroxyl radical and shifted the mechanism of peroxidation away from the Haber-Weiss reaction. Several reports have indicated that the ferrous iron alone could initiate peroxidation (103,104). The rate of hydroperoxide-dependent peroxidation promoted by ferrous iron was found to be greater than the EDTA-ferrous iron (90). This result suggested that chelation by EDTA does not impart a special characteristic to ferrous iron in catalyzing the hydroperoxide-dependent peroxidation. Kornburst and Davis (105) reported that, when ferrous or ferric solutions were freshly prepared, no enahanced stimulation of NADPH-dependent peroxidation was noted upon the addition of EDTA or ADP.

The nature of the iron chelate dictates the mechanism by which iron participates in initiation of peroxidation. The redox potential of the ferric/ferrous pair can vary quite dramatically depending on the nature of the chelate (106). Iron chelates with low redox potential, such as

EDTA, have a rapid reaction with hydrogen peroxide and produce oxidative products (107). In contrast, iron chelates with high redox potentials are characterized by slow decomposition of hydrogen peroxide and low yield of oxidative products. As seen above the EDTA-chelates allow for the generation of hydroxyl radicals from hydrogen peroxide. However, with the ADP-chelate the generation of hydroxyl radical is inhibited, and peroxidation appeared to occur via an active oxygen-iron intermediate. It has been proposed that the chelation not only facilitated the solubility of iron but also changed the redox potential or iron (65,70). Some chelates such as ADP can donate electrons to the ferrous-oxygen ion and thereby strengthen the iron-oxygen bond. Other chelates such as EDTA will withdraw electrons and reduce the stability of the perferryl complex. In addition certain chelates of iron, EDTA or DTPA, are more suitable electron acceptors from reductases than chelates such as ADP or citrate. Thus, the type of chelation may influence not only the reactive intermediates to be formed and the mechanism of initiation but also the source from which the ferric iron is reduced.

5. CONSTRAINTS ON PEROXIDATION

The controversy surrounding the precise mechanism of lipid peroxidation in biological samples will in most likelihood remain until new sensitive and specific methods are developed to assay for the hydroperoxides and to detect the reactive intermediates. Furthermore, in any experimental system the balance amoung the various mechanisms of peroxidation will depend upon the physical state of the lipid, the concentration of the iron species, the nature of the chelating agents, and the presence of scavenging agents such as tris, xanthine or ethanol. One special difficulty relevant to the experimental system is the predominance of hydroperoxide-dependent peroxidation over the active oxygen initiated mechanisms. Unless special precautions are taken to eliminate or minimize hydroperoxide formation during sample handling, the hydroperoxide-driven peroxidation will mask the possible initiation by any other process and mislead the investigator in conclusions on the mechanism of initiation. The effort to explore peroxidation under simple and well-defined conditions have been successful in discerning the possible role of intermediates such as hydroxyl radicals. However, biological samples are complex in their constituency; and, in the presence of biological chelates such as ADP, the mechanism of lipid peroxidation probably does not involve the hydroxyl radical.

The concentration of inorganic iron in liver cytosol has been estimated in the range of 10–100 μM, and the cytosolic concentration of ADP is reported to be in the millimolar range (56). Since the concentration of both components are at levels compatible with promoting lipid peroxidation, it might be reasonable to consider the mechanism(s) which restricts the occurance of peroxidation. As noted in Section 4, the some complexes of ferric iron cannot be directly reduced by reductases but can be reduced by superoxide. The presence of superoxide dismutase will scavenge the superoxide and reduce it to hydrogen peroxide. Hydrogen peroxide can then be further reduce by catalase or glutathione peroxidase. Glutathione peroxidase and glutathione S-transferases can also diminish the threat of peroxidation by reducing lipid hydroperoxides. These scavenger enzymes represent one aspect of the defense against lipid peroxidation. However, cytosolic enzymes will not inhibit peroxidation occurring in membrane domains. For glutathione peroxidase to reduce hydroperoxides esterified into membranes the addition of a hydrolase such as phospholipase A_2 is necessary to first release the hydroperoxy lipid (108,109). Inhibition of peroxidation in membranes can be achieved with lipid soluble (hydrophobic) radical scavengers such as cholesterol, vitamin E (α-tocopherol), or β-carotene. Vitamin E is frequently used to prevent peroxidation because it forms a stable radical species and thus terminates both the initiation and propagation steps of lipid peroxidation. Vitamin E will reduce superoxide, singlet oxygen, and lipid radicals (110). β-Carotene is an efficient singlet oxygen scavenger (111). A number of other cell components can react with free radical and form products that are less toxic to the cell than the original free radical. The use of only one type of scavenger, either the hydrophilic enzymes or hydrophobic antioxidants, will restrict peroxidation in only one compartment. However, the combination of both vitamin E and the soluble scavenger enzymes may prove most effective against peroxidation.

Most cells contain an appreciable level of scavenger activity to suppress the level of cellular peroxides. Radicals and their precursors, such as superoxide and hydrogen peroxide, only represent a problem when their rate of generation exceeds their rate of removal. The cellular rate for generation of these radicals is modulated by the various flavin-containing and ferrodoxin enzymes whose activity is influenced by substrate concentration, cofactor availability, and oxygen tension. Certain metabolic states such as hyperoxia, ischemia or antibiotic therapy can favor free radical production in excess of basal rates. Lipid hydroperoxide formation can then become appreciable, approaching levels (10^{-6} M) that would have deleterious effects. This situation may also present itself when neutrophils and macrophages arrive at a site of insult. In response to irritation, these phagocytic cells produce large quantitites of active oxygen products including hydroperoxides. These peroxides can act as signals for the production of prostaglandins, and play an important role in the elevated eicosanoid synthesis seen in various disease states, including cancer.

ACKNOWLEDGEMENTS:

The author would like to give special thanks to Drs. W.E.M. Lands and R.J. Kulmacz for their helpful discussion during the preparation of this manuscript. This work was supported in part by a National Research Service Award (AI 06719).

REFERENCES

1. Wynder EL: Nutrition and cancer. Federation Proc (35): 1309-1315, 1976.
2. Carroll KK: Experimental evidence of dietary factors and hormone-dependent cancers. Cancer Res (35): 3374-3383, 1975.
3. Gammal EB, Carroll KK, Plunkett ER: Effects of dietary fat on mammary carcinogenesis by 7,12-dimethylbenz[a]anthracene in rats. Cancer Res (27): 1737-1742, 1967.
4. Carroll KK, Hopkins GJ, Kennedy TG, Davidson MB: Esential fatty acids in relation to mammary carcinogenesis. Prog Lipid Res (29): 685-690, 1981.
5. Hopkins GJ, West CE: Effect of dietary polyunsaturated fat on the growth of a transplantable adenocarcinoma in C3HA^{vy}fB mice. J Natl Cancer Inst (58): 753-756, 1977.
6. Rao GA, Abraham S: Enhanced growth rate of transplanted mammary adenocarcinoma induced by C3H mice by dietary linoleate. J Natl Cancer Inst (56): 431-432, 1976.
7. McCay PB, King MM, Pitha JV: Evidence that the effectiveness of antioxidants as inhibitors of 7,12-dimethylbenz[a]anthracene-induced mammary tumors is a function of dietary fat consumption. Cancer Res (41): 3745-3748, 1981.
8. Bridges JW, Benford DJ, Hubbard SA: Mechanisms of toxic injury. Ann NY Acad Sci (407): 42-63, 1983.

9. Mak IT, Misra HP, Weglicki WB: Temporal relationship of free radical-induced lipid peroxidation and loss of latent enzyme activity in highly enriched hepatic lysosomes. J Biol Chem (258): 13733-13737, 1983.
10. Sivarajah K, Anderson MW, Eling TE: Metabolism of benzo[a]pyrene to reactive intermediate(s) via prostaglandin biosynthesis. Life Sciences (23): 2571-2578, 1978.
11. Dix TA, Marnett LJ: Metabolism of polycyclic aromatic hydrocarbon derivatives to ultimate carcinogens during lipid peroxidation. Science (221): 77-79, 1983.
12. Shamberger RJ, Adreone TL, Willis CE: Antioxidants and cancer IV. Initiating activity of malonaldehyde as a carcinogen. J. Natl Cancer Inst (53): 1771-1773, 1974.
13. Marnett LJ, Tuttle MA: Comparison of mutagenicities of malondialdehyde and the side products formed during its chemical synthesis. Cancer Res (40): 276-282, 1980.
14. Perez HD, Wekster BB, Goldstein IM: Generation of chemotactic lipid from arachidonic acid by exposure to a superoxide-generating system. Inflammation (4): 313-328, 1980.
15. Hemler ME, Lands WEM: Evidence for a peroxide-initiated free radical mechanism of prostaglandin biosynthesis. J Biol Chem (255): 6253-6261, 1980.
16. Lands WEM, Kulmacz RJ, Marshall PJ: Role of hydroperoxides in prostaglandin synthesis. In: Pryor WA (ed) Free radicals in biology. Vol. VI. Academic Press, New York, 1984, pp. 37-61.
17. Lupulescue A: Enhancement of carcinogenesis by prostaglandins. Nature (272): 634-636, 1978.
18. Bennett A: Prostanglandins and cancer. In: Karim SMM (ed) Advances in Prostaglandin Research: practical applications of prostaglandins and their synthesis inhibitors. University park Press, Baltimore, 1979, pp. 149-188.
19. Karmali RA: Prostaglandins and cancer. Prostaglandins and Medicine (5): 11-28, 180.
20. Goodwin JS, Webb DR: Regulation of the immune response by prostaglandins. Clin Immunol Immunopathol (15): 106-122, 1980.
21. Slater TF: Free radical as reactive intermediates in tissue injury. Adv Exp Med Biol (136): 575-589, 1981.
22. Freeman BA, Crapo JD: Biology of disease, free radicals and tissue injury. Lab Invest (47): 412-426, 1982.
23. Aust SD, Svingen BA: The role of iron in enzymatic lipid peroxidation. In: Pryor WA (ed) Free radicals in biology. Vol 5. Academic Press, New York, 1982, pp. 1-28.
24. Porter NA: Prostaglandin endoperoxides. In: Pryor WA (ed) Free radicals in biology. Vol. 4. Academic Press, New York, 1980, pp 261-294.
25. Hochstein P, Ernster L: ADP-activated lipid peroxidation coupled to the TPNH oxidase system of microsomes. Biochem Biophys Res Comm (12): 388-394, 1963.
26. Ernster L, Nordenbrand K, Orrenius S: Microsomal lipid peroxidation: mechanism and some biomedical implications. In: Yagi K (ed) Lipid peroxides in biology and medicine. Academic Press, New York, 1982, pp 55-79.
27. Fridovich I: The biology of oxygen radicals. Science (201): 875-879, 1978.

28. Tien M, Svingen BA, Aust SD: Superoxide dependent lipid peroxidation. Federation Proc (40): 179-182, 1981.
29. Schoenberg MH, Younes M, Haglund U, Fredholm B, Sellin D, Jung H, Schildberg FW: The participation of the hydroxanthine-xanthine oxidase system in the generation of free radicals after intestinal ischemia. In Bors W, Saran M, Tait D (eds) Oxygen Radicals in Chemistry and Biology. W. de Gruyter, Berlin, 1984, pp. 851-857.
30. Weiss SJ, LoBuglio AF: Phagocyte-generated oxygen metabolites and cellular injury. Lab Invest (47): 5-18, 1982.
31. Mason RP, Chignell CF: Free radicals in pharmacology and toxicology. Pharmacol Rev (33): 189-211, 1981.
32. Noguchi T, Nakano M: Effect of ferrous ions on micorosomal phospholipid peroxidation and related light emission. Biochim Biophys Acta (368): 446-455, 1974.
33. Bachur NR, Gordon SL, Gee MV, Kon H: NADPH cytochrome P-450 reductase activation of quinone anticancer agents to free radicals. Proc Natl Acad Sci USA (76): 954-957, 1979.
34. Wendel A, Fuerstein S: Drug-induced lipid peroxidation in mice. I. modulation of monoxygenase, glutathione, and selenium status. Biochem Pharmacol (30): 2513-2520, 1980.
35. Levine WG: Glutathione, lipid peroxidation and regulation of cytochrome P-450 activity. Life Sci (31): 779-784, 1982.
36. Koster JF, van Berkel TJC: The effect of diethyldithiocarbamate on the lipid peroxidation of rat liver micorsomes and intact hepatocytes. Biochem Pharmacol (32): 3307-3310, 1983.
37. Ottolenghi A: Interaction of ascorbic acid and mitochondrial lipids. Arch Biochem Biophys (79): 355-363, 1959.
38. Bernheim F, Bernheim MLC, Wilber KM: Reaction between thiobarbituric acid and the oxidation products of certain lipids. J Biol Chem (174): 257-264, 1948.
39. Yagi M: A simple fluorometric assay for lipoperoxide in blood plasma. Biochem Med (15): 212-216, 1976.
40. Asakawa T, Matsushita S: Coloring condition of thiobarbituric acid test for detecting lipid hydroperoxides. Lipids (15): 137-140, 1980.
41. Buege JA, Aust SD: Microsomal lipid peroxidation. In: Fleischer S, Packer L (eds) Methods of enzymology (52). Academic Press, New York, 1978, pp 302-310.
42. Donato H: Lipid peroxidation cross-linking reactios and aging. In: Schal RW (ed) Age Pigments, Elsevier, Amsterdam, 1981, pp 63-83.
43. Asakawa T, Matsushita S: Thiobarbituric acid test for detecting lipid peroxides. Lipids (14): 401-406, 1979.
44. Kellogg EM, Fridovich I: Superoxide, hydrogen peroxide, and singlet oxygen in lipid peroxidation by a xanthine oxidase system. J Biol Chem (250): 8812-8817, 1975.
45. Dillard CJ, Tappel AL: Volatile hydrocarbon and carbonyl products of lipid peroxidation: a comparison of pentane, ethane, hexanal, and acetone as in vivo indices. Lipids (14): 989-995, 1979.
46. Riely CA, Cohen G, Lieberman M: Ethane evolution: a new index of lipid peroxidation. Science (183): 208-210, 1974.
47. Wright JR, Rumbaugh RC, Colby HD, Miles PR: The relationship between chemiluminescence and lipid peroxidation in rat hepatic microsomes. Arch Biochem Biophys (192): 344-351, 1979.
48. Fletcher BL, Dillard CJ, Tappel AL: Measurement of fluorescent lipid peroxidation products in biological systems and tissues. Anal Biochem (52): 1-9, 1979.

49. Hicks M, Gebicki JM: A spectrophotometric method for the determination of lipid hydroperoxides. Anal Biochem (99): 249-253, 1979.
50. Cathcart R, Schwiers E, Ames BN: Detection of picomole levels of hydroperoxides using a fluorescent dichlorofluorescin assay. Anal Biochem (134): 111-116, 1983.
51. Marshall PJ, Kulmacz RJ, Lands WEM: Hydroperoxides, free radicals and prostaglandin synthesis. In Bors W, Saran M, Tait D (eds) Oxygen Radicals in Chemistry and Biology. W. de Gruyter, Berlin, 1984, pp. 299-307.
52. Lai CS, Piette LH: Spin-trapping studies of hydroxyl radical production involved in lipid peroxidation. Arch Biochem Biophys (190): 27-38, 1978.
53. Floyd RA: Direct demonstration that ferrous ion complexes of di and triphosphate nucleotides catalyze hydroxyl free radical formation from hydrogen peroxide. Arch Biochem Biophys (225): 263-270, 1983.
54. Finkelstein E, Rosen GM, Rauckman EJ: Spin trapping of superoxide and hydroxyl radical practical aspects. Arch Biochem Biophys (200): 1-16, 1980.
55. Raleigh JA, Kremers W: DMSO does not protect against hydroxyl radical induced peroxidation in model membranes. Int J Radiat Biol (39): 441-444, 1981.
56. Fong KL, McCay PB, Poyer JL, Misra HP, Keele BB: Evidence for superoxide-dependent reduction of Fe^{3+} and its role in enzyme-generated hydroxyl radical formation. Chem Biol Interactions (15): 77-89, 1976.
57. Sugioka K, Nakano H, Nakano M, Tero-Kubota S, Ikegami Y: Generation of hydroxyl radicals during the enzymatic reductions of the Fe^{3+}-ADP-phosphate-adriamycin and Fe^{3+}-ADP-EDTA systems. Biochim Biophys Acta (753): 411-421, 1983.
58. Pryor WA: Free radical reactions and their importance in biochemical systems. Federation Proc (32): 1862-1869, 1973.
59. Mason RP: Free-radical intermediates in the metabolism of toxic chemicals. In: Pryor WA (ed) Free Radicals in Biology. Vol. 5. Academic Press, New York, 1982, pp. 161-222.
60. Root RK, Metcalf J, Oshino N, Chance B: H_2O_2 release from human granulocytes during phagocytosis. J Clin Invest (55): 945-955, 1975.
61. Czapski G, Ilan Y: On the generation of the hydroxylation agent from superoxide radical. Can the Haber-Weiss reaction be the source of ^1OH radicals. Photochem Photobiol (28): 651-653, 1978.
62. Kellogg EW, Fridovich I: Liposome oxidation and erythrocyte lysis by enzymatically generated superoxide and hydrogen peroxide. J Biol Chem (252): 6721-6728, 1977.
63. Dixit R, Mukhtar H, and Bickers DR: Evidence that lipid peroxidation in microsomal membranes of epidermis is associated with generation of hydrogen peroxide and singlet oxygen. Biochem Biophys Res Comm (105): 546-552, 1982.
64. King MM, Lai EK, McCay PB: Singlet oxygen production associated with enzyme-catalyzed lipid peroxidation in liver microsomes. J. Biol Chem (250): 6496-6502, 1975.
65. Svingen BH, O'Neal FO, Aust SD: The role of superoxide and singlet oxygen in lipid peroxidation. Photochem Photobiol (28) 803-809, 1978.

66. Fong KL, McCay PB, Poyer JL, Keele BB, Misra H: Evidence that peroxidation of lysomal membranes is initiated by hydroxyl free radicals produced during flavin enzyme activity. J Biol Chem (248): 7792-7797, 1973.
67. McCord JM, Day ED: Superoxide-dependent production of hydroxyl radical catalyzed by iron-EDTA complex. FEBS Letts (86): 139-142, 1978.
68. Koster JF and Slee RG: Lipid peroxidation of rat liver microsomes. Biochim Biophys Acta (620): 489-499, 1980.
69. Piette LH, Hagadorn-Baxley L, Grover TA and Harwood PJ: A comparative kinetic study of the initiation of lipid peroxidation with OH radicals and ferrous iron. In: Bors W, Saran M, and Tait D (eds) Oxygen Radicals in Chemistry and Biology. W. de Gruyter, Berlin, 1984, pp 137-145.
70. Tien M, Svingen BA, Aust SD: An investigation into the role of hydroxyl radical in xanthine oxidase-dependent lipid peroxidation. Arch Bioch Biophys (216): 142-151, 1982.
71. Hochstein P, Nordenbrand K, Ernster L: Evidence for the involvement of iron in the ADP-activated peroxidation of lipids in microsomes and mitochrondia. Biochem Biophys Res Comm (14): 323-328, 1964.
72. Heaton FW, Uri N: The aerobic oxidation of unsaturated fatty acids and their esters: cobalt stearate-catalyzed oxidation of linoleic acid. J Lipid Res (2): 152-160, 1961.
73. Pederson TC, Buege JA, Aust SD: Microsomal electron transport. J Biol Chem (248): 7134-7141, 1973.
74. Pederson TC, Aust SD: The mechanism of liver microsomal lipid peroxidation. Biochim Biophys Acta (385): 232-241, 1975.
75. Poyer JL, McCay PB: Reduced triphosphopyridine nucleotide oxidase-catalyzed alterations of membrane phospholipids. J Biol chem (246): 263-296, 1971.
76. Chain-Beloff A, Catanzaro R, Serlupi-Crescenzi G: Oxidation of reduced nicotinamide-adenine dinucleotide phosphate in rat liver microsomes. Nature (198): 351-354, 1963.
77. Svingen, BA, Buege JA, O'Neal FO, Aust SD: The mechanism of NADPH-dependent lipid peroxidation. J Biol Chem. (254): 5892-5899, 1979.
78. Sugioka K, Nakano M: Mechanism of phospholipid induced ferric ion-ADP-adriamycin-coordination complex. Biochim Biophys Acta (713): 333-343, 1982.
79. Sugioka K, Nakano H, Noguchi T, Tsuchiya J, Nakano M: Decomposition of unsaturated phospholipid by iron-ADP-adriamycin coordination complex. Biochem Biophys Res Comm (100): 1251-1258, 1981.
80. Koppenol WH: Thermodynamics of the Fenton-driven Haber-Weiss and related reaction. In: Cohen G, Greenwald RA (eds) Oxy-Radicals and their Scavenger Systems. Vol I: Molecular Aspects. Elsevier Sci, Amsterdam, 1983, pp. 84-88.
81. Tam BK, McCay PB: Reduced triphosphopyridine nucleotide oxidase-catalyzed alterations of membrane phospholipids. J Biol Chem (245): 2295-2300, 1970.
82. Bucher JR, Tien M, Aust SD: The requirement for ferric in the initiation of lipid peroxidation by chelated ferrous iron. Biochem Biophys Res Comm (111): 777-784.

83. Aust SD, Buchner JR, Tien M: Evidence for the initiation of lipid peroxidation by a ferrous-dioxygen-ferric chelate complex. In: Bors W, Saran M, Tait D (eds) Oxygen Radicals in Chemistry and Biology. W. de Gruyter, Berlin, 1984, pp 147–154.
84. Morehouse LA, Thomas CE, Aust SD: Superoxide generation by NADPH-cytochrome P-450 reductase: The effect of iron chelators and the role of superoxide in microsomal lipid peroxidation. Arch Biochem Biophys (232): 366–377, 1984.
85. Thomas MJ, Mehl KS, Pryor WA: The role of the superoxide anion in the xanthine oxidase-induced autoxidation of linoleic acid. J Biol Chem (257): 8343–8347, 1982.
86. Samumi A, Chevion M, Czapski G: Unusual copper-induced sensitization of the biological damage due to superoxide radicals. J Biol Chem (256): 12632–12635, 1981.
87. Wills ED: Lipid peroxide formation in microsomes. Biochem J (113): 315–324, 1969.
88. May HE, McCay PB: Reduced triphosphopyridine nucleotide oxidase-catalyzed alterations of membrane phospholipids. J Biol Chem (243): 2288–2295, 1968.
89. Winterbourn CC: Hydroxyl radical production in body fluids. Biochem J (198): 125–131, 1981.
90. Tien M, Morehouse LA, Bucher JR, Aust SD: The multiple effects of ethylenediaminetetracetate in several models of lipid peroxidation systems. Arch Biochem Biophys (218): 450–458, 1982.
91. Lai CS, Piette LH: Hydroxyl radical production involved in lipid peroxidation of rat liver microsomes. Biochem Biophys Res Comm (78): 51–59, 1977.
92. Morehouse LA, Thomas CE, Aust SD: The effect of EDTA-Fe^{3+} on NADPH cytochrome P450 reductase-dependent superoxide generation and lipid peroxidation. Federation Proc (42): 2062, 1983.
93. Winston GW, Feierman DE, Cederbaum AI: The role of iron chelates in hydroxyl radical production by rat liver microsomes, NADPH-cytochrome P-450 reductase and xanthine oxidase. Arch Bioch Biophys (232): 378–390, 1984.
94. Winterbourn CC: Comparison of superoxide with other reducing agents in the biological production of hydroxyl radicals. Biochem J (1982): 625–628, 1979.
95. Tien M, Bucher JR, Aust SR: Thiol-dependent lipid peroxidation. Biochem Biophys Res Comm (107): 279–285, 1982.
96. Laurie SH: Transport and storage of metals. J Inher Metab Dis (6) (suppl. 1): 9–14, 1983.
97. Jacobs A: An intracellular transit iron pool. In: Iron metabolism. Ciba Found Sympo (51): 91–100, 1977.
98. Crichton RR, Roman F, Roland F: Iron mobilization from ferritin by chelating agents. J Inorganic Bioch (13): 305–316, 1980.
99. Primosigh JV, Thomas ED: Studies on the partition of iron in bone marrow cells. J Clin Invest (47): 1473–1482, 1968.
100. Wills ED: Lipid peroxide formation in microsomes: the role of non-haem iron. Biochem J (113): 325–332, 1969.
101. Tappel AL, Brown WD, Zalkin H, Maier VP: Unsaturated lipid peroxidation catalyzed by hematin compounds and its inhibition by vitamin E. J Am Oil Chem Soc (38): 5–9, 1961.
102. Tyler DD: Role of superoxide radicals in the lipid peroxidation of intracellular membranes. FEBS Lett (51): 180–183, 1975.

103. Wills ED: Mechanisms of lipid peroxide formation in tissues: role of metals and haematin proteins in the catalysis of the oxidation of unsaturated fatty acids. Biochim Biophys Acta (98): 238-251, 1965.
104. Bernheim F: Biochemical implications of pro-oxidants and antioxidants. Radiat Res Suppl (3): 17-32, 1963.
105. Kornburst DJ, Davis RD: Microsomal lipid peroxidation: 1. Characterization of the role of iron and NADPH. Mol Pharmacol (17): 400-407, 1980.
106. Richter HW, Wadell WH: Nonclassical Fenton systems: chain decomposition of hydrogen peroxide catalyzed by reduced 5-methylphenazinium and metal loss. In: Cohen G, Greenwald RA (eds) Oxy Radicals and Their Scavenger Systems. Vol I, Elseveir, Amsterdam and New York, 1983, pp 89-94.
107. Schwarzenbach G, Heller J: Komplexone XVIII. Die Eisen (II)-und Eisen (III)-komplexe der athylendiamin-tetraessigsaure und inr redoxgleichgewicht. Helv Chim Acta (34): 576-591, 1951.
108. Sevanian A, Muakkassah-Kelly SF, Montestruque S: The influence of phospholipase A_2 and glutathione peroxidase on the elimination of membrane lipid peroxides. Arch Biochem Biophys (223): 441-452, 1983.
109. Grossmann A, Wendel A: Non-reactivity of the selenoenzyme glutathione peroxidase with enzymatically hydroperoxized phospholipid. Eur J Biochem (135): 549-552, 1983.
110. Nishikimi M, Yamada H, Yagi K: Oxidation by superoxide of tocopherols dispersed in aqueous media with deoxycholates. Biochim Biophys Acta (627): 101-108, 1980.
111. Kearns DR: Physical and chemical properties of singlet oxygen. Chem Rev (71): 395, 1971.

7

PEROXIDATIC ACTIVATION OF PROCARCINOGENS: A ROLE FOR PROSTAGLANDIN H SYNTHASE IN INITIATION OF CHEMICAL CARCINOGENESIS

TERRY V. ZENSER and BERNARD B. DAVIS

1. INTRODUCTION

Chemical carcinogenesis is thought to be initiated by the enzymatic generation of reactive electrophilic intermediates. Enzymatic activation of procarcinogens to carcinogens by the mixed-function oxidase system has been clearly documented to be involved in the initiation of cancer by certain chemicals (1, 2). However, due to the apparent lack of metabolism of certain procarcinogens by mixed-function oxidases and/or negligible enzyme activity in certain tissues, additional pathways for enzymatic activation of procarcinogens have been investigated.

Peroxidatic activation of procarcinogens to electrophiles which covalently bind macromolecules has been demonstrated. This has been proposed as an additional pathway for initiation of chemical carcinogenesis. This chapter is designed to provide a basic background to peroxidases and their possible involvement in chemical carcinogenesis. A more detailed description of some specific peroxidatic oxidations related to chemical carcinogenesis are provided in volume 3 of this series with more extensive reviews of peroxidases in the literature (3-5).

Peroxidases (Donor: H_2O_2 oxidoreductase; EC 1.11.1.7.) are enzymes which reduce peroxides. The peroxidase reaction is characterized by a two-electron oxidation-reduction reaction as follows:

$$P + ROOH \longrightarrow P(I) + ROH + H_2O \quad 1)$$

$$P(I) + AH_2 \longrightarrow P(II) + AH^{\cdot} \quad 2)$$

$$P(II) + AH_2 \longrightarrow P + AH^{\cdot} \quad 3)$$

P represents the native peroxidase enzyme; P(I) is the two-electron oxidized enzyme; and P(II) is the one-electron oxidized form of the native enzyme. In this reaction, ROOH is either hydrogen peroxide or an organic hydroperoxide. Oxidized forms of the enzyme are reduced by single electron donors such as AH_2 with formation of the oxidized free radical cation species ($AH^{\cdot}$). An alternative mechanism utilizes a two-electron donor (BH_2) to reduce the oxidized enzyme to its native state. This reaction is the following:

$$P(I) + BH_2 \longrightarrow P + B \quad 4)$$

Many procarcinogens are suitable electron donors for peroxidases and, as such, may be considered cosubstrates in the peroxidase reaction. Peroxidase metabolism of the bladder procarcinogen benzidine results in the production of a free radical cation (6) and the covalent binding of activated benzidine to nucleic acids (7). Peroxidase activation of this aromatic amine in intact tissue has been demonstrated (8). Successive one-electron reduction of oxidized peroxidase enzymes with subsequent formation of free radicals is an attractive concept for the initiation of chemical carcinogenesis. However, a one step two-electron oxidation of certain procarcinogens can not be ruled out at this time. In addition to the classical dehydrogenation reactions described above, peroxidases can also catalyze hydroperoxide reduction by transfer of the hydroperoxide oxygen to an acceptor (A) as follows:

$$ROOH + A \longrightarrow ROH + AO \quad 5)$$

Thus, the mechanism of peroxidase metabolism of chemicals is complex and multiple reaction pathways exist.

2. CHARACTERISTICS OF HEMOPROTEIN PEROXIDASES

Mammalian peroxidases have different hemoprotein prosthetic groups associated with their apoproteins. Lactoperoxidase and eosinophil peroxidase are both thought to contain hemin (ferriprotoporphyrin IX) (9). The hemoprotein in myeloperoxidase is not known but is thought to be similar to heme a, the heme found in cytochrome c oxidase (9, 10). The porphyrin iron atom in all hemoprotein peroxidases is in the Fe(III) rather than Fe(II) state. The four equitorial coordinate positions of iron are occupied by porphyrin nitrogen atoms. A fifth axial position is occupied with a protein group and for cytochrome c peroxidase this has been shown to be histidine. The remaining axial position appears to comprise part of the active site of the enzyme. In the absence of substrate, this position is occupied by a weakly coordinating water molecule. During the catalytic cycle, this aquo ligand is displaced by a portion of the peroxide substrate, which dissociates during enzymatic reduction and product formation. This coordination position may also be irreversibly occupied by a strongly coordinating ligand such as cyanide, fluoride, or azide, which explains the ability of these agents to inhibit peroxidatic reactions (3, 11). The amino acid composition of human eosinophil peroxidase, human myeloperoxidase, bovine milk lactoperoxidase, and pig intestinal peroxidase are all similar in having a high content of arginine, leucine and aspartic acid (9). Their isoelectric points are generally basic with that for eosinophil peroxidase being greater than pH 11. While the properties of the prosthetic groups and amino acid analyses of human eosinophil peroxidase and lactoperoxidase are similar, eosinophil peroxidase but not lactoperoxidase can use a chloride anion as an electron donor. Therefore, differences in the protein environment of the hemin prosthetic group can affect the overall catalytic activity of the peroxidase.

Peroxidases are often distinguished from one another by their substrate, inhibitor and pH profiles (12, 13). A wide range of organic hydroperoxide specificity is observed with higher activities usually observed with comparable concentrations of H_2O_2. Because some peroxidases are capable of one-electron oxidation of donors while others are capable of both one- and two-electron oxidation, donor specificity provides a means of characterizing some peroxidases. Guaiacol is a common one-electron donor used to measure peroxidase activity. In contrast, peroxidase-catalyzed halogenation is a property of only the few enzymes capable of two-electron oxidations (i.e. thyroid peroxidase) (5). The affinity for cyanide and azide as reflected by decreased enzymatic activity also varies with the peroxidase. Optimum pH profiles for mammalian peroxidases are usually between pH 7 and 8 with lactoperoxidase having a higher optimum. Plant peroxidases such as horseradish peroxidase have a broad pH range with considerable activity at pH's as low as 4.0.

Spectrally distinct complexes of peroxidases are readily formed by the peroxide substrate in the absence of an electron donor. Chance (14, 15) and George (16, 17) have used the spectral changes observed with peroxidase and H_2O_2 to determine basic characteristics of this reaction mechanism. The two-electron oxidized form of horseradish peroxidase described as P(I) in equation 1 has a distinct absorption maxima at 658 and 407 nm and is referred to as compound I (green). P(II), equation 2, the one-electron oxidized form of the enzyme is referred to as compound II (red) and has absorption maxima at 561, 530, and 417 nm (3, 11). Although other suggestions have been made, it now appears that one electron of the resting state enzyme is lost from the iron atom, yielding Fe(IV), and a second electron is lost from the π-system of the porphyrin ring. Reduction of compound I to compound II results from reduction of the porphyrin radical, and reduction of compound II occurs by reduction of Fe(IV) to Fe(III) (18). Similar results have been reported for other hemoprotein peroxidases.

The molecular weight of peroxidases varies over a wide range. Purified eosinophil peroxidase has a molecular weight of 71,000 ± 1,000 being composed of two subunits of 58,000 and 14,000 in a 1:1 stoichiometry (9). Myeloperoxidase is thought to be composed of a tetramer containing subunits similar to that of eosinophil peroxidase. Bovine lactoperoxidase has a molecular weight of 77,500 (19). Two rat uterus peroxidases with molecular weights of 92,000 and 40,000 which are designated peroxidase I and II, respectively (13). Rat uterus peroxidases have been separated and characterized in terms of their substrate specificities, kinetics, pH optima, inhibitors, and stabilities. Both enzymes are hemoprotein peroxidases. The origin of these peroxidases has been a point of considerable controversy. Recently, Olsen and Little have shown that antibodies against rat uterine peroxidase II strongly inhibit eosinophil peroxidase (20). Both enzymes also have similar electrophoretic mobilities, and spectral properties and could be purified by essentially identical techniques. Thus, uterine peroxidase II and eosinophil peroxidase may be one and the same enzyme. Wagai and Hosoyo (12) purified rat uterus peroxidase using 4% cetyltrimethylammonium bromide to solubilize the enzyme. Those investigators report a peroxidase with molecular weight of 77,000 and a smaller molecular weight peroxidase not characterized. Part of the discrepancy in assigning molecular weights to uterine peroxidases is due to the localization of mammalian peroxidases to membranes (lactoperoxidase and salivary gland peroxidase are notable exceptions) and the different solubilization procedures employed by the various investigators during their purification procedures. Aggregation is also a problem encountered during purification of these solubilized enzymes.

Prostaglandin H synthase (PHS) is a hemoprotein peroxidase consisting of two separate activities: fatty acid cyclooxygenase and prostaglandin hydroperoxidase (21). Fatty acid cyclooxygenase is responsible for the initial bis-dioxygenation of the unsaturated fatty acid (Figure 1).

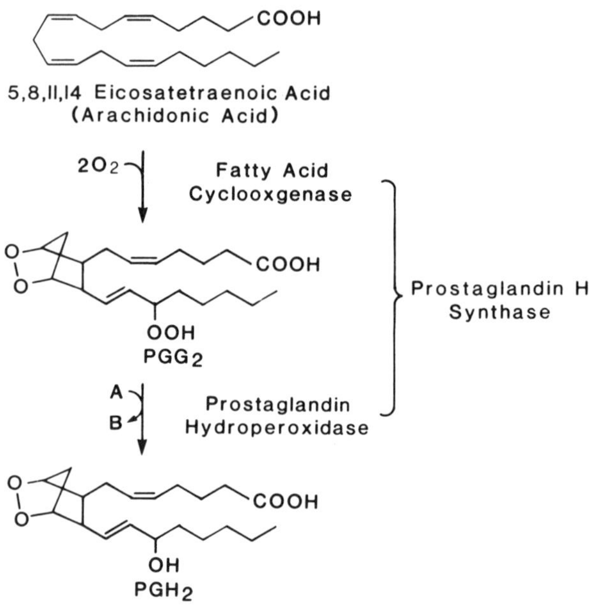

FIGURE 1. Reactions catalyzed by prostaglandin H synthase. "A" denotes an electron donating cosubstrate which is oxidized to "B".

The hydroperoxidase activity is responsible for the subsequent reduction of the lipid peroxide prostaglandin G_2. During this reduction, suitable electron donors become oxidized (see equations 1 to 5 above). This is illustrated in Figure 1 by the conversion of A to B. The substrate requirements of fatty acid cyclooxygenase are very specific. Unsaturated fatty acid substrates have in common the location of a triene system which terminates at the sixth carbon from the alkyl terminus. In contrast, prostaglandin hydroperoxidase can reduce a variety of lipid peroxides. Several procarcinogens and protoxins have been shown to be oxidized by the hydroperoxidase activity of PHS to reactive intermediates which bind macromolecules (for review, 22-24). Aspirin irreversibly inhibits the cyclooxygenase by acetylation of a serine moiety (25). Other nonsteroidal anti-inflammatory drugs also inhibit cyclooxygenase. These drugs do not inhibit peroxidase activity. Prostaglandin hydroperoxidase appears to have a much weaker affinity for cyanide and azide (less inhibition) than horseradish peroxidase (26). PHS has been shown to exist as a dimer with each individual subunit having a molecular weight estimated to be from 70,000 to 79,000 (25, 27, 28). The hemin prosthetic group of PHS is weakly bound to the enzyme and becomes dissociated during isolation (29). In fact, the completely purified enzyme is entirely devoid of enzymatic activity unless a source of hemin is provided.

Other mammalian hemoproteins which have been extensively studied are catalase, cytochrome P450 and hemoglobin. The former is well known for its ability to regulate cellular H_2O_2 content. Cytochrome P450 is a necessary component of the mixed-function oxidase system. Hemoglobin has the vital function of an oxygen carrier in red cells. Under certain conditions, each of these hemoproteins has peroxidatic activity. In contrast to mammalian peroxidases, plant peroxidases have been studied in more detail. The peroxidatic activity of the yeast cytochrome c peroxidase is unusual in that it preferentially oxidizes ferrocytochrome c. The proposed stereochemical mechanism of the cytochrome c peroxidase reaction has been reported (30).

Peroxidases thus far implicated in the direct activation of chemical procarcinogens are hemoprotein peroxidases as opposed to peroxidases such as the selenium requiring glutathione peroxidase. The latter cannot be converted from its oxidized to native state by a procarcinogenic electron donating cosubstrate.

Glutathione peroxidase will only use reduced glutathione (GSH) as an electron donor yielding oxidized glutathione (GSSG) as illustrated by the following equation:

$$ROOH + 2\ GSH \longrightarrow ROH + GSSG + H_2O \qquad 6)$$

Glutathione peroxidase will reduce a variety of hydroperoxide (ROOH) substrates. Thus, like catalase, it functions to keep tissue peroxide tone at a low level (31). The availability of peroxide substrate will determine the extent of tissue hemoprotein peroxidatic activation of procarcinogens. Thus, peroxidases may both initiate and prevent cancer.

3. SOME HEMOPROTEIN PEROXIDASES INVOLVED IN THE ACTIVATION OF PROCARCINOGENS

From a historical point of view, myeloperoxidase was one of the earliest recognized peroxidases. In 1863, the oxidation of organic compounds by pus was observed (32). More than thirty years later, this reaction was attributed to myeloperoxidase (33). Neutrophils and monocytes contain myeloperoxidase activity. This enzyme may constitute as much as 5% of the dry weight of these cells and is located in the azurophilic primary lysosomal granules. The phagocytizing potential of these cells has been attributed in part to their myeloperoxidase activity. During phagocytosis, up to 5-10% of a cells myeloperoxidase can be released. H_2O_2 produced during phagocytosis is available to this peroxidase which catalyzes H_2O_2-dependent oxidation of Cl^- to hypochlorous acid. The latter contributes to intracellular bactericidal activity. Myeloperoxidase oxidizes the potent anti-inflammatory agent sulindac sulfide to its sulfoxide (34). A similar reaction is not catalyzed by lactoperoxidase, horseradish peroxidase or hemin (35). Sulindac sulfide was oxidized by human polymorphonuclear leukocytes stimulated with phagocytizing opsonized zymosan or phorbol myristate. This reaction was inhibited by catalase but not superoxide dismutase.

Polymorphonuclear leukocyte oxidation of the sulfide in each case was attributed to myeloperoxidase (34).

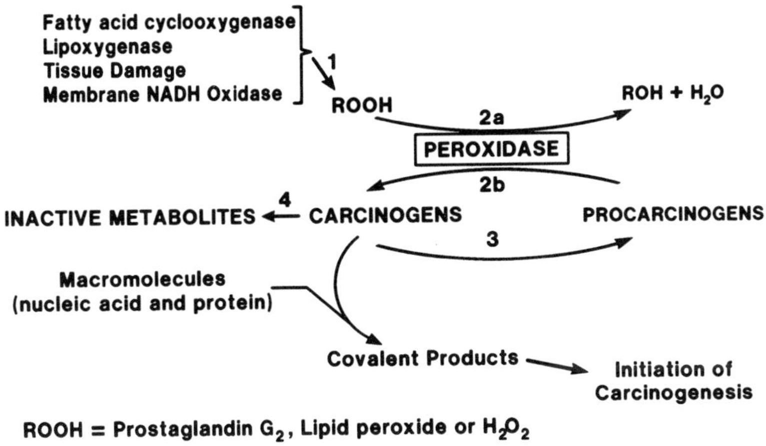

FIGURE 2. Proposed model for peroxidatic activation of procarcinogens. Peroxides are produced by several in vivo processes (1) and are reduced by peroxidases (2a). Procarcinogenic electron donating cosubstrates are then activated by peroxidases to carcinogens (2b). Carcinogens can be reduced back to their procarcinogens (3), further metabolized to inactive metabolites (4) or bind covalently to macromolecules and initiate carcinogenesis. Activation may be prevented as indicated at the numbered sites: 1) prevent synthesis or decrease availability of peroxides (e.g. glutathione peroxidase); 2a) binding of certain agents (e.g. cyanide) to active site; 2b) competition for oxidation by alternative electron donors (e.g. propylthiouracil); 3) reduction of carcinogens back to procarcinogens (e.g. ascorbate); and 4) inactivation of carcinogens (e.g. conjugate formation).

Pulmonary macrophages play an important role in host defense by virtue of their capacity to phagocytize inhaled foreign material. Human pulmonary macrophages have been shown to activate (±)-trans-7,8-dihydroxy-7,8-dihydrobenzo(a)pyrene to mutagens that were detected in co-cultivated Chinese hamster V79 cells (36). This mutagenic activity was ascribed to mixed-function oxidases. However, peroxidatic activation of this proximal carcinogen has been demonstrated (37-39) and macrophage myeloperoxidase activation is also possible.

Pulmonary macrophages have been hypothesized to carry particles of carcinogens absorbed from tobacco smoke in the respiratory tract and to interact with respiratory epithelium in the causation of respiratory cancer (36). A general scheme illustrating peroxidase involvement in chemical carcinogenesis is shown in Figure 2.

Uterus peroxidase activity can be induced by estrogen. The physiological role for this induction is not clear. Attempts have been made to use peroxidase activity as a marker for tissue growth regulated by estrogen. However, these results have been conflicting. It has also been hypothesized that diethylstilbestrol-induced carcinogenicity may involve activation of this synthetic estrogen by target tissue peroxidase(s) (40). The transformation frequency induced in Syrian hamster embryo fibroblast cells by diethylstilbestrol and its analogues suggested that peroxidatic activation may be involved. This conclusion was strengthened by subsequent studies demonstrating PHS-catalyzed metabolism of diethylstilbestrol to cis,cis-dienestrol (41). This product has previously been shown to result from horseradish peroxidase and mouse uterine peroxidase metabolism of diethylstilbestrol (40) and is a product of in vivo metabolism. It is important to note that molecular weights for PHS and the high molecular weight uterine peroxidase are similar (13, 25, 27, 28) and that high levels of PHS are reported in the uterus of cycling female rats (42). The relationship between the uterine peroxidase and PHS is not known.

Hemoglobin is a hemoprotein with a molecular weight of 67,000 containing four hemes per molecule with iron in the ferrous, Fe(II), state (43). This protein is responsible for the transport of oxygen in red blood cells. It is estimated that there are about 750 gm of hemoglobin in the total circulating blood of a 70 kg man. In contrast to the cytochromes, which undergo reversible changes between Fe(II) and Fe(III) forms during electron transport, these globular hemoproteins do not change valence as oxygen is bound and lost. Thus, hemoglobin remains in the Fe(II) state. Oxidized hemoglobin, Fe(III), methemoglobin, does not function as a reversible oxygen carrier. Eyer has proposed that oxidation of arylamines by red cells may have important toxic and carcinogenic effects (44). According to this proposal, activated metabolites, like aminophenols, occasionally escape the liver and become further activated by oxyhemoglobin in blood. This was demonstrated by an i.v. infusion of dimethylaminophenol which produced a maximal rise in ferrihemoglobin in 5 to 10 min. Dimethylaminophenol is thought to be oxidized to the phenoxyl radical and N,N-dimethylquinonimine. Hemoglobin has also been shown to N-oxidize 4-chloroaniline in the presence of NADPH-methemoglobin reductase (45). In the latter case, there is a close resemblance between hemoglobin and the microsomal monoxygenases, where cytochrome b_5 functions as an electron donor to cytochrome P450. Substrate specificity and reactivity by this mechanism has not been widely studied. The apparent K_m value (5.9 mM) for 4-chloroaniline was quite high compared to concentrations of xenobiotics which may occur in vivo. There may be intrinsic chemical properties which may influence a compound to react by this pathway.

Leukotrienes are important products of certain unsaturated fatty acid hydroperoxides. The initial product in this pathway is an epoxide (i.e. leukotriene A_4) which decomposes nonenzymatically to epoxy alcohols and triols. Isolation of the enzyme catalyzing this epoxidation reaction has proven difficult. Activity is observed in rat lung cytosol. However, this activity is not abolished by heating and has been difficult to purify (46). Several laboratories have reported evidence for hemoglobin and other hemoproteins catalyzing the conversion of a fatty acid hydroperoxide to an epoxide. Dix and Marnett (47) have proposed a mechanism for this reaction in which hematin reduces the hydroperoxide by one electron generating a fatty acid alkoxyl radical and a ferryl-hydroxo complex. The alkoxyl radical cyclizes to an epoxide-containing allylic radical. The hydroxyl radical (coordinated to hematin) can then be incorporated into the epoxy allylic radical to form the epoxy alcohols and triols. Alternatively, the hydroxyl radical coordinated to hematin is also available for insertion into procarcinogens. If hemoglobin and/or hemoproteins play such a role in leukotriene synthesis in vivo, a considerable number of activated hemoglobin molecules may be available for the activation of procarcinogens. Hematin-catalyzed epoxidation of 7,8-dihydroxy-7,8-dihydrobenzo(a)pyrene by unsaturated fatty acid hydroperoxides has been demonstrated (48).

The role of hemoglobin in procarcinogen activation and xenobiotic metabolism is for the most part interesting speculation at this time. However, due to the abundance of hemoglobin in in vivo systems care must be taken to prevent artifactual results due to hemoglobin contamination of samples. This may be particularly true for analyses of peroxidatic products in whole blood.

4. INVOLVEMENT OF PROSTAGLANDIN H SYNTHASE IN THE INITIATION OF CHEMICAL CARCINOGENESIS

Although peroxidatic activation of certain procarcinogens has been shown to result in covalent binding to nucleic acids in vitro, the involvement of these events in the initiation of chemical carcinogenesis has not been demonstrated in vivo. Confirmation of such a cause/effect relationship will be facilitated by the development of treatment strategies for prevention of chemically-induced cancers.

In vivo metabolism of procarcinogens by PHS has not been directly assessed. Retrograde perfusion of dog kidneys with a solution containing benzidine and arachidonic acid demonstrated that less aqueous soluble material (metabolites) was present if dogs were pretreated with meclofenamic acid, a PHS inhibitor (49). While this suggests PHS catalyzed metabolism, no products of PHS-catalyzed benzidine metabolism were identified. Experiments with intact tissue preparations are consistent with the occurrence of PHS-dependent activation in vivo. Eling and collaborators have used cultured cells (50) and enriched populations of certain lung cells (51) to demonstrate PHS activation. With these cells, PHS-catalyzed metabolism of diethylstilbestrol and ($\pm$)-trans-7,8-dihydroxy-7,8-dihydrobenzo(a)pyrene was demonstrated. Renal inner medullary slice PHS has been shown to catalyze binding of ^{14}C-benzidine to macromolecules (8). In vivo studies would be facilitated with the availability of an easily detectable specific product of PHS peroxidatic oxidation.

Benzidine is a very good peroxidase cosubstrate. Several peroxidases have been shown to use benzidine as an electron donor (52). This metabolism can result in the formation of benzidine free radical cations (6). Benzidine, in contrast to its mono and diacetylated analogues, is not a substrate for mixed-function oxidases (53). In addition, flavin-containing monoxygenases do not metabolize benzidine (54). PHS-catalyzed benzidine metabolism has been reported with microsomes prepared from several organs in different species. A glutathione conjugate has been a demonstrated product of this reaction (55) and may be useful for assessing in vivo peroxidatic metabolism of benzidine or other aromatic amines. However, because benzidine is a substrate for many, if not all peroxidases (see Figure 3, type 2 procarcinogen), the involvement of PHS may be better assessed with a more specific substrate (see Figure 3, type 1 procarcinogen).

The peroxidase or peroxidases involved in in vivo activation of procarcinogens need to be identified. As indicated above, there are many different kinds of peroxidases. One tissue may have several different peroxidases all of which may be capable of metabolizing

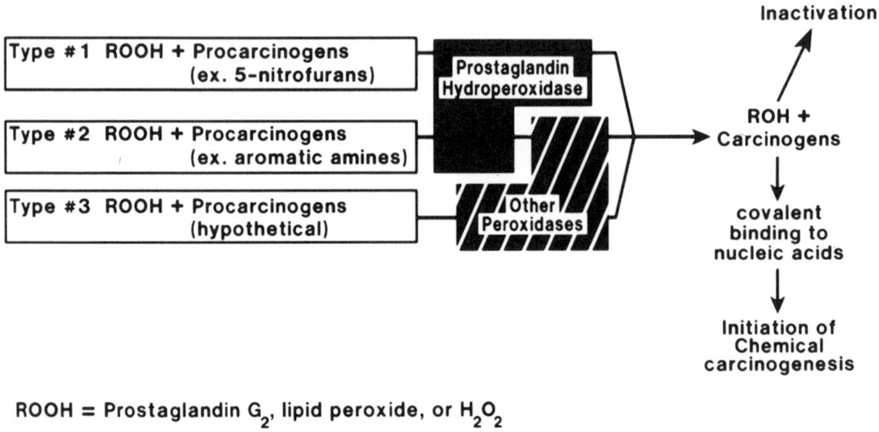

FIGURE 3. Multiple peroxidase involvement in the activation of various procarcinogens.

the chemical of interest (Figure 3). This problem has been addressed with microsomes prepared from ram seminal vesicles. Monoclonal antibodies against PHS were used to precipitate peroxidase activity in 1% Tween 20 solubilized seminal vesicle preparations (56). Using this technique, the hydroperoxidase activity of PHS was shown to be the major peroxidase catalyzing oxidation of diphenyliso-benzofuran, epinephrine and phenylbutazone. Because seminal vesicles contain the highest levels of PHS of any mammalian tissue and negligible levels of other peroxidases such as cytochrome P450, these findings may not be extrapolated to other tissues. This technique may be applicable to in vivo experiments if one assumes that activation by target tissues is important in initiation. Extracts of target tissue could be assessed for metabolism of relevant carcinogens in the presence and absence of monoclonal antibodies. While theoretically possible, the application of this technique may prove difficult due to

(a) unavailability of antibody, (b) practical considerations associated with the analysis of oxidation, and (c) difficulty in reaching unequivocal conclusions. There have been instances where a chemical is only metabolized by a specific peroxidase or in which the product of metabolism by one peroxidase is unique. A procarcinogen with either of these characteristics would identify the peroxidase responsible for its in vivo activation.

The 5-nitrofuran bladder carcinogen 2-amino-4-(5-nitro-2-furyl)thiazole (ANFT) may be useful for in vivo studies assessing PHS activation. ANFT is thought to be the more proximal carcinogen in N-[4-(5-nitro-2-furyl)-2-thiazolyl]formamide (FANFT)-induced bladder cancer (57). The only oxidative metabolism of ANFT demonstrated to date has been PHS catalyzed (58). Mixed-function oxidase metabolism of ANFT has not been demonstrated. Horseradish peroxidase, lactoperoxidase and chloroperoxidase do not metabolize ANFT (52). Thus, ANFT oxidation appears specific for the prostaglandin hydroperoxidase portion of PHS. A glutathione conjugate of PHS activated ANFT has recently been identified using spectral and electrochemical techniques (59). Preliminary results suggest that this conjugate is produced by renal medullary slices. Thus, a thioether conjugate of PHS activated ANFT may be a suitable biological marker for PHS metabolism in vivo.

In vivo PHS activation is thought to be involved in FANFT-induced rat bladder cancer. Co-administration of 0.5% aspirin with 0.2% FANFT reduced the number of early morphological bladder lesions (60) and reduced the incidence of bladder cancer from 87% to 37% in rats (61). Aspirin treatment was shown to significantly reduce bladder prostaglandin E_2 synthesis during the 12 weeks that FANFT was administered (60). Because prostaglandins can contribute to promotion (62), aspirin's effect on FANFT carcinogenesis could be due to inhibition of promotion as well as initiation.

Demonstration of an ANFT glutathione conjugate in vivo and its inhibition with selective PHS inhibitors would strongly support conclusions from the feeding studies described above that FANFT-induced cancer is initiated by PHS. Such a completely defined system for enzymatic activation of any procarcinogen would allow development of an in vivo screening procedure. This screening procedure would test for inhibitors of in vivo product formation in accessible body fluids such as blood, bile and urine. Following dose response inhibition and toxicology studies, suitable inhibitors might be chosen for the more expensive animal feeding tumorigenesis studies.

A scheme for PHS activation of procarcinogens has been proposed for the purpose of developing chemoprevention strategies (Figure 4). In this model, benzidine, a representative procarcinogen, is activated by the prostaglandin hydroperoxidase component of PHS (63). One biologically relevant reaction of activated benzidine metabolite(s) may be covalent binding to critical cellular macromolecules such as protein and DNA. Such binding is thought to be involved in the initiation of the carcinogenic process (64). According to this model, prevention of benzidine binding to macromolecules can occur by 1) inhibiting the generation of peroxide cosubstrate (e.g. aspirin at site 1a); 2) inhibiting hydroperoxidase-catalyzed activation (e.g. propylthiouracil); 3) reduction of oxidized intermediate(s) back to the parent (procarcinogenic) compound (e.g. vitamin C); and 4) conjugation of activated intermediate(s) (e.g. glutathione). Each of these steps has been specifically demonstrated for PHS metabolism of benzidine in vitro (63). Site 2 is of particular interest. Agents acting at this site are thought to inhibit the necessary interaction between the procarcinogen, e.g. benzidine and prostaglandin hydroperoxidase. They would function as competitive inhibitors of procarcinogen metabolism. Competition by

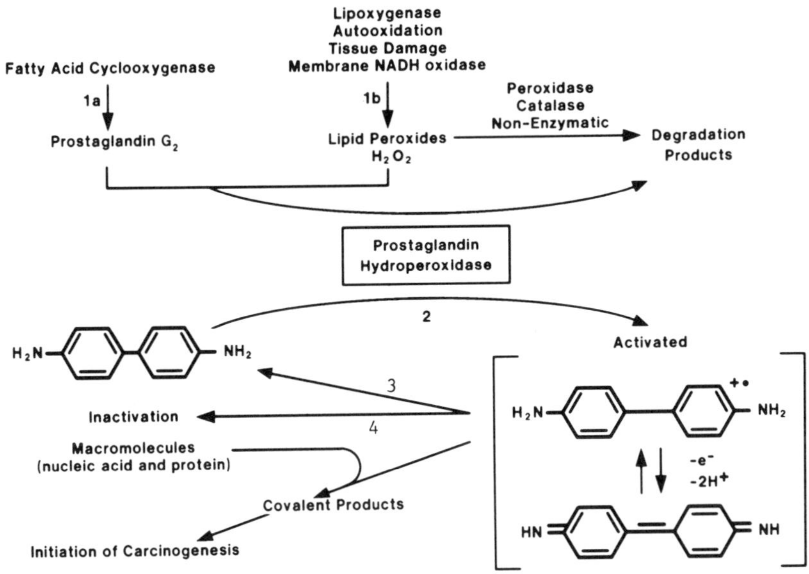

FIGURE 4. A general model for the activation of procarcinogens by prostaglandin hydroperoxidase. Benzidine is used as a representative procarcinogen. Both the one-electron (free radical cation) and the two-electron (benzidine diimine) oxidation products have been detected during peroxidatic activation of benzidine. Possible sites at which strategies may be developed to prevent prostaglandin hydroperoxidase initiation of carcinogenesis are numbered and discussed in the text (63).

cosubstrates for PHS metabolism has been demonstrated with phenidone (65). Similar in vivo results appear to explain the antithyroid effects of propylthiouracil and methimazole in the treatment of hyperthyroidism (66). These two drugs are metabolized by PHS (67) and prevent benzidine activation by PHS (63). Inhibitors at site 2 may inhibit activation and prevent tumorigenesis but not reduce prostaglandin and thromboxane syntheses. This would be an advantageous outcome because the latter have many important physiologic functions. ANFT and other procarcinogens may fit into this scheme. Although an ANFT radical has not been observed, an activated intermediate is formed which produces a unique glutathione conjugate (59).

Prevention of initiation is an important goal of studies investigating enzymatic activation of procarcinogens. Once the steps in the initiation process are known, experiments designed to prevent this process can be more rationally approached. Criteria for relating activation of a procarcinogen by a specific enzyme to initiation of carcinogenesis should include the following: 1) enzymatic activation in vitro; 2) enzymatic activation in vivo; 3) selective inhibitors reduce activation both in vitro and in vivo; 4) selective inhibitors reduce tumorigenicity; and 5) corresponding amounts of inhibitors do not alter other steps in the carcinogenic process (e.g. promotion).

REFERENCES

1. Gelboin HV: Benzo[a]pyrene metabolism, activation, and carcinogenesis: role and regulation of mixed-function oxidases and related enzymes. Physiol Rev (60): 1107-1166, 1980.
2. Conney AH: Induction of microsomal enzymes by foreign chemicals and carcinogenesis by polycyclic aromatic hydrocarbons: G.H.A. Clowes memorial lecture. Cancer Res (42): 4875-4917, 1982.
3. Saunders BC, Holmes-Siedle AG, Stark BP: Peroxidase. Butterworths, Washington, 1964.
4. Yamazaki I: Peroxidase. In: Hayaishi O (ed) Molecular mechanism of oxygen activation. Academic Press, New York, 1974, pp 535-555.
5. Morrison M, Schonbaum GR: Peroxidase-catalyzed halogenation. Annu Rev Biochem (45): 861-888, 1976.
6. Wise RW, Zenser TV, Davis BB: Prostaglandin H synthase metabolism of the urinary bladder carcinogens benzidine and ANFT. Carcinogenesis (4): 285-289, 1983.
7. Zenser TV, Mattammal MB, Armbrecht HJ, Davis BB: Benzidine binding to nucleic acids mediated by the peroxidative activity of prostaglandin endoperoxide synthetase. Cancer Res (40): 2839-2845, 1980.
8. Rapp NS, Zenser TV, Brown WW, Davis BB: Metabolism of benzidine by a prostaglandin-mediated process in renal inner medullary slices. J Pharmacol Exp Ther (215): 401-406, 1980.
9. Bolscher BGJM, Plat H, Wever R. Some properties of human eosinophil peroxidase, a comparison with other peroxidases. Biochim Biophys Acta (784): 177-186, 1984.

10. Olsen RL, Little C: Purification and some properties of myeloperoxidase and eosinophil peroxidase from human blood. Biochem J (209): 781-787, 1983.
11. Saunders BC: Peroxidases and catalases. In: Eichhorn GL (ed) Inorganic biochemistry. Elsevier Science Publishers, New York, 1973, pp 988-1021.
12. Wagai N, Hosoya T: Partial purification of estrogen-dependent peroxidase of rat uterus and comparison of the properties with those of other animal peroxidases. J Biochem (91): 1931-1942, 1982.
13. Olsen RL, Little C: The peroxidase activity of rat uterus. Eur J Biochem (101): 333-339, 1979.
14. Chance B: The spectra of the enzyme-substrate complexes of catalase and peroxidase. Arch Biochem Biophys (41): 404-415, 1952.
15. Chance B: The kinetics and stoichiometry of the transition from the primary to the secondary peroxidase peroxide complexes. Arch Biochem Biophys (41): 416-424, 1952.
16. George P: The chemical nature of the second hydrogen peroxide compound formed by cytochrome c peroxidase and horseradish peroxidase. Biochem J (54): 267-276, 1953.
17. George P: Chemical nature of the secondary hydrogen peroxide compound formed by cytochrome c peroxidase and horseradish peroxidase. Nature (169): 612-613, 1952.
18. Dolphin D: The electronic configurations of catalases and peroxidases in their high oxidation states: a definitive assessment. Israel J Chem (21): 67-71, 1981.
19. Rombauts WA, Schroeder WA, Morrison M: Bovine lactoperoxidase. Partial characterization of the further purified protein. Biochemistry (6): 2965-2977, 1967.
20. Olsen RL, Little C: Comparative studies on oestrogen-induced rat uterus peroxidase and rat eosinophil peroxidase. Biochem J (207): 613-616, 1982.
21. Ohki S, Ogino N, Yamamoto S, Hayaishi O: Prostaglandin hydroperoxidase, an integral part of prostaglandin endoperoxide synthetase from bovine vesicular gland microsomes. J Biol Chem (254): 829-836, 1979.
22. Marnett LJ, Eling TE: Cooxidation during prostaglandin biosynthesis: a pathway for the metabolic activation of xenobiotics. In: Hodgson E, Bend JB, Philpot RM (eds) Reviews in biochemical toxicology. Elsevier Science Publishers, New York, 1983, pp 135-172.
23. Eling T, Boyd J, Reed G, Mason R, Sivarajah K: Xenobiotic metabolism by prostaglandin endoperoxide synthetase. Drug Metab Rev (14): 1023-1053, 1983.

24. Gale PH, Egan RW: Prostaglandin endoperoxide synthase catalyzed oxidation reactions. In: Pryor WA (ed) Free radicals in biology, vol VI. Academic Press, Orlando, Florida, 1984, pp 1-38.
25. Roth GJ, Siok CJ, Ozols J: Structural characteristics of prostaglandin synthetase from sheep vesicular gland. J Biol Chem (255): 1301-1304, 1980.
26. Zenser TV, Mattammal MB, Davis BB: Mechanism of FANFT cooxidation by prostaglandin endoperoxide synthetase. J Pharmacol Exp Ther (214): 312-317, 1980.
27. Hemler M, Lands WEM, Smith WL: Purification of the cyclooxygenase that forms prostaglandins. Demonstration of two forms of iron in the holoenzyme. J Biol Chem (251): 5575-5579, 1976.
28. Ho PPK, Towner RD, Esterman MA: Purification and characterization of fatty acid cyclooxygenase from human platelets. Prep Biochem (10): 597-613, 1980.
29. Ueno R, Shimizu T, Kondo K, Hayaishi O: Activation mechanism of prostaglandin endoperoxide synthetase by hemoproteins. J Biol Chem (257): 5584-5588, 1982.
30. Poulos TL, Kraut J: The stereochemistry of peroxidase catalysis. J Biol Chem (255): 8199-8205, 1980.
31. Chance B, Sies H, Boveris A: Hydroperoxide metabolism in mammalian organs. Physiol Rev (59): 527-605, 1979.
32. Klebs VE: Die pyrogene substanz. Zentralblatt fur die Medicinischen Wissenschaften (6): 417-418, 1863.
33. Linossier MG: Contribution a Letude des ferments oxydants. Sur La peroxydase du pus. Comptes Rendus Hebdomadaires des Seances et Memoires de la Societe de Biologie (50): 373-375, 1898.
34. Dougherty HW, Hen A, Kuehl FA Jr: The role of polymorphonuclear peroxidase-dependent oxidants in inflammation. Agents Actions (7): 167-173, 1980.
35. Egan RW, Gale PH, Baptista EM, Kennicott KL, VandenHeuvel WJA, Walker RW, Fagerness PE, Kuehl FA Jr: Oxidation reactions by prostaglandin cyclooxygenase-hydroperoxidase. J Biol Chem (256): 7352-7361, 1981.
36. Hsu IC, Harris CC, Yamaguchi M: Induction of ouabain-resistant mutation and sister chromatid exchanges in Chinese hamster cells with chemical carcinogens mediated by human pulmonary macrophages. J Clin Invest (64): 1245-1252, 1979.
37. Marnett LJ, Johnson JT, Bienkowski MJ: Arachidonic acid-dependent metabolism of 7,8-dihydroxy-7,8-dihydrobenzo[a]pyrene by ram seminal vesicles. FEBS Lett (106): 13-16, 1979.
38. Reed GA, Marnett LJ: Metabolism and activation of 7,8-dihydrobenzo[a]pyrene during prostaglandin biosynthesis. J Biol Chem (257): 11368-11376, 1982.
39. Sivarajah K, Lasker JM, Eling TE: Prostaglandin synthetase-dependent cooxidation of (±)-benzo[a]-pyrene-7,8-dihydrodiol by human lung and other mammalian tissues. Cancer Res (41): 1834-1839, 1981.

40. Metzler M, McLachlan JA: Peroxidase-mediated oxidation, a possible pathway for metabolic activation of diethylstilbestrol. Biochem Biophys Res Commun (85): 874-884, 1978.
41. Degen GH, Eling TE, McLachlan JA: Oxidative metabolism of diethylstilbestrol by prostaglandin synthetase. Cancer Res (42): 919-923, 1982.
42. Ham EA, Cirillo VJ, Zanetti ME, Kuehl FA Jr: Estrogen-directed synthesis of specific prostaglandins in uterus. Proc Natl Acad Sci USA (72): 1420-1424, 1975.
43. Harper HA, Rodwell VW, Mayes PA: Review of physiological chemistry. Lange Medical Publications, Los Altos, California, 1969, p 201-203.
44. Eyer P: The red cell as a sensitive target for activated toxic arylamines. Arch Toxicol Suppl (6): 3-12, 1983.
45. Golly I, Hlavica P: The role of hemoglobin in the N-oxidation of 4-chloroaniline. Biochim Biophys Acta (760): 69-76, 1983.
46. Pace-Asciak CR, Mizuno K, Yamamoto S: Resolution by DEAE-cellulose chromatography of the enzymatic steps in the transformation of arachidonic acid into 8,11,12- and 10,11,12-trihydroxy-eicosatrienoic acid by the rat lung. Prostaglandins (25): 79-84, 1983.
47. Dix TA, Marnett LJ: Hematin-catalyzed rearrangement of hydroperoxylinoleic acid to epoxy alcohols via an oxygen rebound. J Am Chem Soc (105): 7001-7002, 1983.
48. Dix TA, Marnett LJ: Free radical epoxidation of 7,8-dihydroxy-7,8-dihydrobenzo[a]pyrene by hematin and polyunsaturated fatty acid hydroperoxides. J Am Chem Soc (103): 6744-6746, 1981.
49. Zenser TV, Mattammal MB, Brown WW, Davis BB: Cooxygenation by prostaglandin cyclooxygenase from rabbit inner medulla. Kidney Int (16): 688-694, 1979.
50. Degen GH, Wong A, Eling TE, Barrett JC, McLachlan JA: Involvement of prostaglandin synthetase in the peroxidative metabolism of diethylstilbestrol in Syrian hamster embryo fibroblast cell cultures. Cancer Res (43): 992-996, 1983.
51. Sivarajah K, Jones KG, Fouts JR, Devereux T, Shirley JE, Eling TE: Prostaglandin synthetase and cytochrome P-450-dependent metabolism of (±)benzo(a)pyrene 7,8-dihydrodiol by enriched populations of rat clara cells and alveolar type II cells. Cancer Res (43): 2632-2636, 1983.
52. Wise RW, Zenser TV, Davis BB: Peroxidase metabolism of the urinary bladder carcinogen 2-amino-4-(5-nitro-2-furyl)thiazole. Cancer Res (43): 1518-1522, 1983.
53. Zenser TV, Mattammal MB, Davis BB: Cooxidation of benzidine by renal medullary prostaglandin cyclooxygenase. J Pharmacol Exp Ther (211): 460-464, 1979.

54. Ziegler DM, Poulsen LL: Hepatic microsomal mixed-function amine oxidase. In: Fleischer S, Packer L (eds) Methods in enzymology, vol LII Biomembranes, Part C: Biological oxidations microsomal, cytochrome P-450, and other hemoprotein systems. Academic Press, New York, 1978, pp 142-151.
55. Wise RW, Zenser TV, Davis BB: Oxidation of the carcinogen benzidine to free radical cations may initiate the carcinogenic process. Fed Proc (43): 589 (abstract), 1984.
56. Pagels WR, Sachs RJ, Marnett LJ, Dewitt DL, Day JS, Smith WL: Immunochemical evidence for the involvement of prostaglandin H synthase in hydroperoxide-dependent oxidations by ram seminal vesicle microsomes. J Biol Chem (258): 6517-6523, 1983.
57. Klemencic JM, Wang CY: Mutagenicity of nitrofurans. In: Bryan GT (ed) Carcinogenesis, a comprehensive survey. Raven Press, New York, 1978, pp 99-130.
58. Mattammal MB, Zenser TV, Davis BB: Prostaglandin hydroperoxidase-mediated 2-amino-4-(5-nitro-2-furyl)-^{14}C-thiazole metabolism and nucleic acid binding. Cancer Res (41): 4961-4966, 1981.
59. Rice JR, Zenser TV, Davis BB: Structural characterization of thioether conjugates of 2-amino-4-(5-nitro-2-furyl)thiazole catalyzed by prostaglandin H synthase. Fed Proc (43): 346 (abstract), 1984.
60. Cohen SM, Zenser TV, Murasaki G, Fukushima S, Mattammal MB, Rapp NS, Davis BB: Aspirin inhibition of N-[4-(5-nitro-2-furyl)-2-thiazolyl]formamide-induced lesions of the urinary bladder correlated with inhibition of metabolism by bladder prostaglandin endoperoxide synthetase. Cancer Res (41): 3355-3359, 1981.
61. Murasaki G, Zenser TV, Davis BB, Cohen SM: Inhibition by aspirin of N-[4-(5-nitro-2-furyl)-2-thiazolyl]-formamide induced bladder carcinogenesis and enhancement of forestomach carcinogenesis. Carcinogenesis (5): 53-55, 1984.
62. Fischer SM, Slaga TJ: Modulation of prostaglandin synthesis and tumor promotion. In: Powles TJ, Bockman RS, Honn KV, Ramwell P (eds) Prostaglandins and cancer: first international conference. Alan R Liss Inc, New York, 1982, pp 255-264.
63. Zenser TV, Mattammal MB, Wise RW, Rice JR, Davis BB: Prostaglandin H synthase-catalyzed activation of benzidine: a model to assess pharmacologic intervention of the initiation of chemical carcinogenesis. J Pharmacol Exp Ther (227): 545-550, 1983.
64. Miller JA: Carcinogenesis by chemicals: an overview-G.H.A. Clowes memorial lecture. Cancer Res (30): 559-576, 1970.

65. Marnett LJ, Siedlik PH, Fung LWM: Oxidation of phenidone and BW755C by prostaglandin endoperoxide synthetase. J Biol Chem (257): 6957-6964, 1982.

66. Gilman AG, Murad F: Thyroid and antithyroid drugs. In: Goodman LS, Gilman A (eds) The pharmacological basis of therapeutics. MacMillan Publishing Co, New York, 1975, pp 1398-1422.

67. Zelman SJ, Rapp NS, Zenser TV, Mattammal MB, Davis BB: Antithyroid drugs interact with renal medullary prostaglandin H synthase. J Lab Clin Med (104): 185-192, 1984.

ACKNOWLEDGEMENTS

This work was supported by the Veterans Administration, U.S. Public Health Service Grant CA-28015 from the National Cancer Institute through the National Bladder Cancer Project, and the American Cancer Society, Missouri Chapter. The authors wish to thank Mrs. Sandy Melliere for skillful assistance in the preparation of this manuscript.

8

SELENIUM-DEPENDENT GLUTATHIONE PEROXIDASE AND EICOSANOID PRODUCTION

NORBERTA W. SCHOENE

1. INTRODUCTION

In 1971, Lands et al. (1) reported that a functioning glutathione peroxidase (GSHPx, EC 1.11.1.9) system inhibited two dioxygenases, prostaglandin (PG) synthase from sheep vesicular glands and soybean lipoxygenase, suggesting that product activation is an important feature in the mechanism for both enzymes. These initial experiments were followed by a series of studies (See Chapters 1 and 2 this book and Ref. 2) which demonstrated the peroxide requirement for activation and maintenance of the catalytic function of PG synthase in vitro. The synthase is inhibited at peroxide concentrations 1 µmole/l or greater. This property of duality, stimulation at low and inhibition at high concentrations, has led to the hypothesis that "peroxide tone" can regulate the production of eicosanoids, oxygenated metabolites of arachidonic acid (20:4). The provocative question that arises from these enzymatic studies is "Does GSHPx function as a modulator of dioxygenases by regulating the hydroperoxide level within the cell?" A survey of the literature indicates that evidence is accumulating to support a positive response to this query and it is this evidence which is the subject of this chapter.

2. SELENIUM-DEPENDENT GLUTATHIONE PEROXIDASE - FUNCTION AND CHARACTERISTICS

An extensive discussion of the discovery and subsequent characterization of GSHPx is beyond the purview of this report. For more detailed information the reader is referred to recent reviews (3,4). Properties of GSHPx which are pertinent to interactions with eicosanoid production will be discussed.

The selenoprotein GSHPx (5) and the hemeprotein catalase, function in a concerted manner with superoxide dismutase(s) to protect the cell from damage by toxic oxygen derivatives (6). Cytosolic superoxide dismutase(s) reduce superoxide to H_2O_2 which can then be removed either by GSHPx or catalase. GSHPx, found in the cytosol and mitochondrial matrix, reduces both hydrogen peroxide (H_2O_2) and hydroperoxides (organic hydroperoxy acids) to H_2O and hydroxy fatty acids, respectively. Catalase, present in peroxisomes, acts only on H_2O_2. In addition to these protective enzymes, there are the GSH-S-transferases which also possess peroxide-reducing activity, but they act only on lipid hydroperoxides and not on H_2O_2. These non-selenium enzymes are found in cytosolic and microsomal fractions (7) and are often designated "selenium-independent peroxidase" activity (8,9). Localization and specificity as well as substrate availability are important characteristics to be noted when considering the function of these enzymes in vivo. Until recently, it was unclear whether GSHPx could reduce esterified hydroperoxides. Sevanian et al. (10) and Grossmann and Wendel (11) have clarified this matter by demonstrating that GSHPx requires its hydroperoxide substrate to be unesterified. For optimal activity, GSHPx also requires a constant supply of GSH and thus relies on the hexose monophosphate (HMP) shunt in tandem with GSH-reductase to furnish this obligatory substrate (4). Diet supplies the selenium which is another absolute requirement for this enzyme (5). In fact, it is this requirement that investigators have exploited to study the relationship between GSHPx activity and 20:4 metabolism since this activity can be manipulated by the amount of selenium in the diet (3,12-14). Activities of some of the isomeric GSH-S-transferases increase when dietary selenium is inadequate, most likely as a compensatory mechanism for maintenance of non-lethal levels of hydroperoxides within the cell (8,9). However, some tissues such as platelets, heart, lung, and skin possess little or no "selenium-independent peroxidase" activity and therefore would be more susceptible to alteration in hydroperoxide levels when GSHPx activity is either low or absent (15). The GSH-S-transferase which catalyzes the conjugation of GSH with leukotriene A to form leukotriene

C is reported to have different properties compared to the transferases of liver and kidney (16) and may represent only one of many different types of GSH transferase activity.

3. DIETARY SELENIUM, SELENIUM-DEPENDENT GLUTATHIONE PEROXIDASE, AND EICOSANOID PRODUCTION

Bryant et al. (12) fed rats a diet deficient in selenium for 5 to 6 weeks and demonstrated that GSHPx activity in platelets decreased to 13% of the activity found in the platelets from control rats supplemented with selenium. With the addition of 1-^{14}C-20:4 to washed, aspirin-treated platelets, these investigators showed that the platelets with decreased GSHPx activity produced significant amounts of isomeric trihydroxy fatty acids (THETE) compared to the control platelets in which only trace amounts of these fatty acids could be found. They suggested that the limited presence of GSHPx in selenium-deficient platelets decreased the conversion of L-12-hydroperoxy-5,8,10,14-eicosatetraenoic acid (12-HPETE) to L-12-hydroxy-5,8,10,14-eicosatetraenoic acid (12-HETE) and resulted in the molecular rearrangement of 12-HPETE to THETE. It was concluded that GSHPx could function as the peroxidase in the platelet lipoxygenase pathway. Studies with human platelets by this group (17) indicated that 20:4 metabolism by the 12-lipoxygenase pathway is linked to the HMP shunt through GSHPx. Recently, these same investigators (18) demonstrated an accumulation of 12-HPETE with a concomitant reduction in the HMP shunt in platelets from selenium-deficient rats compared to platelets from selenium-supplemented rats. This is considered further evidence for enzymatic coupling of the lipoxygenase activity with GSHPx activity. In addition to altering product profiles as described above, GSHPx may also suppress lipoxygenase activity as it does PG synthase activity. For example, a 12-lipoxygenase has been partially purified from rat lung and has been shown to be sensitive to GSHPx (19). Also, the 5-lipoxygenase, the dioxygenase which produces the leukotrienes, is inhibited by GSHPx (20).

Dietary studies with selenium to investigate interactions of altered GSHPx activity on PG synthase have also been conducted. The production of two different end products resulting from PG synthase activity, TXA_2

thromboxane A_2 (TXA_2) and prostacyclin (PGI_2), have been studied in platelets and aortic tissue, respectively (13,21-24). These two products have opposing effects on platelet function, TXA_2 stimulates platelet aggregation and PGI_2 inhibits this response. An imbalance in the PGI_2/TXA_2 may contribute to the development of vascular diseases (25).

Kawaguchi et al. (21) isolated microsomal fractions from platelets of hypercholesterolemic rabbits to determine the mechanism for the increase of TXB_2 (TXA_2 metabolite) biosynthesis seen earlier in this dietary model. They observed decreases in glutathione and GSHPx activity compared to the values found in controls. These investigators suggested that the decrease in GSHPx activity was large enough to account for the increased synthesis of TXB_2. These experiments indicate that other dietary factors such as cholesterol could affect GSHPx activity by yet unknown mechanisms. Again, with selenium-deficient rodents, Masukawa et al. (22) reported increased aggregation to ADP, collagen, and 20:4 in platelets from selenium-deficient rats compared to platelets from selenium-adequate rats. They also found that 20:4-induced respiratory distress in mice was enhanced in a selenium-deficient group compared to a selenium-supplemented one. This type of induced respiratory distress is caused by platelet thrombi formation in the lungs and is most probably due to increased synthesis of TXA_2. Preliminary results from our laboratory (23) also indicate that PG synthase in platelets is enhanced by a deficiency in GSHPx. Collagen stimulated platelets from rats fed a selenium-deficient diet for 12 weeks produced increased amounts of TXB_2 as measured by radio-immunoassay compared to platelets from control rats. Selenium-deficient, gel-filtered platelets also aggregated to a greater extent when stimulated with ADP and collagen than the control platelets. Supplementing rats with amounts of selenium above the known nutritional requirement produced anomalous results (13,24). GSHPx activity was increased both in platelets and aortas of rats fed sodium selenite. The increased production of malondialdehyde (MDA) observed led these investigators to conclude that GSHPx stimulates both the production of TXA_2 in platelets and PGI_2 in aortas. However, the use of MDA as a measure of TXA_2 and PGI_2 formation is subject to question and these results should be interpreted with caution (2).

From the above investigations, it can be concluded that a link does exist between GSHPx and the in vivo oxidation of 20:4 to eicosanoids by both PG synthase and lipoxygenase(s). Further studies will clarify the subtle control that the intracellular "peroxide tone" exerts on the amount and type of eicosanoid produced. The source of substrate for the dioxygenases may influence the relative amount of eicosanoid formed. In platelets, exogenous 20:4 is preferentially utilized by the 12-lipoxygenase whereas endogenous 20:4 is preferentially utilized by PG synthase (26). This result is indicative of coupling between the release of 20:4 and subsequent oxidation by PG synthase. Recently, Kent et al. (27) infused 15-HPETE into rabbit aortas and demonstrated that PG and PGI_2 synthases were differentially inhibited in a dose dependent manner by the hydroperoxide. PG synthase was stimulated by low doses of exogenous 15-HPETE whereas PGI_2 synthase was inhibited. In vitro inhibition of PGI_2 synthase by hydroperoxides had been demonstrated earlier (28). Aortas from selenium-deficient rats produced significantly less PGI_2 compared to controls as determined by a bioassay based on the inhibition of platelet aggregation (22) probably as a result of the sensitivity of the PGI_2 synthase to inhibition by hydroperoxides. The experiment with perfused aortas and other experiments (29,30) utilizing mixtures of platelets and leukocytes illustrate how eicosanoid products from different types of cells interact to modulate physiological functions associated with hemostatic and inflammatory mechanisms.

4. HYDROPEROXIDES AND PATHOPHYSIOLOGICAL CONSEQUENCES

Elevated levels of hydroperoxides could unfavorably alter the balance among eicosanoids produced by various cells resulting in aberrant responses to noxious stimuli. For example, because of the different effect of hydroperoxides on the two synthases, TXA_2 biosynthesis in simulated platelets coupled with a decreased PGI_2 production by endothelial cells could potentiate the development of vascular diseases (25) or enhance carcinogenic (31) and metastatic processes (32). Thus, it appears that hydroperoxides and TXA_2 could act in a concerted fashion to promote chemotaxis, adhesion, and other metastatic mechanisms, while PGI_2, if not depressed by the presence of hydroperoxides, could counteract these deleterious responses to circulating tumor cells. Studies on influences of dietary selenium on the development of neoplasia have recently been initiated. Mammary carcinogenesis was increased in rats fed selenium-deficient diets compared to control rats only when the diet also contained high levels of polyunsaturated fat (33). Formation of hydroperoxides would be enhanced by such diets. On the other hand, selenium supplementation above nutritional requirements appears to protect laboratory animals from cancer by mechanism(s) independent of GSHPx activity (31,34).

Products of the 12-lipoxygenase pathway have been implicated in platelet-platelet interactions (35) and have been shown to act as chemotactic agents (36). These products may be involved in promotion of skin tumors in mice (37) and may also participate in the metabolic activation of carcinogens (38). Interestingly enough, tumor cell metastasis is thought to be a consequence of adherence of circulating tumor cells to vessel walls (39). Chemotactic agents have been demonstrated to induce hyperadherence of tumor cells to endothelial cells (40). The relative amounts of TXA_2, PGI_2 and hydroperoxides could produce an environment either promoting or inhibiting carcinogenesis and metastasis. Honn et al. (32) have proposed that PGI_2/TXA_2 determines the interactions among platelets, tumor cells, and vessel walls. Experimental evidence to support this hypothesis has been reported (41-43). Mice pretreated with 15-HPETE prior to tail vein injection of B16a tumor cells develop 300-500% more metastatic lesions than untreated controls (41). Donati et al. (42) have shown that TXA_2 is increased in highly metastatic cells and that there is an inverse correlation between PGI_2 synthesis and metastatic potential. Increased levels of TXB_2 have also been found in breast tumor tissue and these increased levels are correlated with tumor size and enhanced metastasis (43).

5. SUMMARY

Because of its ability to reduce hydroperoxides, GSHPx functions as a regulator of the dioxygenases which produce the eicosanoids. If for any reason the activity of GSHPx is compromised in a cell, e.g., by inadequate intake of selenium, hydroperoxide levels would increase and thereby cause alterations in the amount and type of eicosanoids that are produced in response to stimuli. Pathophysiological responses due to these alterations may favor the inflammatory, proliferative, and invasive aspects of cancer.

REFERENCES

1. Lands WEM, Lee RE, Smith WL: Factors regulating the biosynthesis of various prostaglandins. Ann NY Acad Sci(180):107-122, 1971.

2. Warso MA, Lands WEM: Lipid peroxidation in relation to prostacyclin and thromboxane physiology and pathophysiology. Brit Med Bull(39):277-280, 1983.

3. Sunde RA, Hoekstra WG: Structure, synthesis and function of glutathione peroxidase. Nutr Rev(38):265-273, 1980.

4. Flohe L: Glutathione peroxidase brought into focus. In: Pryor WA(ed) Free radicals in biology Vol V. Academic Press, New York, 1982, pp 223-254.

5. Rotruck JT, Pope AL, Ganther HE, Swanson AB, Hafeman DG, Hoekstra WG: Selenium: biochemical role as a component of glutathione peroxidase. Science(179):588-590, 1973.

6. Diplock AT: The role of vitamin E and selenium in the prevention of oxygen-induced tissue damage. In: Spallholz JE, Martin JL, Ganther HE(eds) Selenium in biology and medicine. AVI Publishing Co, Westport, CT, 1981, pp 303-316.

7. Friedberg T, Bentley P, Stasieki P, Glatt HR, Raphael D, Oesch F: The identification, solubilization, and characterization of microsome-associated glutathione S-transferases. J Biol Chem(254): 12028-12033, 1979.

8. Lawrence RA, Burk RF: Glutathione peroxidase activity in selenium-deficient rat liver. Biochem Biophys Res Commun(71):952-958, 1976.

9. Prohaska JR, Ganther HE: Glutathione peroxidase activity of glutathione-S-transferases from rat liver. Biochim Biophys Res Commun(76):437-445, 1977.

10. Sevanian A, Muakkassah-Kelly SF, Montestruque S: The influence of phospholipase A_2 and glutathione peroxidase on the elimination of membrane lipid peroxides. Arch Biochem Biophys(223):441-452, 1983.

11. Grossmann A, Wendel A: Non-reactivity of the selenoenzyme glutathione peroxidase with enzymatically hydroperoxidized phospholipids. Eur J Biochem(135):549-552, 1983.

12. Bryant RW, Bailey JM: Altered lipoxygenase metabolism and decreased glutathione peroxidase activity in platelets from selenium-deficient rats. Biochem Biophys Res Commun(92):268-276, 1980.

13. Doni MG, Avventi GL, Bonadiman L, Bonaccorso G: Glutathione peroxidase, selenium, and prostaglandin synthesis in platelets. Am J Physiol(240):H800-H803, 1981.

14. Levander OA, DeLoach DP, Morris VC, Moser PB: Platelet glutathione peroxidase activity as an index of selenium status in rats. J Nutr(113):55-63, 1983.

15. Carmagnol R, Sinet PM, Jerome H: Selenium-dependent and non-selenium-dependent glutathione peroxidases in human tissue extracts. Biochim Biophys Acta(759):49-57, 1983.

16. Jakschik BA, Harper T, Murphy RC: Leukotriene C_4 and D_4 formation by particulate enzymes. J Biol Chem(257):5346-5349, 1982.

17. Bryant RW, Simon TC, Bailey JM: Role of glutathione peroxidase and hexose monophosphate shunt in the platelet lipoxygenase pathway. J Biol Chem(257):14937-14943, 1982.

18. Bryant RW, Simon TC, Bailey JM: Hydroperoxy fatty acid formation in selenium deficient rat platelets: coupling of glutathione peroxidase to the lipoxygenase pathway. Biochem Biophys Res Commun(117):183-189, 1983.

19. Yokoyama C, Mizuno K, Mitachi H, Yoshimoto T, Yamanoto S, Pace-Asciak CR: Partial purification and characterization of arachidonate 12-lipoxygenase from rat lung. Biochim Biophys Acta(750):237-243, 1983.

20. Egan RW, Tischler AN, Baptista EM, Ham EA, Soderman DD, Gale PH: Specific inhibition and oxidative regulation of 5-lipoxygenase. In: Samuelsson B, Paoletti R, Ramwell, P(eds) Advances in prostaglandin, thromboxane, and leukotriene research Vol 11. Raven Press, New York, 1983, pp 151-157.

21. Kawaguchi H, Ishibashi T, Imai Y: Increased thromboxane B_2 biosynthesis in platelets. Lipids(17):577-584, 1982.

22. Masukawa T, Goto J, Iwata H: Impaired metabolism of arachidonate in selenium deficient animals. Experientia(39):405-406, 1983.

23. Schoene NW, Morris VC, Levander OA: Effects of selenium deficiency on aggregation and thromboxane formation in rat platelets. Fed Proc(43):477, 1984.

24. Doni MG, Bonaccorso G, Piva E: High glutathione peroxidase and prostacyclin-like activity generation in rat aorta. Haemostasis(13):248-253, 1983.

25. Moncada S, Vane JR: Arachidonic acid metabolites and the interactions between platelets and blood-vessel walls. N Engl J Med(300):1142-1147, 1979.

26. Sautebin L, Caruso D, Galli G, Paoletti R: Preferential utilization of endogenous arachidonate by cyclooxygenase in incubations of human platelets. FEBS Lett(157):173-178, 1983.

27. Kent RS, Diedrich SL, Whorton AR: Regulation of vascular prostaglandin synthesis by metabolites of arachidonic acid in perfused rabbit aorta. J Clin Invest(72):455-465, 1983.

28. Salmon JA, Smith DR, Flower RJ, Moncada S, Vane JR: Further studies on the enzymatic conversion of prostaglandin endoperoxides into prostacyclin by porcine aorta microsomes. Biochim Biophys Acta(523):250-262, 1978.

29. Maclouf J, Fruteau de Laclos B, Borgeat P: Stimulation of leukotriene biosynthesis in human blood leukocytes by platelet-derived 12-hydroperoxy-icosatetraenoic acid. Proc Natl Acad Sci USA(79): 6042-6046, 1982.

30. Marcus AJ, Broekman MJ, Safier LB, Ullman HL, Islam N: Formation of leukotrienes and other hydroxy acids during platelet-neutrophil interactions in vitro. Biochem Biophys Res Commun(109):130-137, 1982.

31. Griffin AC: The chemopreventive role of selenium in carcinogenesis. In: Arnott MS, Van Eys J, Wang Y-M(eds) Molecular interrelations of nutrition and cancer. Raven Press, New York, 1982, pp 401-408.

32. Honn KV, Busse WD, Sloane BF: Prostacyclin and thromboxanes-implications for their role in tumor cell metastasis. Biochem Pharmacol(32):1-11, 1983.

33. Ip C, Sinha DK: Enhancement of mammary tumorigenesis by dietary selenium deficiency in rats with a high polyunsaturated fat intake. Cancer Res(41):31-34, 1981.

34. Ip C: Selenium-mediated inhibition of mammary carcinogenesis. Biol Tr Elem Res(5):317-330, 1983.

35. Dutilh CE, Haddeman E, ten Hoor F: Role of the arachidonate lipoxygenase pathway in blood platelet aggregation. In: Samuelsson B, Ramwell PW, Paoletti R(eds) Advances in prostaglandin and thromboxane research Vol 6. Raven Press, New York, 1980, pp 101-105.

36. Nakao J, Ooyama T, Ito H, Chang WC, Murota ST: Comparative effects of lipoxygenase products of arachidonic acid on rat aortic smooth muscle cells. Atherosclerosis(44):339-342, 1982.

37. Fischer SM, Mills GD, Slaga TJ: Inhibition of mouse skin tumor promotion by several inhibitors of arachidonic acid metabolism. Carcinogenesis(3):1243-1245, 1982.

38. O'Brien PJ: Peroxide-mediated metabolic activation of carcinogens. In: Yagi K(ed) Lipid peroxides in biology and medicine. Academic Press, New York, 1982, pp 317-338.

39. Fantone JC, Elgas LJ, Weinberger L, Varani J: Modulation of tumor cell adherence by prostaglandins. Oncology(40):421-426, 1983.

40. Hayashi H, Yoshida K, Ozak T, Ushijima K: Chemotactic factor associated with invasion of cancer cells. Nature(226):174-175, 1970.

41. Honn KV, Cicone B, Skoff A: Prostacyclin: a potent antimetastatic agent. Science(212):1270-1272, 1981.

42. Donati MB, Rotilio D, Delaini F, Giavazze A, Mantovani A, Poggi A: Animal models for the study of platelet-tumor cell interaction. In: Jamieson GJA(ed) Interaction of platelets and tumor cells. Alan R Liss, New York, 1982, pp 159-176.

43. Karmali RA, Welt S, Thaler HT, Lefevre F: Prostaglandins in breast cancer: relationship to disease stage and hormone status. Br J Cancer(48):689-696, 1983.

9

RADIATION EFFECTS ON EICOSANOID FORMATION

PETER POLGAR, GERALD HAHN AND LINDA TAYLOR

INTRODUCTION

Over the past few years a number of reports have come out describing alterations in prostaglandin (PG) synthesis consequent to irradiation of tissue. This includes the various ranges of ultraviolet (UV) and the ionizing radiation, gamma and X-ray. The literature indicates that all these forms of radiation potentially alter the anabolism and perhaps the catabolism of the prostaglandins. These alterations manifest as either increases or decreases in the content of PGs within the whole tissue, in body fluids or culture medium. In certain cases following radiation there appear to be temporal changes in PG synthesis such as an initial increase in the production of PGs followed by a decrease in the production of PGs after radiation. As we develop this topic it will become apparent that an understanding of the relationship of eicosanoid metabolism to irradiation may be important toward the understanding of the role of radiation in malignancy. This is particularly so with respect to the production of prostacyclin (PGI_2) which is a potent antiaggregatory substance and which may act to inhibit metastasis via the circulatory system (1).

TYPES OF RADIATION

Ultraviolet Radiation

Ultraviolet radiation can be divided into 3 types, according to wavelength: UVA at 315-400nm, UVB at 280-315nm, and UVC at 100-280nm. Unlike ionizing radiation, ultraviolet radiation is selectively absorbed by proteins and nucleic acids. This results in the excitation of valence electrons within the absorbing molecules to higher energy levels. These molecules then have a

W.E.M. Lands (ed.), BIOCHEMISTRY OF ARACHIDONIC ACID METABOLISM.

greater probability of undergoing a chemical reaction. Cysteine and cystine at neutral and alkaline pH absorb UV at 230 nm and above in appreciable amounts. In addition the probability that the absorbed light will result in a chemical change is relatively high. This makes the cystine S-S bond when it is present in a protein a labile target (2). UV radiation also results in alterations in cellular DNA (3). These changes may be occurring through several means, one of them being through formation of breaks in the chromosomal DNA. These breaks are probably generated through the formation of oxygen radicals (4,5). UV radiation also causes the peroxidation of membrane lipids (6,7,8).

Ionizing Radiation

X-ray and gamma radiation are termed ionizing radiation because this extremely short (X-ray: 1.5×10^{-5} - 1×10^{-9} cm, gamma: 1×10^{-8} cm and less), highly energetic radiation ionizes atoms by shearing off some of their electrons (9,10). The amount of energy absorbed by a tissue is measured in units called rads or grays (Gy). A rad is defined as 100 ergs of absorbed energy/gm of tissue. One hundred grays equals one rad. Ionizing radiation may affect a number of metabolic and structural events within the cell. Perhaps the most critical being 1) scission of nucleotide bonds within the DNA structure within the chromosome (11) and 2) an increase in lipid peroxidation which results in the release of oxygen radicals. These events are interrelated and, in fact, much of the damage produced by ionizing radiation may be achieved through the generation of free radicals (12,13). The generation of oxygen radicals could be particularly important for the synthesis of PGs since some of the enzyme systems involved in converting arachidonate to the PGs appear to be sensitive to hydroxyl radicals and are activated or inactivated depending on the existing concentration of these substances within the cell (14). Interestingly, experiments on the peroxidation of unsaturated fatty acids in micelle or liposome forms show that arachidonate is extremely vulnerable to peroxidation when exposed to ionizing radiation under conditions resembling the fatty acid, as it is in biomembranes (15).

EFFECTS OF IRRADIATION ON PG METABOLISM

The effect of gamma and X-radiation on cells in culture

There is no doubt that gamma or X-radiation alters the synthesis of PGs by certain mammalian cells in culture. For example, the endothelial cell is one cell type which appears to be very sensitive to this type of radiation. At least three laboratories have demonstrated that radiation above 200 rads has immediate as well as long term effects on the synthesis of prostacyclin (PGI_2) by these cells. Beatty et al. (16) demonstrated a dose dependant increase in PGI_2 synthesis by human umbilical vein endothelial cells. PGI_2 was measured as 6-keto- PGF_{1a} by radioimmunoassay 48 hours after irradiation with 1000, 5,000 and 10,000 rads. Our own laboratory showed that exposure of endothelial cells, obtained from calf pulmonary artery, to a 60Co source emitting gamma rays at 200-2,000 rads increases the synthesis of PGI_2 by these cells 2-3 fold above that of non- irradiated cultures (17,18). This increase in prostacyclin production by endothelial cells occurred in response to the vasoactive hormone, bradykinin, or the addition of free arachidonate to the cultures. The increase was observed at 4 hours past exposure and continued for at least 24 hours. It is also important to stress that in these experiments the cells were washed and incubated in fresh culture medium to determine PG production. This was done to eliminate the contribution to PG synthesis by any dead, floating cells and also to establish a realistic rate of PG synthesis (30 minutes or less). However, analysis of the medium in which the cells were irradiated showed no difference in PGI_2 content from control medium. The increase in PGI_2 production was not associated with cell death. However, all cell division ceased. More recent experiments by us suggest that not all mammalian cells are affected by gamma radiation in the same way. For example, we find no change in the synthesis of PGs by human lung fibroblasts after they are exposed to 60Co at 200-5,000 rads. These cells, after treatment, remain viable but, are nondividing (unpublished results).

Eldor and coworkers (19) reported an effect of X-ray (100 rads/minute for radiation below 1,000 rads) and 60Co, gamma radiation (doses above 1,000 rads) on PGI_2 synthesis by calf aortic endothelial cells. Their results show that at short intervals following irradiation (3-48 hours), endothelial cells increase their production of PGI_2. They suggest that the initial increase in PGI_2 synthesis by endothelial cells following irradiation is due to cell death. The medium was not changed after irradiation and some cells detached from the culture dish. These cells were suggested as the source of the increased PGI_2 production. At longer intervals after radiation (2-21 days), PGI_2 synthesis in the irradiated cultures falls below that of the control cultures. PGI_2 synthesis by these cells was determined using washed cells incubated with mellitin, or ionophore A23187, arachidonate, or PGH_2. The decline in PGI_2 synthesis after 2-21 days may have clinical significance as will be discussed below.

The effect of gamma irradiation on intact tissue

As with cells in culture there is a consensus in the literature that ionizing radiation alters the synthesis and perhaps the metabolism of prostaglandins in intact tissues. Generally an increase in PG synthesis has been reported following various doses of irradiation. However, decreases in synthesis have also been reported.

Sinzinger et al. reported an increase in PGI_2 synthesis in the rabbit aorta (20) following gamma irradiation of the rabbit in a $10cm^2$ area of the abdominal aorta. The increase was dose dependent (1,000 to 5,000 rads) and peaked between 2 and 4 hours after irradiation. By 48 hours PGI_2 production in the irradiated and control tissues were equal. This was followed by a long-lasting depression in PGI_2 production. These results are similar to those obtained with endothelial cells in culture. These authors suggest that this long-lasting depression might be contributing to the formation of lesions found at the sight of irradiation. On the other hand, Allen et al. reported a decrease in PGI_2 synthesis after a single dose of 200 rads gamma radiation to an area of excised human umbilical artery (21). Irradiation of human platelets, however, resulted in no change in TXB_2 formation even at doses of radiation above 2,000 rads.

Scheidkraut and coworkers observed an increase in urinary

TXB_2 and PGF_{2a} levels following whole body gamma irradiation of 200 and 1,000 rads (22). They postulated the increase to be due, at least in part, to alterations in the metabolism of PGs, either anabolic or catabolic. Donlon and coworkers (23) looked at urinary PGs following whole body irradiation of rats(100-900 rads). They reported an increase in PGE_2 and TXB_2 in the urine following irradiation. They suggested measurement of PGs as a noninvasive measure of radiation exposure. An increase in the presence of PGs in humans following radiation was reported by Tanner and coworkers (24). They reported an increase in PG like molecules in the plasma of patients with head and neck cancer undergoing irradiation. This increase in PGs was associated with the development of mucositis by these patients.

Much lower doses of ionizing radiation also affect PG production. Steel and coworkers (25) reported an increase in PGE_2 and PGF_{2a} concentrations in bronchial airways following exposure of guinea pigs to a 60Co source of 3 Gy. At three hours post radiation PGE concentrations went up. At 48 hours a transient increase in PGF_{2a} was observed. TXB_2 concentrations did not increase at any time. By 72 hours all values returned to normal. Steel and Cantravas in an earlier publication (26) reported on changes in PG production by parenchymal lung tissue of the guinea pig following whole body irradiation (a single dose of 0.5-3.0 Gy of gamma radiation). At 1-3 hours post exposure PGE_2, PGF_{2a} and TXB_2 were all elevated in parenchymal lung tissue of animals receiving 3.0 Gy. By 24 hours tissue PG levels returned to control levels. PG synthesis in response to histamine showed the same response. Irradiated tissue showed an increase in sensitivity to ionophore stimulated TXB_2 production at 24 hours post irradiation and PGF_{2a} production at 72 hours post radiation.

Bito and Klein (27) reported that X-irradiation (10 Gy) of the rabbit eye results in ocular inflammation and a rise in intraocular pressure. This was inhibited by intravenous pretreatment of the rabbit with flurbiprofen or indomethacin, inhibitors of PG synthesis. Vitamin E, a sink for oxygen radicals (28), also exerted some protective effect on the inflammation. These results suggested that an increase in PG synthesis upon irradiation of the eye produced the ocular inflammation.

Not all reports show a rapid increase in PG synthesis at low radiation doses. Ts'Ao et al. (29) reported a decrease in PGI_2 production by the rat lung on the first day after irradiation of the rat with 25 Gy of gamma rays to the right hemithorax. From 3 to 14 days after irradiation, PG production by strips of lung from the irradiated portion of the lung equaled those of either the shielded portion or strips from sham irradiated controls. However, following that time an increase in PGI_2 synthesis was observed for 2-6 months following irradiation. PGI_2 production and arterial perfusion were inversely related.

Effects of ultraviolet radiation on PG synthesis

UV radiation also has an important influence on PG synthesis. The effect of the different types of UV radiation on prostaglandin production by skin and the relationship of this production to the inflammatory response to UV has been well studied (30). In 1970 Sondergaard and Greaves (31) isolated an ethyl acetate extractable smooth muscle contracting agent from human skin exposed to UVB. They implicated PGs, present in the dermal perfusates, as the mediators in UV induced inflammation in human skin. Mathur and Gandhi (32) identified PGs in homogenates of UV irradiated skin. Suction blister fluid (33) has been widely used to investigate relationships between prostaglandin activities and UV irradiation. Arachidonic acid, PGD_2, PGE_2, PGF_{2a} and 6-keto-PGF_{1a} have been demonstrated by combined gas chromatography-mass spectrometry in suction blister exudates from human skin subjected to UVB irradiation (34,35). Arachidonate and metabolites increased with time in parallel with the increase in inflammation reaching a peak at 24 hours. However, inflammation persisted to 48 hours at which time levels of arachidonic acid and metabolites had returned to normal.

With UVC irradiation the time course of PG production correlates better with the observed inflammation than with UVB. Suction blister exudates from human skin irradiated with UVC show an increase in arachidonic acid, PGE_2 and PGF_{2a} which peaks at 18 hours (36). The levels return to normal by 48 hours at which time the inflammation is also fading.

With UVA, a 1,000 times greater radiation dose is needed to get a visible inflammation than with UVB or UVC. Suction blister fluid from UVA irradiated skin shows a rise in arachidonic acid and PGE_2 which peaks earlier (5-9 hours) as opposed to 18-24 hours for UVB or UVC. PGD_2 and 6-keto-PGF_{1a} also rise earlier(5 hours). The inflammatory response to UVA also develops earlier than with UVB or UVC indicating that prostaglandin may be involved in the early stages of inflammation in response to UVA irradiation. However, at 48 hours prostaglandin levels are down while inflammation is still up (37).

Additional evidence for the involvement of prostaglandins in the early stages of UV stimulated inflammmation comes from studies with nonsteroidal anti-inflammatory drugs (NSAID). Intradermally injected NSAID in guinea pigs at the time of UVB irradiation delayed the development of the inflammatory response (38). Topical application of indomethacin to guinea pig skin before irradiation with UVB decreased PGE_2 content and UVB induced redness to near normal (39). Oral administration of indomethacin inhibited PG production by human skin in response to UVC (36). However the inflammatory response was incompletely inhibited.

EFFECT OF RADIATION ON PG RECEPTORS

Johnson et al. (40) have conducted studies on the effects of UV light on the prostaglandin receptor. They exposed the isolated rat uterus to UV at 253.7 nm and found an inhibition of the contractile response to PGE_1 and PGF_{2a}. The response to acetylcholine was not affected. The inhibition could be reversed with the sulfhydryl-oxidizing agent, 5,5-dithiobis-(2-nitrobenzoic acid) (DTNB). Addition of DTNB at the time of UV irradiation prevented the inhibition of PG induced contractions. The effect of UV on the binding of PGE_2 to membrane preparations from human skin has been studied by Lord and Ziboh (41). They found that exposure of the membrane preparation to UVB irradiation inhibited subsequent binding of ^3H-PGE_2 and that the exposure of the bound ^3H- PGE_2 membrane complex to UVB results in rapid release of ^3H-PGE_2. Protection of disulfide groups in the membrane preparation prevented the effects of UVB irradiation. In addition a-tocopherol also prevented the UVB-induced inhibition of ^3H-PGE_2 binding, suggesting that membrane lipids are also affected by UVB.

POSSIBLE MECHANISMS LEADING TO ALTERATIONS IN PG METABOLISM FOLLOWING IRRADIATION

Synthesis of Prostaglandin

Our results with endothelial cells in culture show that gamma irradiation has little if any effect on the release of arachidonate from lipid stores (17,18). The effect of irradiation appears to be on the enzyme system(s) responsible for the conversion of arachidonate to the PGs. It is not clear at this time whether it is the endoperoxide synthetase or prostacyclin synthetase which is affected. Treatment of the cells with cycloheximide simultaneously with the radiation, thus preventing the formation of new enzyme, still resulted in an increase in PGI_2 synthesis (18). This suggests a direct activation of the enzyme system. With regard to the inhibition of PGI_2 synthesis observed by Eldor and coworkers at 2-21 days following irradiation of endothelial cells, their results suggest inhibition of prostacyclin synthetase. The addition of PGH_2 to the cells did not stimulate PGI_2 synthesis (19).

UV radiation stimulates both the activity of cyclooxygenase and phospholipase in vitro. Irradiation of a high speed particulate fraction of human skin stimulated the activity of phospholipase 4-fold when measured by incubation with 14C lecithin and cofactors (42). Arachidonate conversion to prostaglandins by cyclooxygenase isolated from fetal calf skin is also stimulated by UV irradiation of the enzyme (43).

Catabolism of prostaglandin

Ionizing radiation also affects the activity of PG 15-OH dehydrogenase. For example, Walker and Eisen (44) demonstrated that X-radiation of mice (200-1,000 rads, 230kV, 15mA, 140R/min) results in extensive falls in PGDH activity in the spleen within four hours of irradiation. A transient recovery in activity between 4-72 hours was followed by a second reduction in PGDH activity which lasted for at least seven days after irradiation. In the jejenum and kidney this reduction in PGDH activity was smaller. In the lung radiation appeared to actually increase PGDH activity in response to high doses (100 rads).

POSSIBLE PHYSIOLOGIC IMPORTANCE OF RADIATION-ALTERED PG SYNTHESIS

Protective effect of PGs on physiology during radiation

Radiation injury to the gastrointestinal tract has been documented as a major medical problem. Prostaglandins are physiologically active in the intestine (45,46). Hanson and Thomas (47) reported that the treatment of C_3H/HeJ mice with prostaglandins may have a protective effect on the survival of intestinal stem cells following irradiation. They looked at whole body irradiation (12-16 Gy from ^{137}Cs gamma ray source) and the survival of intestinal stem cells from the irradiated mice. Radioprotection was observed when 16,16- dimethyl PGE_2 was injected subcutaneously (10ug/mouse) 1 hour prior to irradiation. They suggest that tumor cells which secrete PGE_2 may protect tumor cells from destruction. PGE_2 was also demonstrated to have a cytoprotective effect on irradiated ileum (48). Rats were pretreated with PGE_2 1 hour prior to irradiation of an exteriorized segment of ileum with 137 Cs (10 and 15 Gy). The rats were sacrificed three and five days following radiation. Morphometric measurements of the ileum suggested that the prostaglandin had a cytoprotective effect on the irradiated ileum. PGs also have a cytoprotective effect on cells in culture. PGE_1 (10 ng/ml) added to cultures of CHO cells before irradiation (505 rad) increased by 2 fold the ability of the cells to form colonies (49).

Possible relevance of PG metabolism to radiation and metastasis

One of the major requirements for the metastasis of a solid tumor is that its cells have the ability to travel, via the circulatory system, to new sites. Honn and coworkers have proposed an interesting hypothesis which places the PGs in an important role in the mechanism of metastasis (1). They propose that PGI_2 acts as an antimetastatic agent by inhibiting the association of the tumor cell with either the platelet or the blood vessel wall. According to this hypothesis radiation with its reported alterations in PG synthesis and turnover could have a significant effect on the dissemination of tumor. Local irradiation of a tumor would not only destroy some of the cells and prevent their growth but could conceivably either inhibit or permit the dissemination of the tumor by affecting the PGI_2 concentrations in the local blood vessels. At this time it appears that the PGI_2 concentration in blood vessels increases, for at least the first two days, following radiation. This phenomenon could perhaps prove important in preventing cell dissemination during a crucial time of cell death and cell sluffing following irradiation. On the other hand, the reported longer term reduction of PGI_2 synthesis by endothelial cells following irradiation could mean an eventual promotion of metastasis by radiation. If the presence of PGI_2 in the blood vessel indeed proves significant in the metastasis of tumors, then supplementation with stable PGI_2 analogs or a long term stimulation of PGI_2 synthesis within the blood vessel (1) could become an important aspect of cancer therapy.

REFERENCES

1. Honn KV, Busse WD and Sloane BF: Prostacyclin and Thromboxanes: Implications of their role intumor cell metastasis. Biochem. Pharmacol. (32):1-11, 1983.
2. Smith KC and Hanawalt PC:Molecular Photobiology, Academic Press, N.Y., 1969, p 85-95.
3. Kornhauser A: Molecular aspects of cytotoxicity. Ann. N.Y. Acad. Sci. (346):398-414, 1980.
4. Cleaver JE: Relationship between rate of DNA synthesis and its inhibition by ultraviolet light in mammalian cell Rad. Res. (30):795-810, 1967.
5. Erickson LC, Matthews OB and Kohn KW: Mechanisms for the production of DNA damage in cultured human and hamster cel irradiated with light from fluorescent lamps, sunlamps, and t sun. Biochim. Biophys. Acta (610:)105-115, 1980.
6. Mandal TK and Chatarjee SNL: Ultraviolet and sunlight-induced lipid peroxidation in liposomal membrane. Rad. Res. (83):29 302, 1980.
7. Wilbur KM, Bernheim F and Shapiro OW: The thiobarbituric reagent as atest for the oxidation of unsaturated fatt acids by various agents. Arch. Biochem. Biophys. (24):305-313, 1949.
8. Mandal TK, Ghosh S, Sur P and Chatterjee SN: Effect of ultraviolet radiation on the liposomal membrane. Int.J.Radiat. Biol. (33):75-79, 1978
9. Little JB: Environmental Hazards, Ionizing Radiation. New Engl. J. Med. (275):929-937, 1966
10. Brancazio PJ: The Nature of Physics. MacMillan Publishing Co, 1975 p353.

11. Wolff S and Atwood KC: Independent X-ray effect on chromosome breakage and reunion. Proc. Natl. Acad. Sci. (40):187-192, 1954.
12. Graslund A, Ehrenberg A and Rupprecht A: Free radical formation in gamma-irradiated oriented DNA containing electron affinic radiosensitizers. J. Radiat. Biol. (31):145-152, 1977.
13. Little JB: Cellular effects of Ionizing Radiation. New Engl. J. Med. (278):308-315, 1968.
14. Taylor L, Menconi M and Polgar P: The participation of hydroperoxides and oxygen radicals in the control of prostaglandin synthesis. J. Biol. Chem. (258):6855-6857, 1983.
15. Mooibroek J, Trieling WB and Konings WT: Comparison of the radiosensitivity of unsaturated fatty acids, structured as micelles or liposomes, under different experimental conditions. Int. J. Radiat. Biol. (42):601-609, 1982.
16. Beatty BA, Hoak JC, Smith JB and DeGowin RL: Effect of irradiation on prostacyclin (PGI_2) production by cultured endothelial cells. Clin. Res. (27):714A, 1979.
17. Hahn GL, Menconi M and Polgar P: The effect of gamma radiation on prostacyclin production in cultured pulmonary artery endothelium. Prostaglandins and Cancer, First International Conference, 1982 p381-384.
18. Hahn GL, Menconi MJ, Cahill M and Polgar P: The influence of radiation on arachidonic acid release and prostacyclin synthesis. Prostaglandins (25):783-791, 1983.
19. Eldor A, Vlodavsky I, HyAm E, Atzmon R and Fuks Z: The effect of radiation on prostacyclin (PGI_2) production by cultured endothelial cells. Prostaglandins (25:)263-280, 1983.
20. Sinzinger H, Firbas, W and Cromwell M: Radiation induced alterations in rabbit aortic prostacyclin formation. Prostaglandins 24:323-329, 1982.
21. Allen JB, Sagerman, RH and Stuart MJ: Irradiation decreases vascular prostacyclin formation with no concomitant effect on platelet thromboxane production.The Lancet (2):1193-1195, 1981.
22. Schneidkraut MJ, Kot PA and Ramwell PW: Urinary prostacyclin and thromboxane levels after whole body gamma irradiation. Adv. Prostaglandin, Thromboxane and Leukotriene Res. (12):107-111, 1983.
23. Donlon M, Steel L, Helgeson EA, Shipp A and Catravas GN: Radiation induced alterations in prostaglandin excretion in the rat. Life. Sci. (32):2631-2639, 1983.
24. Tanner NS, Stamford IF and Bennet A: Plasma prostaglandins in mucositis due to radiotherapy and chemotherapy for head and neck cancer. Br. J. Cancer (43):767-771, 1981.
25. Steel LK, Sweedler IK and Catravas GN: Effects of 60Co radiation on synthesis of prostaglandins F_{2a}, E, and thromboxane B_2 in lung airways of the guinea pigs. Rad. Res. (94):156-165, 1983.
26. Steel LK and Cantravas, GN: Radiation induced changes in production of prostaglandins F_{2a}, E, and thromboxane B_2 in guinea pig parenchymal lung tissue. Int. J. Radiat. Biol. (42):517-530, 1982.

27. Bito LZ and Klein EM: The role of arachidonic acid cascade in the species- specific X-ray induced inflammation of the rabbit eye. Invest. Opthalmol. Vis. Sci. (22):579-587, 1982.
28. Burton GW, Cheeseman KH, Doba T and Slater TF: Vitamin E as an antioxidant in vitro and in vivo. Ciba-Found. Symp. (101):4-18, 1983.
29. Ts'Ao CH, Ward WF and Port CD: Radiation injury in rat lung: I. Prostacyclin (PGI_2) production, arterial perfusion and ultrastructure. Rad. Res. (96):284-293, 1983.
30. Warin AP: The ultraviolet erythemas in man. Br. J. Dermatol. (98):473-477, 1978.
31. Sondergaard J and Greaves MW: Pharmacological studies in inflammation due to exposure to ultraviolet radiation. J. Path. (101):93-97, 1970.
32. Mathur GP and Gandhi V: Prostaglandin in human and albino rat skin. J. Invest. Dermatol. (58):291-295, 1972.
33. Black AK, Greaves MW, Hensby CN, Plummmer NA and Eady RA: A new method for recovery of exudates from normal and inflammed human skin. Clin. Exp. Dermatol. (2):209-216, 1977.
34. Black AK, Fincham N, Greaves MW and Hensby CN: Time course changes in levels of arachidonic acid and prostaglandins D_2, E_2 and F_{2a} in human skin following ultraviolet B irradiation. Br. J. Clin. Pharmacol. (10):453-457, 1980.
35. Black AK, Hensby, CN and Greaves MW: Increased levels of 6-oxo-PGF_{1a} in human skin following ultraviolet B irradiation. Br. J. Clin. Pharmacol. (13):351-354, 1982.
36. Camp RD, Greaves MW, Hensby CN, Plummer NA and Warin AP: Irradiation of human skin by short avelength ultraviolet radiation (100-290 nm) (U.V.C.) increased concentrations of arachidonic acid and prostaglandin E_2 and F_{2a}. Br. J. Clin. Pharmacol. (6):145-148. 1978.
37. Hawk JL, Black A, Jaenicke KF, Barr RM, Soter NA, Mallett, AI, Gilchrest BA, Hensby CN, Parrish JA and Greaves MW: Increased concentrations of arachidonic acid,prostaglandins E_2, D_2 and 6-oxo-PGF_{1a}, and histamine in human skin following UVA irradiation. J. Inv. Derm. (80):469-499, 1983.
38. Snyder DS and Eaglestein WA: Intradermal anti-prostaglandin agents and sunburn J. Invest. Dermatol. (62):47-50, 1974.
39. Snyder DS: Effect of topical indomethacin on UVR-induced redness and PGE levels in sunburned guinea pig skin. Prostaglandins (11):631-643, 1976.
40. Johnson M, Jessup R and Ramwell P: Ultraviolet light modification of prostaglandin receptor. Prostaglandins (4):593-605, 1973.
41. Lord J, and Ziboh VA: Specific binding of prostaglandin E_2 to membrane preparations from human skin. Receptor modulation by UVB-irradiation and chemical agents. J. Invest. Dermatol. (73):73-77, 1979.
42. Ziboh VA, Lord JT, Uematsu S and Blick G: Activation of phospholipase A_2 and increased release of prostaglandin precursor from skin by ultraviolet radiation. J. Invest. Dermatol. (70):211, 1978.
43. Lord JT, Ziboh VA, Poitier J, Legget G and Penneys NS: The effects of photosensitizers and ultraviolet irradiation on the biosynthesis and metabolism of prostaglandins. Br. J.

Dermatol. (95):397-406, 1976.
44. Walker DI and Eisen V: Effect of ionizing radiation on 15-hydroxy prostaglandin dehydrogenase (PGDH) activity in tissue. Int. J. Radiat. Biol. (36):399-407, 1979.
45. Robert A: Antisecretory, antiulcer, cytoprotective and diarrheogenic properties of prostaglandins. Adv. Prostaglandin, Thromboxane Res. (2:)507-520, 1976.
46. Bennett A, Stamford IF and Unger WG: Prostaglandin E_2 and gastric acid secretion in man. J. Physiol. (229):349-360, 1973.
47. Hanson WR and Thomas C: 16,16-dimethyl prostaglandin E_2 increases survival of murine intestinal stem cells given before photon radiation. Rad. Res. (96):393-398, 1983.
48. Tomas-de la Vega JE, Banner BF, Hubbard M, Boston DL, Thomas CW, Straus AK and Roseman DL: Cytoprotective effect of prostaglandin E_2 in irradiated rat ileum. Surg. Gynecol. Obstetr. (158):39-45, 1984.
49. Prasad KN: Radioprotective effect of prostaglandin and an inhibitor of cyclic nucleotide phosphodiesterase on mammalian cells in culture. Int. J. Rad. Biol. (22):187-189, 1972.

10

PHOSPHOLIPASES

LARRY W. DANIEL

1. INTRODUCTION

The enzymes which degrade phospholipids are ubiquitous in nature and have been described in almost every cell type examined. These phospholipases have a major role in the catabolism of dietary lipids for energy sources and also are important in the metabolism of membrane structural phospholipids. This latter function contributes to the control of each cell type's unique phospholipid composition and undoubtedly has a major role in regulating cell function (1).

The specificity of phospholipases for a generalized diacylphospholipid molecule is shown below:

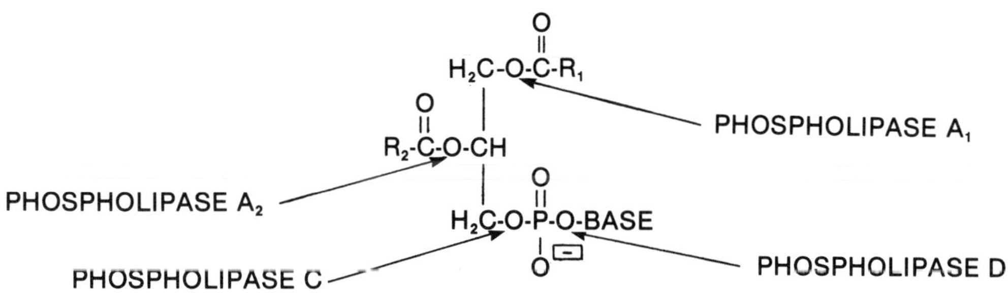

SCHEME 1

Phospholipases A_1 remove the fatty acid (R_1 as shown above) from the <u>sn</u>-1 position of diacylglycerophospholipids. The rat liver enzyme has been most thoroughly studied; however the enzyme is found in a variety of prokaryotic and eukaryotic cells (2). Phospholipases A_2 are also widespread in nature and are a major component of the venoms of many species of snakes. These enzymes specifically hydrolyze the acyl moiety (designated R_2 above) from the sn-2 position of phospholipids. Phospholipases A_2 are involved in controlling synthesis of a variety of bioactive lipids including the eicosanoids (3) and platelet activating factor (1-alkyl-2-acetyl-sn-glycero-3-phosphocholine) (4). Phospholipases C are enzymes which hydrolyze the glycerophosphate ester bond of phospholipids to yield diacylglycerol and a phosphobase moiety. This activity was first described as a bacterial toxin with broad substrate specificity (3). However the mammalian phospholipase C enzymes specifically degrade inositol phospholipids and are involved in many stimulus mediated cellular responses including growth control and eicosanoid synthesis (5,6,7). Phospholipases D which degrade a variety of phospholipids to yield phosphatidic acid were first discovered in carrots and were subsequently found in a variety of plant species (3).

Since the entire subject of phospholipases is beyond the scope of this review it is fortunate that several excellent reviews on the phospholipases in general are available (2,3). In addition, there are more specialized reviews on the purification and characterization of phospholipases A_1 (8), and A_2 (9,10); on the control of the enzymes which release arachidonic acid (2,3,11); and on the enzymes which degrade phosphatidylinositol (5,12). The present review concentrates on the phospholipases which may be involved in the phenotypic changes observed in tumor cells. The enzymes which release arachidonic acid from cellular phospholipids will be considered since products of the arachidonic acid cascade have been implicated in altered cellular growth control and metastasis. The enzymes which degrade phosphatidylinositol will also be considered since much recent work implicates phosphatidylinositol degradation in the control of cellular functions.

2. GENERAL CHARACTERISTICS OF PURIFIED PHOSPHOLIPASES

The characterization of phospholipases is complex due to the water insoluble nature of their substrates. In addition most mammalian phospholipases are membrane bound and their solubilization has added an additional hindrance to purification. Soluble phospholipases A_2 are abundant in several venoms, and are also found at relatively high concentrations in mammalian pancreatic secretions. Due to their abundance, ease of purification and stability, the understanding of the mechanisms of action and structure of the venom and pancreatic phospholipase A_2 enzymes is much more sophisticated than the understanding of membrane bound phospholipases A_2 (9,10).

Although phospholipase A_2 can hydrolyze monomeric phospholipids, optimal activity is only achieved above the critical micellar concentration of the substrate. Therefore, the physical nature of the substrate appears to be critical to enzyme activity; and it is essential to understand the physical requirements for enzyme-substrate interactions. Studies of phospholipase hydrolysis using liposomal substrates have demonstrated the importance of the physical form of the substrate in determining the effective amount and concentration of phospholipid available to the enzyme. For example if a given phospholipid preparation has 1 µmol of phospholipid per ml of solution the formation of liposomal structures would yield an effective concentration of phospholipid at the particle surface of several molar. In addition to determining the effective concentration of substrate molecules, the physical form of the substrate can also influence the conformation of the phospholipase molecules and the removal of reaction products. Due to these many influences of physical form a variety of different model substrates have been required. The substrate model used depends on the questions being asked. Monolayers, liposomes and micelles have been used as substrates and although they are far less complex than the natural substrates for phospholipases in vivo they have proven to be important model systems for the basic characterization of phospholipases.

2.1 Physical characteristics of phospholipid substrates

The effects of molecular packing or surface pressure have been measured using monomolecular films of phospholipids as model substrates. These studies have shown that increased surface pressure lowers phospholipase activity presumably by preventing the enzyme access to portions of the phospholipid other than the polar head group. The apparatus for these studies has been described by Verger et al. (13).

Liposomes are a commonly used substrate for phospholipases. Different methods of preparation can yield liposomes which vary widely in size and can form single-shell vesicles or large multi-lamellar vesicles. Mixed liposomes with varying charge have been used to assess the effect of surface charge on phospholipase activity (14). In these studies the zeta potential of mixed amphiphile-phospholipid liposomes is measured by microelectrophoresis (15). These types of studies have shown that surface charge as well as surface concentration are critical for optimal enzymatic activity.

Recent evidence, obtained by nuclear magnetic resonance, has shown that pure phosphatidylethanolamine does not form bilayer structures but forms structures similar to the hexagonal type II phase described by Cullis and de Kruijff (16). However, when phosphatidylethanolamine is mixed with phosphatidylcholine, bilayer structures are formed. The functional significance of the hexagonal II type structures is unknown, but may be a potentially important control mechanism regulating phospholipid hydrolysis.

For soluble phospholipases the kinetic parameters are determined by binding of the enzyme to the surface of the substrate particle as well as the effective surface concentration (9,10). The importance of surface concentration has been demonstrated by lowering the substrate surface concentration with increased detergent concentrations while maintaining a constant bulk concentration of substrate.

In summary, although much less complex than the natural substrates for phospholipases, liposomes, emulsions and monomolecular films have been used to determine the basic characteristics of phospholipase-substrate interactions. In addition, these studies have shown the importance of the physical form of the substrate in determining kinetic properties of the phospholipases.

2.2 Phospholipases A_2

Phospholipases A_2 are relatively easily purified from a variety of venoms and pancreatic secretions since they comprise 1-10% of the total protein present. In addition, these enzymes are relatively stable to variations in temperature, pH and denaturing conditions. Because of their ease of isolation these enzymes have been purified from a number of sources and found to have similar properties. The stability of the enzymes is probably due to the multiple disulfide bonding which is a common feature among venom and pancreatic enzymes (9). Both the venom and pancreatic phospholipases are small, water soluble enzymes of about 14,000 daltons and require Ca^{2+} for activity. Pancreatic phospholipases A_2 appear to be controlled by activation of a proenzyme form by proteolytic cleavage; however, no other phospholipases A_2 have been shown to exist as a proenzyme. The pancreatic and venom phospholipases have been studied in detail and have been the subject of many mechanistic studies in addition to studies with chemically modified enzymes (for reviews see 9,10).

Relatively few phospholipases A_2 have been purified from sources other than venoms and pancreas. One reason for the difficulty, as pointed out by van den Bosch (3), is the relatively small amount of enzyme in cells other than pancreas. For example homogeneous preparations of pancreatic phospholipase A_2 have been obtained after 210 fold purification; however, a 13,000 fold purification was required to obtain a homogeneous preparation of phospholipase A_2 from rat hepatoma cells.

Membrane bound phospholipases A_2 have been purified from a variety of other sources including rabbit polymorphonuclear leukocytes (17), erythrocytes (18) and platelets (19). In general these phospholipases are similar to the soluble phospholipases in size (12,000-18,000 daltons) and Ca^{2+} dependence. Most of the mammalian phospholipases A_2 have optimum activity between pH 7.0 and pH 9.5 except those from lysosomes which have an acidic pH optimum (20).

Future attempts to purify membrane-bound phospholipases should benefit from the affinity chromatography technique of Rock and Snyder (21) which uses a nonhydrolyzable alkyl phospholipid bound to AH-Sepharose 4B. This technique has been used successfully for the purification of phospholipase A_2 from erythrocytes (22).

2.3 Phospholipases C

Bacterial phospholipases C degrade a variety of phospholipids to yield diacylglycerols and a phosphate monoester (see Scheme 1) (3). However, mammalian phospholipases C have been identified which exhibit a rigid specificity for phosphatidylinositol (PI) and its phosphorylated derivatives phosphatidylinositol 4-phosphate (DPI) and phosphatidylinositol 4,5-bisphosphate (TPI). Phospholipase C is the key to the well-recognized PI cycle, represented in Scheme 2, which is involved in many stimulus coupled responses involving Ca^{2+} (5).

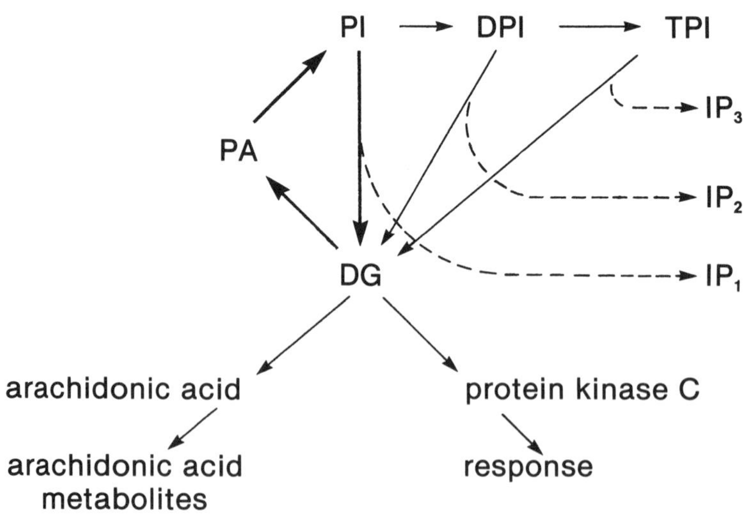

SCHEME 2

This cycle has been implicated in the release of arachidonic acid as discussed above and more recently in the production of a variety of bioactive molecules which may function as intracellular messengers (see section 3.2).

One of the early reports of mammalian phospholipase C came from Sloane-Stanley who found the enzyme in brain tissue (23). In rat brain (24) and in human platelets (25) phospholipase C appears to be predominately soluble but exists in both soluble forms and membrane bound forms in other tissues (12).

The specificity of phospholipases C for phosphatidylinositol and its phosphorylated derivatives is a matter of current controversy. It appears that some phospholipases C degrade all three species. However, a recent report by Graff et al. (26) found that the degradation of DPI by human platelet phospholipase C was inhibited by TPI. These authors speculate that the synthesis of TPI may be a negative regulatory signal for the PI cycle. In a contrasting report, Irvine et al. (27) have shown that TPI is preferentially degraded by phospholipase C from rat brain. These authors found that, in the presence of phosphatidylethanolamine, TPI was degraded 10 times faster than PI even when PI was in 10 fold molar excess. Thus, species and tissue specific differences may exist in the specificity of phospholipase C.

The regulation of phospholipase C in both the crude and purified state appears to be complex but current information suggests that its activity may be regulated by the membrane structure. The products of the reaction, diacylglycerol and products derived from diacylglycerol, including polyunsaturated fatty acids and phosphatidic acid (PA), see Scheme 2, are potential stimulators of PI hydrolysis (28,29). This observation points to the intriguing possibility of a self-potentiating system. Further evidence that the membrane composition affects phospholipase C activity is the observation that hydrolysis of PI in a mixture of phospholipid resembling the inner membrane leaflet is stimulated by PA. No stimulation was observed in a mixture of phospholipids resembling the outer leaflet.

In summary, phospholipases C from a variety of sources are stimulated by Ca^{2+} in vitro although they may not be regulated by Ca^{2+} concentration in vivo (29). Further comparative studies are required to determine the relative importance of the hydrolysis of PI versus DPI or TPI. Current studies indicate that the regulation of phospholipase C may be a complex interaction among the relative amounts of phosphatidylinositol species and other phospholipids and the intracellular Ca^{2+} ion concentration.

3. ROLE OF PHOSPHOLIPASES IN CELLULAR FUNCTIONS

3.1 Release of prostaglandin and leukotriene precursors

Prostaglandins and leukotrienes have been implicated in the growth control, metastatic spread and resistance of tumor cells to host immune responses. The mechanisms involved in these biological effects of prostanoids are not clear; however the subject has been well reviewed elsewhere (30,31,32). Several questions remain to be answered. Which phospholipases are required? Which substrates are hydrolyzed? How is the hydrolysis controlled? These questions are of importance since control of prostanoid synthesis is generally accepted to be at the level of arachidonic acid release from cellular phospholipids.

Three pathways have been proposed for the release of arachidonic acid. All these pathways have been suggested to occur in platelets (6,25,33); however their relative importance is the subject of much debate. A portion of the controversy in the study of platelets may be due to the use of different species of platelet donors and due to the different stimuli used. In early studies with platelets phosphatidylcholine and phosphatidylinositol were found to be the major donors of arachidonic acid in stimulated cells (34). However, later studies have shown that phosphatidylethanolamine is also deacylated in thrombin stimulated horse and rabbit platelets (35). These discrepancies may also be due to differences in the specific activity of radiolabeled arachidonic acid in the various phospholipid classes or in the localization of specific pools of arachidonyl phospholipids within the cells.

It is generally accepted that arachidonic acid is released by the action of a phospholipase A_2 (3). However, platelets also possess a phospholipase C which when stimulated by thrombin results in a 30 fold increase in diacylglycerol production (6). The resultant diacylglycerol may be a donor of arachidonic acid when degraded by lipases or may be converted to phosphatidic acid by diacylglycerol kinase. Platelets have been shown to have a phospholipase A_2 which hydrolyzes phosphatidic acid, and this enzyme has been postulated to be important in releasing arachidonic acid (33). Thus, two pathways for removal of arachidonic acid from phosphatidylinositol following phospholipase C hydrolysis have been suggested. However, the quantitative significance of phosphatidylinositol as a donor of arachidonic acid has not been firmly established and the turnover of phosphatidylinositol may be more important in relation to the production of other bioactive intracellular signals (see section 3.2).

Madin-Darby canine kidney (MDCK) cells have also been used as a model for the study of arachidonic acid release and prostaglandin synthesis. Ohuchi and Levine (36) first demonstrated that the MDCK cells synthesized high levels of prostaglandin in response to the tumor promoter 12-O-tetradecanoyl-phorbol-13-acetate (TPA). This cell has subsequently been used to study the mechanism of action of tumor promoters and the control of prostaglandin synthesis (37). Therefore the MDCK cells will be used herein to illustrate recent work on the control of arachidonic acid release and to illustrate some of the pitfalls which may be encountered in these studies.

Most previous studies to determine the source of arachidonic acid for prostaglandin synthesis have utilized cells preincubated with radiolabeled arachidonic acid, and later stimulated. This approach can provide useful data on the source of arachidonic acid; however, potential problems exist. First, this approach assumes that the radiolabeled arachidonic acid is in equilibrium with the endogenous pools of arachidonic acid. In the MDCK cell, relatively long (16-24 h) labeling periods are required to obtain a constant specific activity among the phospholipid classes (38). Of special interest is the relatively slow incorporation of labeled arachidonic acid into phosphatidylethanolamine compared to phosphatidylcholine. This is significant since phosphatidylethanolamine is a major source of released arachidonate in TPA stimulated MDCK cells. Thus, the release of labeled arachidonic acid may be misleading due to large differences in the specific radioactivity of arachidonate in different phospholipid classes.

Specific pools of arachidonyl phospholipids may also exist in topologically or functionally separated locations within the cell. Humes et al. (39) have given evidence for such pools in macrophages. They found that different stimuli result in the synthesis of different arachidonic acid derived products and suggest that the prostaglandin synthetic enzymes may be spatially segregated from lipoxygenase synthetic enzymes with each utilizing distinct pools of arachidonic acid-containing phospholipids.

Another determinant of phospholipid specificity that is generally ignored is the group at the *sn*-1 position. Few studies have compared the release of arachidonic acid from 1-*O*-alkyl- or 1-alk-1'-enyl-linked phospholipids with the release from the corresponding 1-acyl-linked phospholipid. This has probably been due to the difficulty of separating these molecular species. However, the ethanolamine plasmalogens (1-alk-1'-enyl-2-acyl-*sn*-glycero-3-phosphoethanolamine) have been found to be important donors of arachidonic acid in MDCK cells (37,40) but not in $HSDM_1C_1$ cells (41). Recently developed techniques for separating molecular species of intact phospholipids by high-pressure liquid chromatography provide a powerful tool for assessing the role of ether-linked phospholipids as donors of arachidonic acid. Using these techniques, Swendsen et al. (42) found that both 1-alkyl and

1-acyl phosphatidylcholine species were donors of arachidonic acid in A23187 stimulated neutrophils. This finding is of additional significance since phospholipase A_2 hydrolysis of the 1-alkyl species yields 1-alkyl-2-lyso-sn-glycero-3-phosphocholine which can be converted to platelet activating factor and arachidonic acid which can be converted to bioactive derivatives. Thus, one molecule may provide the substrates for two metabolic pathways producing bioactive lipid mediators of inflammation.

Another open question is whether arachidonoyl-specific phospholipases exist. This problem has been approached by prelabeling MDCK cells with [^{14}C]linoleic acid and [^{3}H]arachidonic acid and measuring the extracellular radio-labeled compounds at 24 h after stimulation with TPA. Using this approach the authors found an increased amount of [^{3}H]arachidonic acid derived products and concluded that the lipase was specific for arachidonic acid (43). However, the dynamic state of intracellular free fatty acids and the deacylation/reacylation balance were not fully considered. Further studies with MDCK cells have shown that [^{14}C]linoleic acid release is also stimulated by TPA treatment, but the linoleic acid is quickly reincorporated into the cells (44). Beaudry et al. (44) have further shown that [1-^{14}C]-Δ8,11,14-eicosatrienoic acid and [1-^{14}C]-Δ5,8,11-eicosatrienoic acid are released at similar rates from TPA stimulated MDCK cells. With both fatty acids there is a transient rise in free fatty acid levels and a decreased level as the fatty acids are reincorporated into the cell (Δ5,8,11-eicosatrienoic acid and Δ8,11,14-eicosatrienoic acid) or converted to prostaglandins (Δ5,8,11-eicosatrienoic acid). These data indicate that the increased release of fatty acids upon stimulation may be a transient alteration in the deacylation/reacylation balance of cellular fatty acids and is not specific for arachidonic acid. As discussed above there may be different functional pools of phospholipids within the cell; further understanding of the specificity of fatty acid release will require purification and subcellular localization of the enzymes and substrates involved.

Many membrane bound phospholipases A_2 have a requirement for Ca^{2+} and this has led to the speculation that they are Ca^{2+} controlled (3). The stimulation of deacylation by the Ca^{2+} ionophore A23187 and other divalent cation ionophores further supports the idea that increased intracellular Ca^{2+} concentrations can stimulate phospholipase A_2. In MDCK cells, A23187 stimulates the deacylation of phospholipids and the pattern of deacylation is similar to that induced by other stimuli (45). However, quantitative differences were observed when platelets were stimulated with thrombin or A23187 (46). This observation demonstrates the need for caution when using A23187. These results also indicate that future studies are required to compare the rates of deacylation from various phospholipid classes after treatment of the cells with different stimuli. These studies must also be combined with the subcellular localization of the substrates and enzymes involved in order to determine the existence and function of topologically isolated pools of substrates.

The membrane bound phospholipases A_2 may also be controlled by the bilayer lipid composition. Dawson et al. (29) have recently shown that diacylglycerol stimulates the hydrolysis of phospholipids in phosphatidylcholine bilayers but does not stimulate the hydrolysis of phosphatidylethanolamine which exists in a hexagonal type II structure. They postulate that the formation of diacylglycerol by phospholipase C after cell stimulation (see Scheme 2) may cause local membrane perturbations which introduce a bilayer-nonbilayer phase transition. These structures may then be more conducive to phospholipase A_2 degradation. Such a mechanism of activation might explain the apparent sequential activation of phospholipase C and phospholipase A_2 which is observed in many systems.

Corticosteroids inhibit phospholipase activity by stimulating the synthesis of inhibitory proteins. The synthesis of these inhibitors has been described in neutrophils and macrophages and is reviewed in Chapter 11 of this volume.

3.2 Turnover of phosphatidylinositol

Phosphatidylinositol degradation is a key event in most cellular responses which involve Ca^{2+} as a second messenger. The first observations of phosphatidylinositol degradation and resynthesis in response to external stimuli was made over 30 years ago (47); however, the number of putative bioactive molecules generated has only recently been appreciated. Bioactivity has been suggested for four molecules as shown in Scheme 2; they are diacylglycerol (DG), myo-inositol 1,4,5-trisphosphate (IP_3), phosphatidic acid (PA) or arachidonic acid and its metabolites.

As previously mentioned (section 2.3) DG, PA and arachidonic acid release may be important in potentiating further degradation of PI by phospholipase C or may also activate phospholipase A_2 and initiate the degradation of other phospholipid classes with the release of arachidonic acid (29). The released arachidonic acid or its metabolites may also have biological activities as discussed in section 3.1.

Recent studies have suggested that intracellular free Ca^{2+} content may be regulated by the level of IP_3. Joseph et al. (48) have shown that the addition of IP_3 to saponin-permeabilized hepatocytes results in an increase in intracellular free Ca^{2+}. Further the return of Ca^{2+} concentrations to their original levels paralleled the degradation of IP_3. These observations may explain one function of the PI cycle; however, studies in the laboratory of Nishizuka have identified another function for the PI cycle. These studies have identified a protein kinase which is activated by DG and requires Ca^{2+} and phosphatidylserine (49).

This protein kinase activity has been termed protein kinase C to differentiate it from the cAMP or cGMP dependent kinases. Subsequent studies have identified protein kinase C in a wide variety of species and tissues (50). Perhaps the most exciting aspect of this enzyme, and the reason for its inclusion in this discussion, is its activation by the tumor promoter TPA (51). Copurification studies indicate that the TPA receptor and protein kinase C are the same molecule (52,53). Thus, the long-sought endogenous molecule which is mimicked by TPA appears to be diacylglycerol generated by the PI cycle. Protein kinase C appears to be important in the response of platelets to thrombin since the degree of shape change, serotonin release and aggregation can be related to the degree of activation of phospholipase C and the phosphorylation of endogenous proteins by protein kinase C (54). The relationship of phospholipase C activation to cellular growth control has been demonstrated by Rozengurt et al. (55). They have shown that the addition of epidermal growth factor to quiescent 3T3 cells stimulates phosphorylation of an 80,000 dalton cellular protein and that treatment of the cells with TPA or exogenous phospholipase C results in phosphorylation of the same protein. Additional evidence that products of the PI cycle may be involved in growth control comes from recent studies with the product of the src oncogene. Sugimoto et al. (56) found that the isolated src kinase phosphorylates PI and DPI in vitro. They also found that transformation of cells with the src containing Rous sarcoma virus results in increased turnover of DPI and TPI.

In summary, the PI cycle results in the generation of a number of bioactive compounds and a cascade of events, the significance of which is only becoming appreciated. Recent studies indicate that a nonphysiological stimulation of this cascade, as by TPA or viral transformation, may result in altered growth control and ultimately lead to tumor formation.

4. CONCLUSIONS

From the previous discussion it is obvious that our understanding of the regulation of phospholipases is in a very early stage. Past studies have used either purified phospholipases and model membrane substrates or intact cells and endogenous substrates. The most significant challenge for future studies is to relate the data from these two approaches. It is also apparent that further studies must determine the physical organization of the substrates and characterize the effects of bilayer to nonbilayer phase transitions in the regulation of phospholipase activity.

Studies on the subcellular localization of the phospholipases and their substrates are of premier importance since an increasing number of studies are identifying physiologically distinct pools of substrates for both phospholipases A_2 and phospholipase C. Further studies on the regulation of the PI cycle are also of importance due to the increasingly recognized importance of this cycle in a variety of cellular functions including growth control.

ACKNOWLEDGEMENTS

This investigation was supported by Grants CA09422, CA12197 and AM11799, awarded by the National Institutes of Health, and by Environmental Protection Agency Grant R807770. I wish to thank Dr. Moseley Waite for many helpful discussions on phospholipases and Gwen Charles for her skillful assistance in preparing this manuscript.

REFERENCES

1. Stubbs CD, Smith AD: The modification of mammalian membrane polyunsaturated fatty acid composition in relation to membrane fluidity and function. Biochim Biophys Acta (779): 89-137, 1984.
2. van den Bosch H: Intracellular phospholipases A. Biochim Biophys Acta (604): 191-246, 1980.
3. van den Bosch H: Phospholipases. In: Hawthorne JN, Ansell GB (ed) Phospholipids. Elsevier Biomedical Press, Amsterdam, 1982, pp 313-357.
4. O'Flaherty JT, Wykle RL: Biology and biochemistry of platelet-activating factor. Clin Rev Allergy (1): 353-367, 1983.
5. Michell RH: Inositol phospholipids and cell surface receptor function. Biochim Biophys Acta (415): 81-147, 1975.
6. Rittenhouse-Simmons S: Production of diglyceride from phosphatidylinositol in activated human platelets. J Clin Invest (63): 580-587, 1979.

7. Bell RL, Kennerly DA, Stanford N, Majerus PW: Diglyceride lipase: a pathway for arachidonate release in platelets. Proc Natl Acad Sci USA (76): 3238-3241, 1979.
8. Waite M, Rao RH, Griffin H, Franson R, Miller C, Sisson P, Fry J: Phospholipases A_1 from lysosomes and plasma membranes of rat liver. Methods Enzymol (71): 674-689, 1981.
9. Dennis EA: Phospholipases. In: Boyer P (ed) The enzymes, 3rd ed., Lipid enzymology. Academic Press, New York, 1983, pp 307-353.
10. Slotboom AJ, Verheij HM, de Haas GH: On the mechanism of phospholipase A_2. In: Hawthorne JN, Ansell GB (ed) Phospholipids. Elsevier Biomedical Press, Amsterdam, 1982, pp 359-434.
11. Irvine RF: How is the level of free arachidonic acid controlled in mammalian cells? Biochem J (204): 3-16, 1982.
12. Hawthorne JN: Inositol phospholipids. In: Hawthorne JN, Ansell GB (ed) Phospholipids. Elsevier Biomedical Press, Amsterdam, 1982, pp 263-278.
13. Verger R, Mieras MCE, de Haas GH: Action of phospholipase A at interfaces. J Biol Chem (248): 4023-4034, 1973.
14. Robinson M, Waite M: Physical-chemical requirements for the catalysis of substrates by lysosomal phospholipase A. J Biol Chem (258): 14371-14378.
15. Bangham AD, Dawson RMC: Electrokinetic requirements for the reaction between Cl. perfringens α-toxin (phospholipase C) and phospholipid substrates. Biochim Biophys Acta (59): 103-115, 1962.
16. Cullis PR, de Kruijff B: Polymorphic phase behavior of lipid mixtures as detected by ^{31}P NMR: evidence that cholesterol may destabilize bilayer structure in membrane systems containing phosphatidylethanolamine. Biochim Biophys Acta (507): 207-218, 1978.
17. Elsbach P, Weiss J, Franson RC, Beckerdite-Quagliata S, Schneider A, Harris L: Separation and purification of a potent bactericidal/permeability increasing protein and a closely associated phospholipase A_2 from rabbit polymorphonuclear leukocytes: observations on their relationship. J Biol Chem (254): 11000-11009, 1979.
18. Jimento-Abendano J, Zahler P: Purified phospholipase A_2 from sheep erythrocyte membrane: preferential hydrolysis according to polar groups and 2-acyl chains. Biochim Biophys Acta (573): 266-275, 1979.
19. Kannagi R, Koizuma K: Effects of different physical states of phospholipid substrates on partially purified platelet phospholipase A_2 activity. Biochim Biophys Acta (556): 423-433, 1979.
20. Franson RC, Waite M, Wegliki W: Phospholipase A activity of lysosomes of rat myocardial tissue. Biochemistry (11): 472-476, 1972.
21. Rock CO, Snyder F: Rapid purification of phospholipase A_2 from Crotalus adamanteus venom by affinity chromatography. J Biol Chem (250): 6564-6566.
22. Kramer RM, Wütrich C, Bollier C, Allegrini PR, Zahler P: Isolation of phospholipase A_2 from sheep erythrocyte

membranes in the presence of detergents. Biochim Biophys Acta (507): 381-394, 1978.
23. Sloane-Stanley GH: Anaerobic reactions of phospholipids in brain suspensions. Biochem J (53): 613-619, 1953.
24. Irvine RF, Dawson RMC: The distribution of calcium-dependent phosphatidylinositol-specific phosphodiesterase in rat brain. J Neurochem (31): 1427-1434, 1978.
25. Billah MM, Lapetina EG, Cuatrecasas P: Phospholipase A_2 and phospholipase C activities of platelets. J Biol Chem (255): 10227-10231, 1980.
26. Graff G, Nahas N, Nikolopolou M, Natarajan V, Schmidt HHO: Possible regulation of phospholipase C activity in human platelets by phosphatidylinositol 4,5-bisphosphate. Arch Biochem Biophys (288): 299-308, 1984.
27. Irvine RF, Letcher AJ, Dawson RMC: Phosphatidylinositol-4,5-bisphosphate phosphodiesterase and phosphomonoesterase activities of rat brain. Biochem J (218): 177-185, 1984.
28. Takenawa T, Yoshitaka N: Effect of unsaturated fatty acids and Ca^{2+} on phosphatidylinositol turnover. J Biochem (91): 793-799, 1982.
29. Dawson RMC, Hemington NL, Irvine RF: Diacylglycerol potentiates phospholipase attack upon phospholipid bilayers: Possible connection with cell stimulation. Biochem Biophys Res Commun (117): 196-201, 1983.
30. Tisdale MJ: Role of prostaglandins in metastatic dissemination of cancer. Exp Cell Biol (51): 250-256, 1983.
31. Honn KV, Bockman RS, Marnett LJ: Prostaglandins and cancer: a review of tumor initiation through tumor metastasis. Prostaglandins (21): 833-864, 1981.
32. Goodwin JS: Prostaglandins and host defense in cancer. Med Clin North Am (65): 829-844, 1981.
33. Billah MM, Lapetina EG, Cuatrecasas P: Phospholipase A_2 activity specific for phosphatidic acid: a possible mechanism for the production of arachidonic acid in platelets. J Biol Chem (256): 5399-5403, 1981.
34. Bills TK, Smith JB, Silver MJ: Metabolism of [^{14}C]-arachidonic acid by human platelets. Biochim Biophys Acta (424): 303-314, 1976.
35. Brockman MJ, Ward JW, Marcus AJ: Phospholipid metabolism in stimulated human platelets: changes in phosphatidyl-inositol phosphatidic acid and lysophospholipids. J Clin Invest (66): 275-283, 1980.
36. Ohuchi K, Levine L: Stimulation of prostaglandin synthesis by tumor promoting phorbol 12,13-diesters in canine kidney (MDCK) cells: cycloheximide inhibits the stimulated prostaglandin synthesis deacylation of lipids and morphological changes. J Biol Chem (253): 4783-4790, 1978.
37. Levine L: Arachidonic acid transformation and tumor promotion. Adv Cancer Res (35): 49-79, 1981.
38. Daniel LW, King L, Waite M: Source of arachidonic acid for prostaglandin synthesis in Madin-Darby canine kidney cells. J Biol Chem (256): 12830-12835, 1981.

39. Humes JL, Sadowski S, Galavage M, Goldenberg M, Subers E, Bonney RJ, Kuehl FA Jr: Evidence for two sources of arachidonic acid for oxidative metabolism by mouse peritoneal macrophages. J Biol Chem (257): 1591-1594, 1982.
40. Beaudry GA, Daniel LW, King L, Waite M: Stimulation of deacylation in Madin-Darby canine kidney cells 12-O-tetradecanoyl-phorbol-13-acetate stimulates rapid phospholipid deacylation. Biochim Biophys Acta (750): 274-281, 1983.
41. Schremmer JM, Blank ML, Wykle RL: Bradykinin-stimulated release of [^{3}H]arachidonic acid from phospholipids of $HSDM_1C_1$ cells: comparison of diacylphospholipids and plasmalogens as sources of prostaglandin precursors. Prostaglandins (18): 491-505, 1979.
42. Swendsen CL, Ellis JM, Chilton FH III, O'Flaherty JT, Wykle RL: 1-O-alkyl-2-acyl-sn-glycero-3-phosphocholine: a novel source of arachidonic acid in neutrophils stimulated by the calcium ionophore A23187. Biochem Biophys Res Commun (113): 72-79, 1983.
43. Ohuchi K, Levine L: Tumor promoting phorbol diesters stimulate release of radioactivity from [^{3}H]arachidonic acid labeled – but not [^{14}C]linoleic acid labeled – cells. Indomethacin inhibits the stimulated release from [^{3}H]arachidonate labeled cells. Prostaglandins Med (1): 421-431, 1978.
44. Beaudry GA, King L, Daniel LW, Waite M: Stimulation of deacylation in Madin-Darby canine kidney cells: specificity of deacylation and prostaglandin production in 12-O-tetradecanoyl-phorbol-13-acetate treated cells. J Biol Chem (257): 10973-10977, 1982.
45. Daniel LW, Beaudry GA, King L, Waite M: Regulation of arachidonic acid metabolism in Madin-Darby canine kidney cells: comparison of A23187 and 12-O-tetradecanoyl-phorbol-13-acetate. Biochim Biophys Acta (792): 33-38, 1984.
46. Rittenhouse-Simmons S: Differential activation of platelet phospholipases by thrombin and ionophore A23187. J Biol Chem (256): 4153-4155, 1981.
47. Hokin MR, Hokin LE: The synthesis of phosphatidic acid from diglyceride and adenosine triphosphate in extracts of brain microsomes. J Biol Chem (234): 1381-1386, 1959.
48. Joseph SK, Thomas AP, Williams RJ, Irvine RF, Williamson JR: myo-Inositol 1,4,5-trisphosphate: a second messenger for the hormonal mobilization of intracellular Ca^{2+} in liver. J Biol Chem (259): 3077-3081, 1984.
49. Kishimoto A, Takai Y, Mori T, Kikkawa U, Nishizuka Y: Activation of calcium and phospholipid-dependent protein kinase by diacylglycerol, its possible relationship to phosphatidylinositol turnover. J Biol Chem (255): 2273-2276, 1980.
50. Kuo JF, Andersson RGG, Wise BC, Mackerlova L, Salomonsson I, Brackett NL, Katoh N, Shoji M, Wrenn RW: Calcium-dependent protein kinase: widespread occurrence in various tissues and phyla of the animal kingdom and comparison of effects of phospholipid, calmodulin and trifluoperazine. Proc Natl Acad Sci USA (77): 7039-7043.

51. Castagna M, Takai Y, Kaibuchi K, Sano K, Kikkawa Y, Nishizuka Y: Direct activation of calcium-activated, phospholipid-dependent protein kinase by tumor-promoting phorbol esters. J Biol Chem (257): 7847-7851, 1982.
52. Niedel JE, Kuhn LJ, Vandenbark GR: Phorbol diester receptor copurifies with protein kinase C. Proc Natl Acad Sci USA (80): 36-40, 1983.
53. Ashendel CL, Staller JM, Boutwell RK: Protein kinase activity associated with a phorbol ester receptor purified from mouse brain. Cancer Res (43): 4333-4337.
54. Siess W, Siegel FL, Lapetina EG: Arachidonic acid stimulates the formation of 1,2-diacylglycerol and phosphatidic acid in human platelets: degree of phospholipase C activation correlates with protein phosphorylation, platelet shape change, serotonin release and aggregation. J Biol Chem (258): 11236-11242, 1983.
55. Rozengurt E, Rodriguez-Pena M, Smith KA: Phorbol esters, phospholipase C and growth factors rapidly stimulate the phosphorylation of a Mr 80,000 protein in intact quiescent 3T3 cells. Proc Natl Acad Sci USA (80): 7244-7248, 1983.
56. Sugimoto Y, Whitman M, Cantley LC, Erickson RL: Evidence that the Rous sarcoma virus transforming gene product phosphorylates phosphatidylinositol and diacylglycerol. Proc Natl Acad Sci USA (81): 2117-2121, 1984.

11

GLUCOCORTICOID-INDUCED ANTI-PHOSPHOLIPASE PROTEINS

L. PARENTE AND R.J. FLOWER

Phospholipase A_2 (PLA_2) is a lipolytic enzyme that catalyzes the hydrolysis of the 2-ester bond of a 1,2-di-acyl-sn-glycerol-3-O-phosphatide to yield a 1-acyl-sn-glycerol-3-O-phosphatide (lysophosphatide) and a free fatty acid. Over the last years, the products of the PLA_2 attack on cellular phosphatides have assumed an important role in many physio-pathological processes. The oxidation of free 20C fatty acids leads to the formation of many highly bioactive metabolites. Prostaglandins (PG), produced by the cyclo-oxygenase pathway, mediate many of the vascular changes which occur in inflammation (1). Other cyclo-oxygenase metabolites such as prostacyclin (PGI_2) and thromboxane A_2 (TXA_2) are homeostatic mediators regulating platelet and leukocyte aggregation and adhesion to endothelial cells (2). Lipoxygenase metabolites have been shown to have dramatic bronchoconstrictor effects (leukotrienes C_4, D_4, E_4) and potent chemotactic actions (leukotriene B_4) (3,4). Lysophosphatides have a wide range of biological activities including tumoricidal effects (5,6) and have also been implicated in several pathological states (7,8). Some lysophosphatides are also the precursors of 1-O-alkyl-2-sn-glycerol-3-phosphocholine (AGEPC), formerly referred to as platelet-activating factor (PAF), a phospholipid that may represent the first of a new class of mediators in anaphylaxis and inflammation (9,10). In this presentation we will review the thesis that this phospholipase A_2 enzyme is an important target for the anti-inflammatory glucocorticoids which, by inducing the synthesis and release of anti-phospholipase proteins, regulate the activity of this enzyme and thereby control the rate of synthesis of these pro-inflammatory lipids.

In 1980, we proposed (11) that the anti-phospholipase action of glucocorticoids was mediated by the synthesis/release of a proteinaceous second messenger. This messenger was released by guinea-pig lungs and rat peritoneal leucocytes (mainly macrophages) after stimulation by glucocorticoids in vitro. The protein was named 'macrocortin': 'cortin' to underline its relationship with

W.E.M. Lands (ed.), BIOCHEMISTRY OF ARACHIDONIC ACID METABOLISM.

steroids and 'macro' because it was a much larger molecule than the steroids, and was released by macrophages.

The release of macrocortin was related to the concentration of hydrocortisone, the number of incubated cells and the incubation time (11). Secretion of the protein could be elicited by concentrations of hydrocortisone which were in the physiological range. Further investigations have established that secretion of macrocortin is complex: some protein is stored preformed in rat macrophages and the steroid-induced release takes about 3 hr in vitro (12). Macrophages depleted of their content of macrocortin can no longer respond to hydrocortisone in the usual way (i.e. by a reduction in PG biosynthesis), until resynthesis of the protein (also stimulated by the steroid) has occurred, but are still sensitive to exogenous macrocortin (13). The steroid-induced secretion of macrocortin from rat isolated peritoneal macrophages is specific for glucocorticoids, as non-glucocorticoid steroids such as aldosterone, testosterone and estradiol as well as non-specific cell activators including cytochalasin, f-MLP, PMA, LPS and A23187 failed to induce the secretion of macrocortin. Thus, in order to induce macrocortin secretion, it is essential that the steroid must interact with specific receptors (14). Macrocortin secretion is inhibited by agents which interfere with microtubule assembly (colchicine, vinblastine and trimethylcholchicinic acid) and by agents which increase the intracellular level of cyclic AMP (e.g. PGI_2 and dibutyryl cyclic AMP) (14).

As expected, glucocorticoids can also release macrocortin in vivo and in fact the process occurs more rapidly than in vitro. Rats injected with dexamethasone (1 mg/kg, s.c.) release maximal amounts of macrocortin in the peritoneal cavity within 60 min of injection (15). During the course of these experiments it was also observed that a basal level of the protein was present in the animals before the injection, and that this background level was much lower in adrenalectomised (ADX) rats. ACTH was effective in generating anti-PLA_2 proteins only in normal or sham operated (SHO) animals, whilst dexamethasone and hydrocortisone acted in both ADX and SHO animals. These data strongly support the idea that the background level of macrocortin observed in normal animals is due to endogenous corticosteroids. In agreement with these results we have recently found that stimulated rat peritoneal macrophages from ADX rats release about three times more eicosanoids than cells from SHO animals (Parente & Flower, manuscript in preparation).

The first reported mol. wt. of macrocortin was 15K daltons (11), but when more sophisticated extraction and purification procedures were employed we were able to demonstrate that the major active species in crude lavage fluid from steroid-treated rats was of around 40K, with smaller amounts of bio-activity eluting at around 200K and 15K (15). Other anti-PLA_2 proteins have been described independently by other two groups of researchers. Hirata and coworkers at NIH (16) have reported a PLA_2-inhibitory protein induced in rabbit neutrophils by glucocorticoids. The main mol. wt. of this protein, named 'lipomodulin' was of 40K with fragments at 25K and 16K. The same author demonstrated that phosphorylation of lipomodulin caused the loss of PLA_2-inhibitory activity (17) (see below). In Paris, Russo-Marie's group (18) reported that dexamethasone is able to induce in cultured renal cells, the synthesis of two polypeptides of 15K and 30K mol. wt., which have been named 'renocortins'. Because of the similarities in the distribution of molecular weights, biological activity (i.e. inhibition of PLA_2), and in the pattern of biosynthesis (i.e. induction by glucocorticoids), the question of the relationship between these proteins was investigated.

Using the anti-lipomodulin antibody, three species of anti-PLA_2 proteins with mol. wt. of 40K, 30K and 16K were identified in the macrocortin-rich peritoneal lavage fluids from steroid-treated rats (19). Most of the immunoreactive material was associated with the 16K fraction, which showed PLA_2-inhibitory activity only after dephosphorylation by alkaline phosphatase treatment. It was then suggested that the 16K protein (macrocortin) was a phosphorylated fragment of the 40K protein (lipomodulin) (19). In other experiments (21) it was demonstrated that renomedullary cells cultured in serum free medium form not only 30K and 15K anti-PLA_2 proteins but also a 45K protein - which was not found when the cells were cultured in medium containing 10% foetal calf serum (18). It seems likely that serum contains a protease which cleaves the 45K peptide into smaller fragments. Again the 15K species of renocortin was active on the isolated PLA_2 only after dephosphorylation, but it was very active in inhibiting PGE_2 production by cultured renal cells (18). As cells contain alkaline phosphatase on their surface (22), it seems likely that the 15K species is activated by dephosphorylation occurring at the cell membrane. In the course of the same experiments (21) the monoclonal anti-macrocortin antibody (20) cross-reacted with the peptides from renal cells and inhibited their biological activity. From all these results it seems likely that macrocortin, lipomodulin and renocortin are very similar if not identical proteins. Therefore, the induction of anti-PLA_2 proteins by glucocorticoids is not confined to inflammatory cells (neutrophils and macrophages) and could represent a phenomenon of much wider physiological relevance. Recently, an anti-PLA_2 protein with a mol. wt. of ≃ 125K has been identified in rat peritoneal macrophages (23). The anti-macrocortin antibody cross-reacts and inhibits the activity of this high mol. wt. protein, suggesting that it could be a precursor, but as yet there is no definitive evidence for this relationship.

The inhibition of PLA_2 may lead to profound alterations in cellular biochemistry. We would now like to review briefly the biological effects caused by macrocortin. As far as lipomodulin is concerned it has been implicated in several immunoregulatory processes. The reader is referred to a comprehensive paper by Hirata (24) which gives an up-to-date review of the subject.

Central to the whole idea that macrocortin is a second messenger of steroid action is the question of whether or not the protein has any anti-inflammatory activity. To examine this problem, we first did a simple experiment in which we injected one group of 5 rats with saline and another similar group with dexamethasone. After 1 hr both groups were killed and the peritoneal lavage fluid was obtained, dialyzed to remove any residual steroid and lyophilized. The lyophilized proteins were redissolved and injected together with carrageenin into the pleural cavity of rats and the ensuing inflammation assessed after 4 hr. This is, of course, a modification of a standard rat pleurisy edema test. We found that the inflammatory response (fluid exudation and cell migration) in rats receiving carrageenin and proteins from saline-treated rats was the same as that in rats receiving saline and carrageenin alone. Those rats that had received proteins from steroid-treated animals, however, showed a much reduced inflammatory response. In other words, there was some macromolecular anti-inflammatory factor in the lavage fluid from dexamethasone-treated animals.

In subsequent experiments we defined more precisely the relationship between anti-PLA_2 and anti-inflammatory effect of this protein. The 40K anti-PLA_2 protein was able to (i) inhibit the carrageenin paw oedema in the rat and (ii) reduce the biosynthesis of PGE_2 and LTB_4 by leucocytes in vitro (25). This was the first link between the anti-inflammatory effect of macrocortin and the decreased level of a potent chemotactic agent such as LTB_4. The inhibition of both eicosanoid formation and inflammatory oedema induced by macrocortin, or steroids themselves, was counteracted by the anti-macrocortin monoclonal antibody (20,25,26).

Two different pools of PLA_2 have been described in macrophages (27,28) one localized in the lysosomes and one associated with cellular membranes. Which PLA_2 does macrocortin inhibit? To investigate the problem we studied the effect of either crude or partially purified macrocortin on PLA_2 activities of rat peritoneal macrophages. We observed that crude macrocortin (i.e. a mixture of species) was able to inhibit both lysosomal and membrane-bound PLA_2 while a selective inhibition was caused by each of the different mol.wt. proteins. In fact proteins with a mol. wt. of 40K selectively inhibited the lysosomal PLA_2, whilst proteins with a mol. wt. of ≃ 125 K selectively inhibited the membrane-bound PLA_2 (29).

In view of these results it is tempting to suggest that the anti-PLA_2 proteins act as endogenous regulators of inflammatory responses. This hypothesis has been further strengthened by very recent investigations on the effect of steroids and macrocortin on the formation of AGEPC and lyso-GEPC (Parente and Flower, manuscript in preparation). AGEPC (acetylglyceryl ether phosphorylcholine) is formed by a two stage process in which the 1-alkyl-2-acyl glyceryl phosphorylcholine ether phosphatide precursor is degraded to lyso-GEPC (1-O-alkyl-2-lyso glyceryl phosphorylcholine) by PLA_2 and then acetylated by a specific Co-A dependent acetyl transferase (30). AGEPC and Lyso-GEPC have many important biological activities and may represent a new type of inflammatory mediator (9,10). To study the effect of glucocorticoids and macrocortin on the formation of these substances, rat peritoneal macrophages were incubated at $37^{o}C$ with either hydrocortisone or partially purified macrocortin. After 90 min zymosan was added to the cells. After further 60 min the AGEPC released by the cells was extracted and half of the extracted sample was chemically acetylated to reveal the lyso-GEPC. The samples were then bioassayed on indomethacin-treated platelets. Hydrocortisone exerted a dose-dependent inhibition on the formation of lyso-GEPC, but had no effect on the acetylation reaction. Macrocortin mimicked the steroid action. This inhibitory effect, besides adding more light to the mechanism of anti-inflammatory effect of PLA_2 inhibitors, can explain the ability of glucocorticoid to suppress the tumoricidal action of macrophages, since lysophosphatides are able to induce destruction of both human leukaemia cells (5) and human solid tumours (6).

CONCLUSION

By inhibiting the catabolism of phospholipids, the glucocorticoid-induced peptides exert a powerful suppressive effect on the release of chemicals involved in the inflammatory response. It is likely that this enzyme is also important for many other membrane phenomena too (cf. refs.) the exact relevance of this enzyme (and its naturally occuring inhibitor) to the genesis of malignancy remains to be settled by further research.

REFERENCES

1. Vane JR: The mode of action of aspirin and similar compounds. J Allergy Clin Immunol (58): 691-712. 1976.
2. Bunting S, Moncada S, Vane JR: The prostacyclin-thromboxane A2 balance: pathophysiological and therapeutic implications. Brit Med Bull (39): 271-276. 1983.
3. Samuelsson B: Leukotrienes: a new class of mediators of immediate hypersensitivity reactions and inflammation. In: Samuelsson B, Paoletti R, Ramwell P (eds) Advances in Prostaglandin, Thromboxane and Leukotriene Research. Raven Press, New York, 1983, pp 1-13.
4. Ford-Hutchinson AW: Leukotrienes as potential mediators of inflammation. In: Otterness I, Capetola R, Wong S (eds) Advances in Inflammation Research, Vol. 7. Raven Press, New York, 1984, pp 29-37.
5. Andreesen R, Modolell M, Weltzien HU, Eibl H, Common HH, Lohr GW, Munder PG: Selective destruction of human leukaemic cells by alkyl-lysophospholipids. Cancer Res (38): 3894-3899. 1978.
6. Runge MH, Andreesen R, Pfleiderer A, Munder PG: Destruction of human solid tumours by alkyl lysophospholipids. JNCI (64): 1301-1306. 1980.
7. Blackwell GJ, Flower RJ: Inhibition of phospholipase. Br Med Bull (39): 260-264. 1983.
8. Vadas P, Pruzanski W: Role of extracellular phospholipase A2 in inflammation. In: Otterness , Capetola R, Wong S (eds) Advances in Inflammation Research, Vol. 7, Raven Press, New York, 1984, pp 51-58.
9. Vargaftig BB, Chignard M, Benveniste J, Lefort J, Wal F: Background and present status of research on platelet-activating factor (PAF-acether). NYAS (370): 119-137. 1981.
10. Pinckard RN, McManus LM, Hanahan DJ: Chemistry and biology of acetyl glyceryl ether phosphorylcholine (platelet-activating factor). In: Weissman G (ed), Advances in Inflammation Research, Vol. 4, Raven Press, New York, 1982, pp 147-180.
11. Blackwell GJ, Carnuccio R, Di Rosa M, Flower RJ, Parente L, Persico P: Macrocortin: a polypeptide causing the anti-phospholipase effect of glucocorticoids. Nature (287): 147-149. 1980.
12. Blackwell GJ, Carnuccio R, Di Rosa M, Flower RJ, Parente L: Storage and steroid-induced release from rat leucocytes of a phospholipase inhibitor. Br J Pharmac (72): 136P-137P. 1981.
13. Carnuccio R, Di Rosa M, RJ Flower, Pinto A: The inhibition by hydrocortisone of prostaglandin biosynthesis in rat peritoneal leucocytes is correlated with intracellular macrocortin levels. Br J Pharmac (74): 322-324. 1981.
14. Blackwell GJ: Specificity and inhibition of glucocorticoid-induced macrocortin secretion from rat peritoneal macrophages. Br J Pharmac (79): 587-594. 1983.
15. Blackwell GJ, Carnuccio R, Di Rosa M, Flower RJ, Langham CSJ, Parente L, Persico P, Russell-Smith NC, Stone D: Glucocorticoids induce the formation and release of anti-inflammatory and anti-phospholipase proteins into the peritoneal cavity of the rat. Br J Pharmac (76): 185-194. 1982.

16. Hirata F, Schiffmann E, Venkatasubramanian K, Salomon D, Axelrod J: A phospholipase A2 inhibitory protein in rabbit neutrophils induced by glucocorticoids. Proc Natl Acad Sci USA (77): 2533-2536. 1980.
17. Hirata F: The regulation of lipomodulin, a phospholipase inhibitory protein, in rabbit neutrophils by phosphorylation. J Biol Chem (256): 7730-7733. (1981).
18. Cloix JF, Colard O, Rothhut B, Russo-Marie F: Characterization and partial purification of 'renocortins': two polypeptides formed in renal cells causing the anti-phospholipase action of glucocorticoids. Br J Pharmac (79): 313-321. 1983.
19. Hirata F, Notsu Y, Iwata M, Parente L, Di Rosa M, Flower RJ: Identification of several species of phospholipase inhibitory protein(s) by radioimmunoassay for lipomodulin. Biochem Biophys Res Commun (109): 223-230. 1982.
20. Flower RJ, Wood JN, Parente L: Macrocortin and the mechanism of action of the glucocorticoids. In: Otterness I, Capetola R, Wong S (eds) Advances in Inflammation Research, Vol. 7, Raven Press, New York, 1984, pp 61-70.
21. Rothhut B, Russo-Marie F, Wood J, Di Rosa M, Flower RJ: Further characterization of the glucocorticoid-induced anti-phospholipase protein "Renocortin". Biochem Biophys Res Commun (117): 878-884. 1983.
22. Neumann H, Klein E, Hauck-Granoth R, Yachnin S, Ben-Bassat H: Comparative study of alkaline phosphatase activity in lymphocytes, mitogen-induced blasts, lymphoblastoid cell lines, acute myeloid leukemia and chronic lymphatic leukemia cells. Proc Natl Acad Sci USA (73): 1432-1436. 1976.
23. Coote PR, Di Rosa M, Flower RJ, Merrett M, Parente L, Wood JN: Detection and isolation of a steroid-induced anti-phospholipase protein of high molecular weight. Br J Pharmac (80): 597P. 1983.
24. Hirata F: Roles of lipomodulin: a phospholipase inhibitory protein in immunoregulation. In: Otterness R, Capetola R, Wong S (eds) Advances in Inflammation Research, Vol. 7, Raven Press, New York, 1984, pp 71-78.
25. Parente L, Di Rosa M, Flower RJ, Ghiara P, Meli R, Persico P, Salmon JA, Wood JN: Relationship between the anti-phospholpase and anti-inflammatory effect of glucocorticoid-induced proteins. Eur J Pharmac (99): 233-239. 1984.
26. Blackwell GJ, Carnuccio R, Di Rosa M, Flower RJ, Ivanyi J, Langham CSJ, Parente L, Persico P, Wood J: Suppression of arachidonate oxidation by glucocorticoid-induced anti-phospholipase peptides. In: Samuelsson B, Paoletti R, Ramwell P, (eds) Advances in Prostaglandin, Thromboxane and Leukotriene Research, Vol. 11, Raven Press, New York, 1983, pp 65-71.
27. Hsueh W, Desai U, Gonzales-Crussi F, Lamb R, Chu A: Two phospholipase pools for prostaglandin synthesis in macrophages. Nature (290): 710-713. 1981.
28. Wightman PD, Humes JL, Davies P, Bonney RJ: Identification and characterization of two phospholipase A2 activities in resident mouse peritoneal macrophages. Biochem J (195): 427-433. 1981.
29. Ghiara P, Meli R, Parente L, Persico P: Distinct inhibition of membrane-bound and lysosomal phospholipase A2 by glucocorticoid-induced proteins. Biochem Pharmac (in the press), 1984.
30. Albert DH, Snyder F: Biosynthesis of 1-alkyl-2-acetyl-sn-glycero-3-phosphocholine (Platelet-Activating Factor) from 1-alkyl-2-acyl-sn-glycero-3-phosphocholine by rat alveolar macrophages. Phospholipase A2 and acetyltransferase activities during phagocytosis and ionophore stimulation. J Biol Chem (258): 97-102. 1983.

12

LIPID NUTRITION, PROSTAGLANDINS AND CANCER

RASHIDA A. KARMALI

1. INTRODUCTION

The Committee on Diet, Nutrition and Cancer (1) concluded that of all dietary components studied the combined epidemiological and experimental evidence is most suggestive of a causal relationship between dietary fat intake and incidence of cancers of the colon and breast. Similar but less consistent correlations have been reported with cancers of the prostate, ovary and endometrium (2,3). This is an attempt to discuss the subject of dietary fat, prostaglandins and breast cancer relationships based on a brief review of both epidemiological and experimental findings.

2. EPIDEMIOLOGICAL EVIDENCE

2.1 Breast cancer

Breast cancer is a leading cause of death among women in industrialized areas such as North America and Western Europe (4,5). The concept that dietary fat plays a major role in the etiology of human breast cancer has evolved over the past 20 years. Dietary fat accounts for approximately 40% of the total calories consumed in the United States (6). This subject has been reviewed extensively (1,7). One of the most prolific contributors in both the experimental field (8-9) as well as in correlation studies based on age-adjusted mortality rates has been Carroll (10).

2.1.1 Eskimos and Japanese. A high breast cancer incidence and mortality has been shown to correlate strongly with increased per capita fat consumption in several populations (7,10). An exception is the low incidence of breast cancer in Greenland Eskimos (11,12). The association between fat intake and breast

W.E.M. Lands (ed.), BIOCHEMISTRY OF ARACHIDONIC ACID METABOLISM.

cancer must therefore be evaluated carefully. Both the quality and quantity of dietary fat may influence the incidence of this disease (10).

Epidemiological evidence indicates that both Japanese and Greenland Eskimo women have a low incidence of breast cancer (11-13). Japanese and Eskimo diets include a large amount of ω3 fatty acids (C20:5 and C22:6) which are unique to the marine lipids. In the past both the Japanese and the Eskimo women were notable for their very low incidence of breast cancer. However, in both populations the situation has been changing in recent years. In Japan the number of annual deaths from breast cancer doubled in the 20-year period from 1955-1975 (13); over the same period the Japanese diet became more akin to that of Western countries, primarily among the younger, more affluent, urban segments of the population (14). This change in dietary practice has particularly affected fat consumption. The typical Japanese diet includes large quantities of fish and some seaweed so that total fat is comprised of an appreciable portion of long chain fatty acids of the ω3 family (15). Epidemiological evidence suggests the change in breast cancer incidence among Eskimo women may be due to an increased exposure to Western influence. At the beginning of this century approximately half the caloric intake of Eskimos consisted of fats of marine origin. One of the consequences of modernization has been a change in dietary habits; imported foods have become more accessible resulting in an increased consumption of saturated fat and unsaturated fatty acids of the ω6 family.

An extremely low incidence of atherosclerotic heart disease in Eskimos has stimulated researchers to study the relationship between intakes of marine oil and incidence of atherosclerosis. This should be translated to studies of human breast carcinogenesis.

3. ANIMAL STUDIES

3.1 Experimental mammary tumor models

In animal studies polyunsaturated fat has been shown to promote development of mammary tumors more effectively than saturated fat. In dimethylbenz(a)anthracene (DMBA)-induced mammary carcinogenesis in rodents, the incidence of mammary carcinomas was greater among animals fed a high polyunsaturated fat diet than among animals fed a high saturated fat diet (9, 10,16). Tumor-enhancing effects of polyunsaturated fatty acids (PUFAs) were also reported in transplantable (17) and spontaneous mammary tumorigenesis (18). Detailed analysis of the various diets used indicates that the principal PUFA studied was C18:2 since corn oil, which contains approximately 60% C18:2, was the main source of fat used.

The above observations should be evaluated in light of the change in the types of dietary fats consumed in the United States. In recent years consumption of animal fat has decreased dramatically whereas intake of PUFAs such as C18:2 ω6 and C18:1 ω9 present in, for instance, margarines, has increased (6,19).

3.2 Mechanism of action of dietary fat

The mechanism by which dietary fat influences mammary carcinogenesis is still unknown. Chan et al. (20) suggested that high levels of dietary fat act by enhancing prolactin secretion. However, in subsequent studies several investigators found that a high fat diet exerts its effect independent of any change in circulatory prolactin or estrogen. This topic has been reviewed by Welsch et al. (21).

It has also been postulated that the effect of dietary fat may be mediated by a change in host immunocompetence. A diet high in PUFAs is known to exert immunosuppressive effects (22).

Recent evidence in the animal tumors suggests that the tumor-promoting effects of a high PUFA diet are mediated through prostaglandins (PGs) and, in particular, products of arachidonic acid (Fig. 1). In DMBA-induced mammary tumorigenesis when indomethacin, a cyclooxygenase inhibitor, was included in a high PUFA diet (15-20% corn oil), the promotional effects of the high PUFA diet were reduced (23,24). Mammary tumor growth was found to be related to C20:4 ω6 content of the tumor (17). Studies reported by a number of investigators suggest that inhibition of arachidonic acid metabolism offers a promising new approach to inhibit mammary tumorigenesis (25-26).

Two ω3 fatty acids, C20:5 and C22:6, found mainly in marine lipids, have been known for some time to inhibit arachidonic acid metabolism (27,28). A brief discussion of the synthetic pathways for the essential fatty acids and prostaglandins will follow.

4. ESSENTIAL FATTY ACIDS AND PROSTAGLANDINS

The term essential fatty acid (EFA) covers a wide range of 18, 20 and 22 carbon chain length PUFAs with 2 to 6 methylene interrupted double bonds with the cis configuration. There are two classes: the linoleic (ω6) and linolenic (ω3). Linolenic acid was considered an EFA by Burr and Burr (29) but since then it has been in and out of the EFA picture. Both Sinclair (30) and Crawford (31) regard the linolenate family to be essential in animals. Both C18:2 ω6 and C18:3 ω3 are required for the normal growth and function of all tissues. These parent EFAs undergo chain elongation and desaturation to produce long chain derivatives of 20 and 22 carbons with 3, 4, 5 and 6 double bonds. The result is two families (ω3 and ω6) of EFAs which are required for cell membrane structures, for prostaglandin synthesis and in the transport and oxidation of cholesterol (Fig. 1).

The third class of PUFAs is the oleic - C18:1 ω9. This can be desaturated and elongated to form C20:3 but this is not an EFA. PUFAs comprise all the EFAs (ω3 and ω6 families), the ω9 family and trans-isomers of EFAs. It would therefore be erroneous to equate EFAs with PUFAs.

5. PROSTAGLANDINS AND BREAST CANCER

There is strong evidence that PG metabolism is linked to the biological behavior of breast cancer. PGE_2 has been implicated in the mediation of breast cancer metastases to bone and hypercalcemia (32) as well as survival (33). It has been demonstrated by Rolland et al. (34) and in our laboratory (35) that PGE_2 (34,35) and thromboxane $(TX)B_2$ (35) are related to metastases in breast cancer patients and may be used as

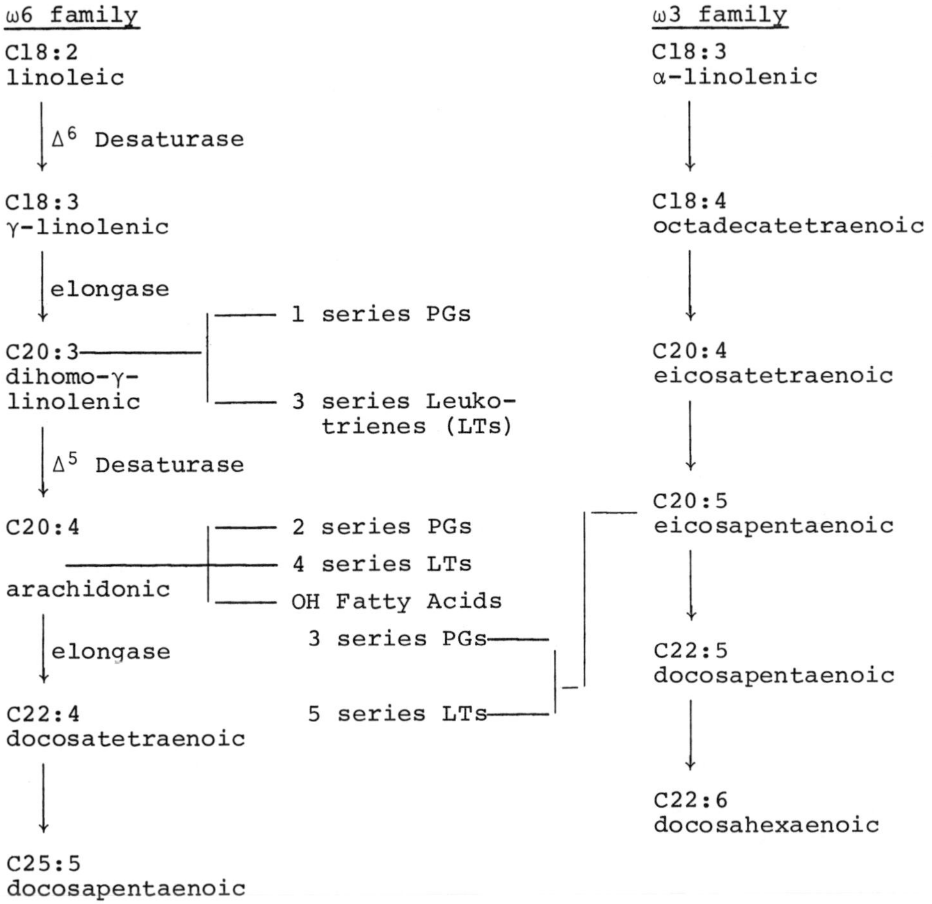

FIGURE 1.

prognostic factors. Thus it is reasonable to hypothesize that PGs may play an important role in the evolution of breast cancer from precursor lesions (Reviews-36,37).

6. ω3 FATTY ACIDS: INHIBITORS OF ARACHIDONIC ACID METABOLISM

Two ω3 fatty acids, C20:5 and C22:6, exert an inhibitory effect on metabolism of C20:4 ω6 and thus inhibit synthesis of dienoic PGs (27,28). C20:5 ω3 is the precursor of the three series PGs and TXA_3. It has been suggested that TXA_3 is less thrombogenic than TXA_2 and PGI_3, like PGI_2, is anti-thrombogenic (38). When C20:5 ω3 is substituted for 20:4 ω6 in the platelet membrane, the two fatty acids compete for cyclooxygenase. Though C20:5 ω3 and C20:4 ω6 have approximately equal affinities for cyclooxygenase, C20:5 ω3 is transferred much more slowly than C20:4 ω6 to cyclic endoperoxides (21,39). Lands et al. (40) have proposed that under conditions where in vivo levels of peroxides are very low, the antithrombotic activity of C20:5 ω3 is due more to its ability to displace C20:4 ω6 in tissue and serum phospholipids and to act as a competitive inhibitor of cyclooxygenase rather than to any ability of C20:5 ω3 to be metabolized to an antithrombotic trenoic PG.

6.1 ω3 Fatty acids: effect on mammary tumor growth

PGE_2 is the major eicosanoid measured in experimental and human tumors (36,37). The hypothesis that ω3 fatty acids may inhibit mammary tumor growth by inhibiting C20:4 ω6 metabolism is consistent with observations that inhibitors of PG synthesis such as indomethacin were found to inhibit tumor growth (26). We have studied the effect of marine oil MaxEPA (containing ≅17.4% C20:5 ω3 and 16.8% C22:6 ω3) on growth of the R3230AC mammary adenocarcinoma transplanted in female Fischer 344 rats. The experimental protocol has been described earlier (41). After three weeks, tumors were excised, measured and used for two PG measurements: Tumor PG content (ng/g wet weight tissue) and in vitro PG synthesis by tumor microsomes (ng/mg protein). Daily dietary supplementation with 0.1, 0.2 and 0.4 ml of MaxEPA significantly inhibited tumor growth. Both tumor content and in vitro synthesis of dienoic PGs - E_2, $F_{2\alpha}$, 6-keto-$PGF_{1\alpha}$ (degradation product of PGI_2) and TXB_2 (degradation product of TXA_2) were inhibited in tumors from rats treated with MaxEPA (42). These findings indicate that the mechanism underlying inhibition of mammary tumorigenesis is linked in part to the inhibitory effect of ω3 fatty acids on C20:4 ω6 metabolism in mammary tumors.

Further evidence supporting the contention that ω3 fatty acids may exert a protective effect on mammary tumorigenesis comes from recent studies. Cohen et al. (43) found that the brown seaweed Laminaria angustata effectively blocked the tumor-promoting effect of a high fat (20% corn oil) diet in mammary tumors induced with N-nitrosomethylurea. When a control diet was supplemented with 5% seaweed, there was a delay in the appearance of tumors induced with DMBA; there was also a reduction in tumor numbers (44). This brown kelp seaweed is frequently consumed in Japan (45). We have found it to contain a large amount of 20:5 ω3 in the choline phospholipid fraction (R. Karmali and L. Cohen, unpublished observations). Whether the chemopreventive effect of this seaweed in two different chemically-induced mammary tumors is related to the C20:5 ω3 content is not determined.

We have discussed preliminary findings on tumor-inhibitory effects of two ω3 fatty acids, C20:5 and C22:6, in the R3230AC mammary adenocarcinoma. It is proposed that the effects of these two macronutrients are mediated in part by modulation of C20:4 ω6 metabolism, resulting in inhibition of synthesis of dienoic PGs in the mammary tumors.

In summary, epidemiological and experimental evidence is suggestive for a causal relationship between fat intake and occurrence of breast cancer. The nature as well as the amount of dietary fat may influence the incidence of mammary cancer. Diets high in C18:2 ω6 enhance development of chemically-induced, transplanted and spontaneous mammary tumors. Animal studies with PG inhibitors suggest that the tumor-promoting effects of a high PUFA (and, in particular, C18:2 ω6) diet are mediated through dienoic PGs. Two ω3 fatty acids, C20:5 and C22:6, are natural inhibitors of C20:4 ω6 metabolism and thus of dienoic PGs. It is proposed that nutritional intervention with these macronutrients be considered in the chemoprevention of breast cancer.

ACKNOWLEDGEMENTS

This investigation was supported by Grant Number CA-32092 awarded by the National Cancer Institute, Department of Health and Human Services, and grant PDT-208 awarded by the American Cancer Society.

The support of Drs. Jerome J. DeCosse, Michael P. Osborne, David W. Kinne, and David Rose is greatly appreciated.

REFERENCES

1. Committee on Diet, Nutrition, and Cancer. Diet, Nutrition, and Cancer. Assembly of Life Sciences, National Research Council, National Academy Press, Washington, D.C., 1982.
2. Armstrong B, Doll R: Environmental factors and cancer incidence and mortality in different countries, with special reference to dietary practices. Int J Cancer (15): 617-631, 1975.
3. Miller AB: Epidemiology of gastrointestinal cancer. Compr. Therapy (7): 53-58, 1981.
4. Segi M, Kurihara M, Matsuyama T: Cancer mortality for selected sites in 24 countries. No. 5 (1964, 1965), Sendai, Japan; Department of Public Health, Tohoku University School of Medicine, 1969.
5. Waterhouse J, Muir C, Correa P: Cancer incidence in five continents. Vol. III. IARC Scientific Publications No. 15, Lyon; International Agency for Research on Cancer, 1976.
6. Rizek RL: Food supply studies and consumption survey statistics on fat in United States diets. Cancer Res (41): 3729-3730, 1981.
7. Wynder EL: Nutrition and Cancer. Fed Proc (35): 1309-1315, 1976.
8. Carroll KK, Gammal EB, Plunkett ER: Dietary fat and mammary cancer. Can Med Assoc J (98): 590-594, 1968.
9. Carroll KK, Khor HT: Effects of level and type of dietary fat on incidence of mammary tumors induced in female Sprague-Dawley rats by 7,12-dimethylbenz(a)-anthracene. Lipids (6): 415-420, 1971.
10. Carroll KK: Experimental evidence of dietary factors and hormone-dependent cancers. Cancer Res (35): 3374-3383, 1975.
11. Stefansson V: Cancer: Disease of civilization? Hill and Wang, New York, 1960.
12. Nielsen NH, Hansen JPH: Breast cancer in Greenland - selected epidemiological, clinical and histological features. J Cancer Res Clin Oncol (98): 287-299, 1980.
13. Hirayama T: Epidemiology of breast cancer with special reference to the role of diet. Prev Med (7): 173-195, 1978.
14. Kagawa Y: Impact of westernization on the nutrition of Japanese: Changes in physique, cancer, longevity and centenarians. Prev Med (7): 205-217, 1978.
15. Bang HO, Dyerberg J, Hjorne N: The composition of food consumed by Greenland Eskimos. Acta Med Scand (200): 69-73, 1976.
16. Carroll KK, Hopkins GI: Dietary polyunsaturated fat versus saturated fat in relation to mammary carcinogenesis. Lipids (14): 155-158, 1979.

17. Rao GA, Abraham S: Brief communication: Enhanced growth rate of transplantable mammary adenocarcinoma induced in C3H mice by dietary linoleate. J Natl Cancer Inst (56): 431-432, 1976.
18. Tinsley IT, Schmitz J, Pierce D: Influence of dietary fatty acids on the incidence of mammary tumors in the C3H mouse. Cancer Res (41): 1460-1465, 1981.
19. Weihrauch JL, Brignoli CA, Reeves JB III, Iverson JL: Fatty acid composition of margarines, processed fats, and oils: a new compilation of data for tables of food composition. Food Technol (31): 80-85, 1977.
20. Chan PC, Didato F, Cohen LA: High dietary fat, elevation of rat serum prolactin and mammary cancer. Proc Soc Exp Biol Med (149): 133-135, 1975.
21. Welsch KW, Aylsworth CF: Enhancement of murine mammary tumorigenesis by feeding high levels of dietary fat: a hormonal mechanism? J Natl Cancer Inst (70): 215-221, 1983.
22. Vitale JJ, Broitman SA: Lipids and immune function. Cancer Res (41): 3706-3710, 1981.
23. Hillyard LA, Abraham S: Effect of dietary polyunsaturated fatty acids on growth of mammary adenocarcinomas in mice and rats. Cancer Res (39): 4430-4437, 1979.
24. Carter CA, Milholland RJ, Shea W, Ip MM: Effect of prostaglandin synthetase inhibitor indomethacin on 7,12-dimethylbenz(a)anthracene-induced mammary tumorigenesis in rats fed different levels of fat. Cancer Res (43): 3559-3562, 1983.
25. Leaper DJ, French B, Bennett A: Reduction by flurbiprofen of primary tumor growth and local metastasis formation in mice. Adv Prostaglandin Thromboxane Res (6): 591-593, 1980.
26. Kollmorgen GM, King MM, Kosanke SD, Do C: Influence of dietary fat and indomethacin on the growth of transplantable mammary tumors in rats. Cancer Res (43): 4714-4719, 1983.
27. Needleman P, Whitaker MO, Wyche A, Watters K, Sprecher H, Raz A: Manipulation of platelet aggregation by prostaglandins and their fatty acid precursors. Pharmacological basis for a therapeutic approach. Prostaglandins (19): 165-181, 1980.
28. Corey RJ, Shih C, Cashman JR: Docosahexaenoic acid is a strong inhibitor of prostaglandin but not leukotriene biosynthesis. Proc Natl Acad Sci USA (80): 3581-3584, 1983.
29. Burr GO, Burr MM: On the nature and role of the fatty acids essential in nutrition. J Biol Chem (86): 587-621, 1930.
30. Sinclair HM: Essential fatty acids and chronic degenerative diseases. Ch. 6. In: Rose J (ed) Nutrition and Killer Diseases. The effects of dietary factors on fatal chronic diseases. Noyes Publ, 1982, pp 69-83.
31. Crawford MA: Background to essential fatty acids and their prostanoid derivatives. Br Med Bull (39): 210-213, 1983.
32. Seyberth HW, Segre GV, Margan LL, Sweetman BJ, Potts JT, Oates JA: Prostaglandin as mediators of hypercalcemia associated with certain types of cancer. N Engl J Med (293): 1278-1283, 1975.

33. Bennett A, Berstock DA, Raja B: Survival time after surgery is inversely related to the amounts of prostaglandins extracted from human breast cancers. Br J Pharmacol (66): 451P, 1979.
34. Rolland PH, Martin PM, Jacquemler J, Rolland AM, Toga M: Prostaglandin in human breast cancer: Evidence suggesting that an elevated prostaglandin production is a marker of metastatic potential for neoplastic cells. J Natl Cancer Inst (64): 1061-1070, 1980.
35. Karmali RA, Welt S, Thaler HT, Lefevre F: Prostaglandins in breast cancer: Relationship to disease stage and hormone status. Br J Cancer (48): 689-696, 1983.
36. Karmali RA: Prostaglandins and cancer. Prostaglandins Leukotriene Med (5): 11-28, 1980.
37. Karmali RA: Prostaglandins and cancer. CA - A Cancer J for Clinicians (33): 322-332, 1983.
38. Dyerberg J, Bang HO, Stoffersen E, Moncada S, Vane JR: Eicosapentaenoic acid and prevention of thrombosis and atherosclerosis. Lancet (ii): 117-119, 1978.
39. Dyerberg J, Bang HO: Lipid metabolism, artherogenesis and haemostasis in Eskimos. The role of the prostaglandin-3 family. Haemostasis (8): 227-233, 1979.
40. Lands WEM, Byrnes MJ: The influence of ambient peroxides on the conversion of 5,8,11,14,17-eicosapentaenoic acid to prostaglandins. Prog Lipid Res (20): 287-290, 1981.
41. Karmali RA: Growth inhibition and prostaglandin metabolism in the R3230AC mammary adenocarcinoma by reduced glutathione. Cancer Biochem Biophys. (In press)
42. Karmali RA, Marsh J, Fuchs C: Effect of omega-3 fatty acids on growth of a rat mammary tumor. (Submitted)
43. Cohen LA, Thompson DO, Teas J: Seaweed blocks the mammary tumor promoting effects of high fat diets. Int Breast Cancer Res Conf, Denver, CO, March 1983, Abstr. 52.
44. Teas J, Harbison ML, Gelman RS: Dietary seaweed as a protective factor in DMBA-induced mammary carcinogenesis. Int Breast Cancer Res Conf, Denver, CO, March 1983, Abstr. 53.
45. Kaneda T, Ando H: Component lipids of purple laver and their antioxygenic activity. In: Nisizawa K (ed) Proceedings of the 7th International Seaweed Symposium, Sapparo, Japan, Aug. 8-12, 1971. Halsted Press, A Division of John Wiley & Sons, New York, pp. 553-555.

13

BIOLOGICAL EFFECTS OF HYDROXY FATTY ACIDS

JACK Y. VANDERHOEK

1. INTRODUCTION

Mammalian cells metabolize arachidonic acid to many different products including a variety of hydroxy fatty acids (1). Most of these hydroxy fatty acids are formed by lipoxygenases. The three most common lipoxygenases are the 5- and 15-lipoxygenases, present in such cells as leukocytes and macrophages and the 12-lipoxygenase which is found in mast cells, platelets, and epidermal cells. These lipoxygenases metabolize arachidonic acid to hydroperoxyeicosatetraenoic acids (HPETEs). The HPETEs are unstable and are either readily reduced by cellular peroxidases to the corresponding stable monohydroxyeicosatetraenoic acids (HETEs) or are converted to other products including leukotrienes, diHETEs or THETEs (2). The nature of the products formed depends on the specificity of the particular lipoxygenase as well as the presence of other metabolizing enzymes such as epoxide synthetase, hydrolases, etc. Usually the HETEs are the predominant lipoxygenase product. Several reports indicate that both 11- and 15-HETE can also be formed via the lipoxygenase activity of the cyclooxygenase enzyme (1). The major hydroxy fatty acid produced by the cyclooxygenase pathway is 12-hydroxyheptadecatrienoic acid (HHT). This chapter will focus on the reported biological activities of these various mono-hydroxy fatty acids.

2. EFFECTS OF HYDROXY FATTY ACIDS ON CELLULAR MIGRATION

The earliest report of a biological function for HETEs was the discovery that the platelet lipoxygenase product 12-HETE was chemotactic for human PMN leukocytes (3). When other HETE

W.E.M. Lands (ed.), BIOCHEMISTRY OF ARACHIDONIC ACID METABOLISM.

isomers (which differed in the position of the hydroxyl group) were tested, they were found to be chemotactic in the concentration range of 2-75 μM (4-8). Furthermore, the following rank order of chemotactic potency for PMN leukocytes and eosinophils was observed: 5-HETE>>8-HETE=9-HETE>>11-HETE=12-HETE. 15-HETE was inactive. However, leukotriene B_4, a 5,12-diHETE, was 100- to 1000-fold more potent (9). The maximal chemotactic responses of the different HETEs did not differ significantly in magnitude. The chemotactically active HETEs also stimulated leukocyte migration in the absence of a concentration gradient, termed chemokinesis. The chemokinetic effects were observed in the 1-10 μM concentration range, but the maximal levels of chemokinetic stimulation were different for different HETE isomers. Maximal chemokinetic stimulation was observed by 5-HETE followed by 8- and 9-HETE which were more potent than 11- or 12-HETE. HHT, a platelet cyclooxygenase pathway product, was about 2- to 3-fold less active than 12-HETE (7). The free carboxyl group is important for the chemotactic and chemokinetic activity of the HETEs since the methyl esters only exhibited about a tenth of the chemotactic activity of the free acid whereas no detectable chemokinetic activity was observed (7,8). Injection of 12-HETE (13 μM) into the guinea pig peritoneal cavity elicited a rapid *in vivo* influx (within 30 min) of eosinophils and a slower influx (after 5 hr) of neutrophils (7). In addition to the HETE-induced mobility of leukocytes, several HETE isomers were reported to stimulate the aggregation of cytochalasin-B-pretreated human neutrophils (10). 12-HETE induced a half-maximal response at 40 nM whereas the half-maximal potency of 5-HETE was 200 nM. The 8-, 9-, 11- and 15-HETEs were ineffective. The aggregating activity of 5- and 12-HETE required extracellular Ca^{+2} and Mg^{+2}. Several reports have described the effects of HETEs on other cell types. 5-HETE has been reported to induce a chemokinetic response by human T lymphocytes but 11-HETE was ineffective (11). 15-HETE suppressed both the random migration of these T-lymphocytes as well as the chemokinetic responses of the cells to concanavalin A. 12-HETE, in a concentration range of 0.02-2 pM, was reported to strongly stimulate the migration of rat aortic smooth muscle cells (12). The locomotion was chemokinetic and was a highly calcium-dependent process (13). 15-HETE was a much less potent chemoattractant (effective at 30-300 nM) whereas 5-HETE was inactive (14). Human platelet aggregation induced by the endoperoxide PGH_2 was equally inhibited by 12- and 15-HETE (I_{50}=6.7 μM) (15). The isomeric 5-HETE and the cyclooxygenase product HHT were three-fold less potent.

3. INTERACTIONS OF HETEs WITH FATTY ACID OXYGENASES AND PHOSPHOLIPASES

Lipoxygenase products are capable of modulating various fatty acid oxygenase activities. We were the first to demonstrate that the 15-lipoxygenase product 15-HETE selectively inhibited the human platelet 12-lipoxygenase (I_{50}=8 μM) since the platelet cyclooxygenase was about 15-fold less sensitive to inhibition by 15-HETE (16). Similarly, 15-HETE was also a more selective inhibitor of the mouse thyroid 12-lipoxygenase (I_{50}=3 μM) (17). However, 12-HETE was a more potent inhibitor of the cyclooxygenase pathway (I_{50}= 3 μM). In addition, 15-HETE has been reported to inhibit the cyclooxygenase in rat pituitary gonadotrophs (18). The effects of various HETEs on different rabbit lipoxygenases is summarized in Table 1. The results

Table 1. Inhibition of the 5- and 15- lipoxygenases in rabbit peritoneal PMN leukocytes and of the 12-lipoxygenase in rabbit platelets by different HETEs.*

	Concentration (μM) of HETE required for half-maximal inhibition of		
HETE isomer	5-Lipoxygenase	12-Lipoxygenase	15-Lipoxygenase
15-HETE	3.7	0.93	20
12-HETE	6.8	>100	9.4
5-HETE	14	21	10

*See reference 19 for detail

show that 15-HETE was about 22-times more potent than 5-HETE in inhibiting the rabbit platelet lipoxygenase (19). The order of inhibitory potencies of these HETEs on the peritoneal PMN 5-lipoxygenase was 15-HETE>12-HETE>5-HETE (19,20). Furthermore, 5-HETE and 12-HETE were equally effective in inhibiting the PMN 15-lipoxygenase but 15-HETE was less potent. 15-HETE was also reported to inhibit the 5-lipoxygenase in human T-lymphocytes but 11-HETE was much less effective (21). In contrast to the inhibitory effects of 15-HETE on the various lipoxygenases just discussed, 15-HETE at micromolar concentrations activated a cryptic 5-lipoxygenase in a mast/basophil cell line to produce leukotriene B_4 and other 5-lipoxygenase metabolites (22). The location of the hydroxyl group in the HETE molecule appears important since the isomeric 5-HETE was less effective than 15-HETE and 12-HETE was inactive (JY Vanderhoek and DH Pluznik, unpublished observations).

In addition to the regulatory effects on fatty acid oxygenases, HETEs can also act as substrates for these and other enzymes. 12-HETE was oxidized by the human and porcine leukocyte 5-lipoxygenase to 5S,12S-diHETE and a trihydroxy product, 5,12,20-trihydroxy-eicosatetraenoic acid (5,12,20-THETE) (23,24). 5S,12S-diHETE was also obtained by the action of the human platelet 12-lipoxygenase on 5-HETE (23). Since human blood neutrophils, eosinophils and rat peritoneal mononuclear cells contain both 5- and 15- lipoxygenases, incubation of these leukocytes with either 5-HETE or 15-HETE produced 5S,15S-diHETE (25-27). Recently human leukocytes were shown to contain an ω-hydroxylase which metabolized 12-HETE (derived from platelets) to 12S,20-diHETE (28,29). Presumably, this enzyme was responsible for the formation of 5,12,20-THETE from 5S,12S-diHETE.

There have been several reports that suggest that HETEs can also affect phospholipases. 15-HETE (0.3-10 μM) generated leukotrienes C_4, D_4 and E_4 from dog mastocytoma cells, presumably via the action of the 15-HETE on both the phospholipase and the 5-lipoxygenase/leukotriene pathways (30). 5-HETE enhanced the chemotactic peptide-induced release of arachidonic acid from endogenous phospholipids of the human promyelocytic leukemia cell line HL 60, the concentration for half-maximal stimulation was 17 μM (31).

4. EFFECTS OF HETEs ON SECRETION OF CELLULAR CONSTITUENTS

5-HETE, 12-HETE and 15-HETE induced degranulation of specific granules of human neutrophils and guinea pig peritoneal PMN leukocytes as measured by lysozyme release (32-34). 5-HETE was the more potent agent and at 10 μM concentration, produced a 4-fold increase over control. Neither 5-, 11- nor 12-HETE was able to release β-glucuronidase, a constituent of azurophilic granules, or lactic dehydrogenase, a cytoplasmic enzyme from human neutrophils and eosinophils (5). However, 5-HETE induced the release of β-N-acetylglucosaminidase, another azurophilic granule enzyme, from guinea pig peritoneal leukocytes (34). In addition to the direct effects of HETEs on degranulation, 5-HETE was also reported to potentiate the degranulating action of platelet-activating factor on human neutrophils (35). 16 nM to 5 μM of 5-L-HETE augmented the release of lysozyme and β-glucuronidase from neutrophils 100- to 1000-fold when the cells were simultaneously exposed to 5-HETE and platelet activating factor. 5-Rac-HETE was also effective but 8-rac-HETE was inactive. 5-L-HETE had no influence on the degranulating actions of C5a, A23187 and N-formylmethionylleucylphenylalanine. Both 5- and 12-HETE (1-10 μM) increased the IgE-mediated release of histamine from rat mast cells (by about 60-90% over control values) whereas 5-HETE was reported to enhance similar quantities of the histamine release from human basophils stimulated with ragweed antigen E (36,37). Neither 5-HETE nor 12-HETE were effective alone in inducing histamine release. Another proinflammatory property of 5-HETE involves the reversal of the inhibition of histamine release by agents that act via adenylate cyclase such as PGE_2 and dimaprit (37,38). The effects of various HETEs on mucous glycoprotein release from cultured human airways has also been reported (39). At 1 and 10 nM, 5-, 8-, 9-, 11-, 12- and 15-HETE were equipotent secretagogues of mucus but at 100 nM, 12- and 15-HETE were the most effective agents. HETE lipoxygenase products appear to modulate glucose-induced insulin secretion from isolated pancreatic islets of rats (40). 5-HETE (10 μM) induced significant insulin secretion in low glucose (3.3 mM) medium. 12-HETE and 15-HETE were ineffective even at 50 μM. However, insulin secretion in 16.7 mM glucose was inhibited by 15-HETE (100 μM caused 60% inhibition) and 12-HETE (100 μM caused 40% inhibition) but not by 5-HETE. Another report indicates that 5-HETE (40 μM) enhanced amylase release by 60% over control values from isolated guinea pig pancreatic acini (41). 5-HETE was also reported to stimulate colonic secretion as measured by increases in short-circuit current in rabbit colonic mucosa (42). A recent report indicated that 5-HETE was the only hydroxy fatty acid that was an effective stimulant of luteinizing hormone release from rat pituitary cells since 11-, 12- and 15-HETE were ineffective (43). Finally, 15-HETE (50-100 μM) was found to elicit superoxide production by guinea pig peritoneal macrophages (44).

5. EFFECTS OF HETEs ON UPTAKE OF CELLULAR CONSTITUENTS

Several HETEs have been shown to affect calcium homeostasis in rabbit peritoneal neutrophils (45). Both 11- and 12-HETE enhanced ^{45}Ca uptake in neutrophils by 250% over control values compared with a 75% increase for 5-HETE and no effect for 15-HETE. Data also has been presented to indicate that 5- and 12-HETE mediate the coupling of membrane receptor stimulation and increased hexose uptake in human PMN leukocytes (46,47). It was found that both isomers induce a 2.4-fold increase in the stereospecific uptake of extracellular hexose by leukocytes and half-maximum uptake was obtained with 176 nM 5-HETE and 823 nM 12-HETE. However, there appear to be mechanistic differences between the 5-HETE and 12-HETE mediated uptake of hexose (47).

6. MODIFICATION OF CELLULAR MEMBRANES BY HYDROXY FATTY ACIDS

Appropriate stimulation of cells results in the release of endogenous arachidonic acid from cellular lipid stores. Arachidonic acid can then be metabolized either via the lipoxygenase or cyclooxygenase pathways (1). Various groups have determined that several of these metabolites were reesterified into cellular lipids whereas other metabolites could not be reacylated. When human neutrophils were stimulated with A23187, 35% of the total

5-HETE, produced from endogenous arachidonic acid, was reesterified into cellular glycerides (48,49). About 25% was reincorporated into phospholipids and 10% into triglycerides. The other metabolite, 5,12-diHETE was not reesterified. Incubation of resting neutrophils with exogenously added 5-HETE resulted in more than 60% esterification into triglycerides and about 20% into phospholipids. When the neutrophils were stimulated with A23187, incorporation of exogenous 5-HETE into phospholipids increased to 33%. Not surprisingly, the presence of albumin trapped the 5-HETE and prevented the reacylation (49). Stimulation of the human leukemic cell line HL60 with A23187 produced 5-HETE, LTB_4 and HHT as the major hydroxylated fatty acids (50). 5-HETE was rapidly esterified into the cellular lipids and 30 min after stimulation, 55% of 5-HETE is incorporated into phospholipids and 35% into triglycerides. No further metabolism of 5-HETE was observed. LTB_4 and the cyclooxygenase product HHT were not reesterified. The predominant process in resting macrophages was further metabolism of the monoHETE products produced from endogenous arachidonic acid rather than reesterification of these monoHETEs (51). About 50% of the 5-, 12- and 15-HETE were metabolized to more polar products and 12-30% of these HETEs were reacylated. The major portion (50-85%) of the reacylated HETEs were incorporated into phospholipids which contrasted with the results obtained with resting neutrophils (49). Neither the more polar metabolites nor the cyclooxygenase products were incorporated into the macrophage lipids. Activated macrophages had a smaller capacity to metabolize HETEs. Furthermore, incorporation of 5-HETE (but not 12- or 15-HETE) into neutral cellular lipids was increased. Somewhat different results were obtained with a mouse macrophage-like tumor line J774.2 (52). Exogenous addition of 5-, 12- or 15-HETE resulted in rapid uptake by the cells and esterification into cellular lipids. 5- and 12-HETE were esterified into both triglycerides and phospholipids while 15-HETE was selectively esterified into phosphatidyl inositol. HHT was also taken up and esterified, though more slowly than the other HETE products. LTB_4 was not taken up at all.

Treatment of the cells with phospholipase A_2 released 5-, 12- and 15-HETE but not HHT, indicating that HHT was not esterified in the 2-position. Mouse thyroid homogenates incorporated 12-HETE, the major lipoxygenase product, primarily into triglycerides (53). However, thyrotropin significantly increased the esterification of 12-HETE into phospholipids. Finally, 15-HETE was reported to specifically bind to mouse lymphocytes and increase the viscosity of the plasma membrane (54).

The above results suggest that the enzymes involved in esterification of HETEs and other hydroxy fatty acids exhibit substrate specificity. Furthermore, the differences in biological activity of these hydroxy fatty acids may be related to the different patterns of intracellular distribution of these acids. Thus, the reacylation of endogenously synthesized products may be a generalized mechanism for controlling membrane functions such as degranulation or chemotaxis. For example, introduction of hydroxy fatty acids into granulocytes may change the fluidity of local sites on membranes allowing fusion to occur leading to degranulation (32).

7. MODULATION OF THE IMMUNE RESPONSE BY HETEs

15-HETE was an effective *in vitro* inhibitor of mouse spleen T-lymphocyte mitogenesis but had very little effect on the B-cell mitogenic response (55). When 15-HETE was injected into C57Bl/6 mice, there was a decreased response of the splenocytes from these animals to lectin stimulation (56). This appeared to be due to an induction of suppressor cells in the 15-HETE-treated mice. In addition, 5-, 11- and 15-HETE have been reported to increase (up to 60%) the expression of I-region-associated antigens by macrophages exposed to lymphokines (57). Similar results were observed with 5-, 9-, 11- and 12-HETE which enhanced the expression of C3b receptors in eosinophils and neutrophils (5,58). Finally, we have observed that 15-HETE inhibited the cytotoxicity of murine natural killer cells (JY Vanderhoek and M Mage, unpublished observations).

8. EFFECTS OF HETEs ON CELLULAR CYCLIC NUCLEOTIDE LEVELS

12-HETE (0.1-50 μM) was reported to increase guanylate cyclase activity in human peripheral blood neutrophils, lymphocytes and psoriatic epidermis (58-60). 5- and 11-HETE were also effective stimulators of the lymphocytic guanylate cyclase but 12-HETE had no effect on the epidermal cyclase activity from normal individuals. In the presence of the lectin mitogen PHA or the tumor promoter PMA, 15-HETE (I_{50}=6 μM) inhibited the membrane-bound lymphocyte guanylate cyclase (61). 15-HETE (100 μM) was also reported to inhibit the thyrotropin-augmented increase in cAMP (by 57%) whereas 12-HETE was inactive (17).

9. CONCLUSIONS

Cells that metabolize arachidonic acid via lipoxygenase pathways produce monohydroxy fatty acids or HETEs as a quantitatively important class of metabolites. These HETEs regulate a wide variety of biological responses with potencies that range from 10^{-14} to 10^{-4} M. The similarities of some of these biological responses to cellular events involved in cancer initiation and/or proliferation suggest that HETEs may also modulate these processes.

Acknowledgement. This work was supported by grants from the National Institutes of Health.

REFERENCES

1. Pace-Asciak CR and Smith WL: Enzymes in the biosynthesis and catabolism of the eicosanoids: Prostaglandins, thromboxanes, leukotrienes and hydroxy fatty acids. In: Boyer PD (ed) The enzymes. Vol 16. Academic Press, New York, 1983, pp 543-603.
2. Samuelsson, B: Leukotrienes: Mediators of immediate hypersensitivity reactions and inflammation. Science (220): 568-575, 1983.
3. Turner SR, Tainer JA and Lynn WS: Biogenesis of chemotactic molecules by the arachidonate lipoxygenase system of platelets. Nature (257): 680-681, 1975.

4. Goetzl EJ, and Sun FF: Generation of unique monohydroxyeicosatetraenoic acids from arachidonic acid by human neutrophils. J Exp Med (150): 406-411, 1979.
5. Goetzl EJ, Weller PF and Sun FF: The regulation of human eosinophil function by endogenous monohydroxyeicosatetraenoic acids (HETES) J Immunol (124): 926-933, 1980.
6. Konig W, Kroegel C, Kunau HW and Borgeat P: Comparison of the eosinophil chemotactic factor (ECF) with endogenous hydroxyeicosatetraenoic acids of leukocytes. Int Archs Allergy Appl Immunol (66, Suppl 1): 168-171, 1981.
7. Goetzl EJ, Valone FH, Reinhold VN and Gorman RR: Specific inhibition of the polymorphonuclear leukocyte chemotactic response to hydroxy-fatty acid metabolites of arachidonic acid by methyl ester derivatives. J Clin Invest (63): 1181-1186, 1979.
8. Goetzl EF, Brash AR, Tauber AI, Oates JA and Hubbard WC: Modulation of human neutrophil function by monohydroxyeicosatetraenoic acids. Immunol (39): 491-501, 1980.
9. Ford-Hutchinson AW, Bray MA, Doig MV, Shipley ME and Smith MJH: Leukotriene B, a potent chemokinetic and aggregating substance released from polymorphonuclear leukocytes. Nature (286): 264-265, 1980.
10. O'Flaherty JT, Thomas MJ, Lees CJ and McCall CE: Neutrophil-aggregating activity of monohydroxyeicosatetraenoic acids. Am J Pathol (104):55-62, 1981.
11. Payan DG and Goetzl EJ: The dependence of human T-lymphocyte migration on the 5-lipoxygenation of endogenous arachidonic acid. J Clin Immunol (1): 266-270, 1981.
12. Nakao J, Ooyama T, Chang WC, Murota S and Orimo H: Platelets stimulate aortic smooth muscle cell migration in vitro. Involvement of 12-L-hydroxy-5,8,10,14-eicosatetraenoic acid. Atherosclerosis (43): 143-150, 1982.
13. Nakao J, Ito H, Ooyama T, Chang WC and Murota S: Calcium dependency of aortic smooth muscle cell migration induced by 12-L-hydroxy-5,8,10,14-eicosatetraenoic acid. Effects of A23187, nicardipine and trifluoroperazine. Atherosclerosis (46): 309-319, 1983.
14. Nakao J, Ooyama T, Ito H, Chang WC and Murota S: Comparative effect of lipoxygenase products of arachidonic acid on rat aortic smooth muscle cell migration. Atherosclerosis (44): 339-342, 1982.
15. Croset M and Lagarde M: Stereospecific inhibition of PGH_2-induced platelet aggregation by lipoxygenase products of icosaenoic acid. Biochem Biophys Res Comm (112): 878-883, 1983.
16. Vanderhoek, JY, Bryant RW and Bailey JM: 15-Hydroxy-5,8,11,13-eicosatetraenoic acid, a potent and selective inhibitor of platelet lipoxygenase. J Biol Chem (255): 5996-5998, 1980.
17. Levasseur S, Sun FF, Friedman Y and Burke G: Arachidonate metabolism in the mouse thyroid. Implication of the lipoxygenase pathway in thyrotropin action. Prostaglandins

(22): 663-677, 1981.

18. Vanderhoek JY, Kiesel L, Naor Z, Bailey JM and Catt KY: Arachidonic acid metabolism in gonadotroph-enriched pituitary cells. Prostaglandins, Leukotrienes and Medicine. In press, 1984.
19. Vanderhoek JY, Bryant RW and Bailey JM: Regulation of leukocyte and platelet lipoxygenases by hydroxyeicosanoids. Biochem Pharmacol (31): 3463-3467, 1982.
20. Vanderhoek JY, Bryant RW and Bailey JM: Inhibition of leukotriene biosynthesis by the leukocyte product 15-hydroxy-5,8,11,13-eicosatetraenoic acid. J Biol Chem (255): 10064-10066, 1980.
21. Goetzl EJ: Selective feedback inhibition of the 5-lipoxygenation of arachidonic acid in human T-lymphocytes. Biochem. Biophys Res Comm (101): 344-350, 1981.
22. Vanderhoek JY, Tare NS, Bailey JM, Goldstein AL and Pluznik DH: New role for 15-hydroxyeicosatetraenoic acid: activation of leukotriene biosynthesis in PT-18 mast/basophil cells. J Biol Chem (257): 12191-12195, 1982.
23. Borgeat P, deLaclos BF, Picard S, Drapeau J, Vallerand P and Corey EJ: Studies on the mechanism of formation of the 5S, 12S-dihydroxy-6,8,10,14 (E,Z,E,Z)-icosatetraenoic acid in leukocytes. Prostaglandins (23): 713-724, 1982.
24. Marcus AJ, Broekman MJ, Safier LB, Ullman HL and Islam N: Formation of leukotrienes and other hydroxy acids during platelet-neutrophil interactions in vitro. Biochem Biophys Res Comm (109): 130-137, 1982.
25. Borgeat P, Picard S, Drapeau J and Vallerand P: Metabolism of arachidonic acid in leukocytes: isolation of a 5,15-dihydroxyeicosatetraenoic acid. Lipids (17): 676-681, 1982.
26. Maas RL, Turk J, Oates JA and Brash AR: Formation of a novel dihydroxyacid from arachidonic acid by lipoxygenase-catalyzed double oxygenation in rat mononuclear cells and human leukocytes. J Biol Chem (257): 7056-7067, 1982.
27. Sok DE, Han CQ, Pai JK and Sih CJ: Inhibition of leukotriene biosynthesis by acetylenic analogs. Biochem Biophys Res Comm (107): 101-108, 1982.
28. Marcus AJ, Safier LB, Ullman HL, Broekman MJ, Islam N, Oglesby TD and Gorman RR: 12S,20-Dihydroxyicosatetraenoic acid: A new icosanoid synthesized by neutrophils from 12S-hydroxyicosatetraenoic acid produced by thrombin or collagen-stimulated platelets. Proc Natl Acad Sci USA (81): 903-907, 1984.
29. Wong PYK, Westlund P, Hamberg M, Granstrom E, Chao PHW and Samuelsson B: w-Hydroxylation of 12-L-hydroxy-5,8,10,14-eicosatetraenoic acid in human polymorphonuclear leukocytes. J Biol Chem (259): 2683-2686, 1984.
30. Goetzl EJ, Phillips MJ and Gold WM: Stimulus specificity of the generation of leukotrienes by dog mastocytoma cells. J Exp Med (158): 731-737, 1983.
31. Siegel MI, McConnell RT, Bonser RW and Cuatrecasas P: The lipoxygenase product, 5-hydroperoxyarachidonic acid, augments chemotactic peptide-stimulated arachidonic acid

release from HL60 granulocytes. Biochem Biophys Res Comm (104): 874-881, 1982.
32. Stenson WF, Parker CW: Monohydroxyeicosatetraenoic acids induce degranulation of human neutrophils. J Immunol (124): 2100-2104, 1980.
33. Goetzl EJ, Pickett WC: The human PMN leukocyte chemotactic activity of complex hydroxy-eicosatetraenoic acids (HETEs). J Immunol (125): 1789-1791, 1980.
34. Bokoch GM and Reed PW: Effect of various lipoxygenase metabolites of arachidonic acid on degranulation of polymorphonuclear leukocytes. J Biol Chem (256): 5317-5320, 1981.
35. O'Flaherty JT, Thomas MJ, Hammett MJ, Carroll C, McCall CE and Wykle RL: 5-L-hydroxy-6,8,11,14-eicosatetraenoate potentiates the human neutrophil degranulating action of platelet-activating factor. Biochem Biophys Res Comm (111): 1-7, 1983.
36. Stenson WF, Parker CW and Sullivan TJ: Augmentation of IgE-mediated release of histamine by 5-hydroxyeicosatetraenoic acid and 12-hydroxyeicosatetraenoic acid. Biochem Biophys Res Comm (96): 1045-1052, 1980.
37. Peters SP, Siegel MI, Kagey-Sobotka A and Lichtenstein LM: Lipoxygenase products modulate histamine release in human basophils. Nature (292): 455-457, 1981.
38. Peters SP, Kagey-Sobotka A, Macglashan DW, Siegel MI and Lichtenstein LM: The modulation of human basophil histamine release by products of the 5-lipoxygenase pathway. J Immunol (129): 797-803, 1982.
39. Marom Z, Shelhamer JH, Sun F and Kaliner M: Human airway monohydroxyeicosatetraenoic acid generation and mucus release. J Clin Invest (72): 122-127, 1983.
40. Yamamoto S, Ishii M, Nakadate T, Nakaki T and Kato R: Modulation of insulin secretion by lipoxygenase products of arachidonic acid. Relation to lipoxygenase activity of pancreatic islets. J Biol Chem (258): 12149-12152, 1983.
41. Stenson WF and Lobos E: Metabolism of exogenous arachidonic acid by isolated guinea pig acini and the effects of arachidonate metabolites on pancreatic secretion. Gastroenterology (80): 1293, 1981.
42. Musch MW, Miller RJ, Field M and Siegel MI: Stimulation of colonic secretion by lipoxygenase metabolites of arachidonic acid. Science (217): 1255-1256, 1982.
43. Naor Z, Vanderhoek JY, Lindner HR and Catt KJ: Arachidonic acid products as possible mediators of the action of gonadotropin-releasing hormone. Adv Prostaglandin Thromboxane and Leukotriene Res (12): 259-263, 1983.
44. Bromberg Y and Pick E: Unsaturated fatty acids as second messengers of superoxide generation by macrophages Cellular Immunol (79): 240-252, 1983.
45. Naccache PH, Shaafi RI, Borgeat P and Goetzl EJ: Mono- and dihydroxyeicosatetraenoic acids alter calcium homeostasis in rabbit neutrophils. J. Clin Invest (67): 1584-1587, 1981.
46. Bass DA, Thomas MJ, Goetzl EJ, DeChatelet LR and McCall CE: Lipoxygenase-derived products of arachidonic acid mediate

stimulation of hexose uptake in human polymorphonuclear leukocytes. Biochem Biophys Res Comm (100): 1-7, 1981.

47. Thomas MJ, O'Flaherty JT, Cousart S and McCall CE: The effect of free radical derived hydroxyicosatetraenoic acids on hexose transport in human polymorphonuclear leukocyte. Prostaglandins (23): 265-272, 1982.

48. Walsh CE, Waite BM, Thomas MJ and DeChatelet LR: Release and metabolism of arachidonic acid in human neutrophils. J Biol Chem (256): 7228-7234, 1981.

49. Stenson WR and Parker CW: Metabolism of arachidonic acid in ionophore-stimulated neutrophils. Esterification of a hydroxylated metabolite into phospholipids. J Clin Invest (64): 1457-1465, 1979.

50. Bonser RW, Siegel MI, Chung SM,, McConnell RT and Cuatrecasas P: Esterification of an endogenously synthesized lipoxygenase product into granulocyte cellular lipids. Biochemistry (20): 5297-5301, 1981.

51. Pawlowski NA, Scott WA, Andreach M and Cohn ZA: Uptake and metabolism of monohydroxyeicosatetraenoic acids by macrophages. J Exp Med (155): 1653-1664, 1982.

52. Stenson WF, Nickells MW and Atkinson JP: Esterification of monohydroxy fatty acids into the lipids of a macrophage cell line. Prostaglandins (26): 253-264, 1983.

53. Levasseur S, Sun F, Friedman Y and Burke G: Esterification of 12-hydroxy-5,8,10,14-eicosatetraenoate into mouse thyroid lipids: Possible physiological significance. Adv. Prostaglandin, Thromboxane and Leukotriene Res (12): 247-251, 1983.

54. Mexmain S, Gualde N, Aldigier JC, Motta C, Chable-Rabinovitch H and Rigaud M: Specific binding of 15-HETE to lymphocytes. Effects on the fluidity of plasmatic membranes. Prostaglandins Leukotrienes and Medicine (13): 93-97, 1984.

55. Bailey JM, Bryant RW, Low CE, Pupillo MB and Vanderhoek JY: Regulation of T-lymphocyte mitogenesis by the leukocyte product 15-hydroxyeicosatetraenoic acid (15-HETE). Cell Immunol (67): 112-120, 1982.

56. Aldigier JC, Gaulde N, Mexmain S, Chable-Rabinovitch H, Ratinaud MH and Rigaud M: Immunosuppression induced in vivo by 15-hydroxyeicosatetraenoic acid (15 HETE) Prostaglandins, Leukotrienes and Medicine (13): 99-107, 1984.

57. Snyder DS, Beller Di and Unanue ER: Prostaglandins modulate macrophage Ia expression. Nature (299): 163-165, 1982.

58. Goetzl EJ, Hill HR and Gorman RR: Unique aspects of the modulation of human neutrophil function by 12-L-hydroperoxy-5,8,10,14-eicosatetraenoic acid. Prostaglandins (19): 71-85, 1980.

59. Coffey RG and Hadden JW: Arachidonate and metabolites in mitogen activation of lymphocyte guanylate cyclase. In: Hadden J, Chedid L, Mullen P and Sprenfico F (eds) Advances in Immunopharmacology, Pergamon Press, Oxford, 1982, pp 365-373.

60. Cantieri JS, Graff G and Goldberg ND: Cyclic GMP metabolism in psoriasis: Activation of soluble epidermal guanylate

cyclase by arachidonic acid and 12-hydroxy-5,8,10,14-eicosatetraenoic acid. J Invest Dermatol (74): 234-237, 1980.

61. Coffey RG and Hadden JW: Phorbol myristate acetate stimulation of lymphocyte guanylate cyclase and cyclic guanosine 3':5'-monophosphate phosphodiesterase and reduction of adenylate cyclase. Cancer Res (43): 150-158, 1983.

14

CHEMOTAXIS

J. VARANI

1. CHEMOTAXIS IN EUKARYOTIC CELLS

Nearly all higher eukaryotic cells have the capacity for active cell movement. Active movement occurs in a variety of normal biological processes and in several pathological conditions. Examples include fertilization, embryonic development, inflammation, wound healing and tumor cell invasion & metastasis. The involvement of cell motility in these complex biological processes insures that cell movement will not occur in an unregulated manner. Rather, the movement of cells in multicellular organisms in closely controlled. A number of different mechanisms appear to be involved. In addition to factors which turn cell movement "on" and "off", there are factors which influence the direction in which cell movement occurs. Contact inhibition, contact guidance, haptotaxis and chemotaxis are all terms that describe the regulation of directional cell movement.

This chapter will focus specifically on the process of chemotaxis and its role in the regulation of cell movement. Chemotaxis can be defined as the directional movement of cells which occurs in response to the presence of a soluble stimulating agent (i.e., the chemotactic factor) in the form of a gradient. Chemotactic responses occur in a broad range of cells including prokaryotic and lower eukaryotic cells as well as the cells of complex, multicellular organisms. In the higher eukaryotic cells, the prototypic, chemotactically-responsive cells are the leukocytes. Of these, the most widely-investigated are the polymorphonuclear neutrophils (PMN) although other granulocytes, monocytes and lymphocytes are also responsive to one or more factors. Most of what we currently know about the cellular and molecular basis of the chemotactic response comes from studies with leukocytes and a number of recent monographs and review articles are available (1-3). In addition to these cells, chemotactic responses have also been identified in platelets, fibroblasts, smooth muscle cells, endothelial cells, tumor cells of diverse histologic type, spermatozoa and cells of neural origin (4). The ability of these cells to carry out chemotactic responses facilitates their participation in the diverse biologicaly processes in vivo which require cellular translocations.

By its definition, the term chemotaxis implies active cell movement. In addition to stimulating directional cell movement, chemotactic factors also induce a number of other functional responses in the appropriate cells. Among these responses are increased random motility (chemokinesis), cell to substrate adhesiveness, cell to cell aggregation, exocytosis of lysozomal enzymes and production of reactive oxygen metabolites (5,6). These functional activities, no doubt, play biologically-important roles in vivo. For example, the increased motility occurring as a result of chemotactic stimulation may have increased cell to substrate adhesiveness at its base. Motility in higher eukaryotic cells (with the exception of spermatozoa) involves the cells "pulling" themselves over the substratum (7,8). Thus, the increased adhesiveness of the stimulated cells may allow them to generate a greater amount of force. Likewise, in leukocytes at least, the release of lysozomal enzymes and reactive oxygen metabolites by the stimulated cells undoubtably contributes both to the bacteriocidal activity and the tissue-destructive activity which these cells possess.

Although chemotactic responses occur in normal, physiological processes, they also occur in and contribute to a number of pathological conditions. The localization of leukocytes at sites of inflammation and the localization of tumor cells at secondary, metastatic, sites are two such examples. Because of the involvement of chemotactic responses in pathological conditions, efforts have been made to delineate how these responses are regulated. It is hoped that once the cellular and molecular basis of these responses are understood, we may be able to develop strategies for interfering with them and thus prevent the pathology associated with them. With both the leukocytes, which have been well-studied, and tumor cells, which have been examined in detail in our labortory, evidence is accumulating which suggests that metabolites of arachidonic acid participate in the endogenous regulation of chemotactic responses. Furthermore, this evidence suggests that arachidonic acid metabolites may be useful for exogenously modifying the response of cells to chemotactic stimulation. The role of arachidonic acid metabolites in chemotactic responses will be described in the following sections of this review.

2. CELLULAR AND MOLECULAR BASIS OF THE CHEMOTACTIC RESPONSE

2.1 Overview. Most of what we currently know about the cellular and molecular basis of the chemotactic response comes from studies on leukocytes. However, now that a number of other cell types are being used in chemotaxis studies, information from cells other than leukocytes is starting to become available. While it is too early to say that no major differences will be identified, it appears that the response to chemotactic factors, at least to the peptide chemotactic factors, is similar in leukocytes and in non-leukocytic cells.

The initial event in the chemotactic response is the binding of the active ligand to cellular receptors. Receptors for chemotactic factors have been identified and characterized on a number of different cell types (9-11). Following the binding of chemotactic factors, there is a rapid membrane depolarization. This is accompanied by alterations in the level of cytoplasmic Ca^{2+}, changes in cyclic nucleotide levels and in the metabolism of arachidonic acid-containing phospholipids (1). There is a very rapid polymerization of actin and a reorganization of cytoskeletal elements (12,13). This is accompanied by the onset of cell polarization (14) and the precipitation of the functional responses associated with the chemotactic event. While the standard in vitro methods for measuring chemotaxis require one or more hours to complete (5), cell orientation can be detected within a few minutes of stimulation (14,15) and other functional responses including cell to cell aggregation and cell to substrate adhesiveness can be detected within seconds after stimulation (5). Thus, the kinetics of the functional responses are in accord with the kinetics of chemotactic factor binding.

2.2 Arachidonic acid metabolism and chemotaxis. The metabolism of arachidonic acid appears to be an important event in the PMN response to many stimuli. Within minutes of exposure to chemotactic factors, ionophores or phagocytic stimuli, PMNs mobilize their phospholipid-bound arachidonic acid and convert it to a number of bioactive metabolites via the cyclooxygenase and lipoxygenase pathways (16,17). Under unstimulated conditions, nearly all of the intracellular arachidonic acid is esterified in phospholipids and very little is free. Considerable evidence suggests that the rate of arachidonic acid metabolism is controlled primarily by the rate of its release from phospholipids. Events which initiate this release may vary from cell to cell but the major pathways seem to be common to all cells.

One major pathway involves the degradation of phospholipids by phospholipase A_2 (PLA_2). This enzyme hydrolyzes the phospholipid almost exclusively at the 2-position releasing equimolar amounts of free fatty acid and lysophospholipids (18). Several lines of evidence suggest that this is a major pathway in PMNs and mast cells (18-19). A second pathway involves the degradation of phospholipids by the sequential actions of a phospholipase C and a diacylglycerol lipase (20). Regardless of the mechanism, the released arachidonic acid is then available for oxidative metabolism through either the lipoxygenase or cyclooxygenase pathways. While these two pathways are known to produce a variety of bioactive intermediates, it should be noted that the other products of phospholipid breakdown including the lysophospholipids could also affect cellular behavior. It should also be noted that the released arachidonic acid need not be oxidatively metabolized, but rather may be reacylated into phospholipids. Acylation into phospholipids is the fate of most of the exogenous arachidonic acid added to various kinds of cells. How the reacylation reactions are controlled is not known and the inhibition of these reactions may be a major role of the stimulants which make arachidonic acid available for metabolism through the lipoxygenase and cyclooxygenase pathways (21,22).

A substantial portion of the arachidonic acid released from the phospholipids of activated cells is converted by specific lipoxygenase enzymes to highly unstable hydroperoxyeicosatetraenoic acids and then to stable monohydroxy-eicosatetranoic acids (mono-HETEs) (23). The nature and quantities of mono-HETEs produced vary from cell type to cell type. The 5-hydroperoxyeicosatetraenoic acid, in addition to being a source of 5-HETE, also serves as the intermediate in the generation of complex HETEs termed leukotrienes (23,24). The mono-HETEs and the more complex leukotrienes have a variety of pro-inflammatory effects (17,23,24).

In addition to metabolism through the lipoxygenase pathway, the arachidonic acid released from phospholipids may also be converted to a variety of cyclooxygenase products (e.g., prostaglandins, prostacyclin and thromboxanes). The first step in the cyclooxygenase path is the formation of unstable endoperoxides catalyzed by the enzyme, prostaglandin synthetase. These intermediates give rise to the more stable endproducts. As with the lipoxygenase products, different cells produce varying kinds and amounts of the different cyclooxygenase metabolites (25,26).

2.2.1 Chemotactic activities of lipoxygenase products. Certain of the lipoxygenase metabolites are potent chemotactic factors for leukocytes and also elicit other responses such as cell to cell aggregation, cell to substrate adhesiveness, chemokinesis and release of lysozomal enzymes. The most potent metabolites are 5-HETE and 5,12-di-HETE (leukotriene B_4) although some of the other metabolites also have activity (27-31). With leukotriene B_4, chemotactic activity can be detected at concentrations as low as 10^{-9} M. In this regard it is approximately as potent as the peptide chemotactic factors. The mono-HETE and leukotriene chemotactic factors are, however, less active than the peptide chemotactic factors in eliciting the degranulation response and induce relatively little oxidative metabolism.

The endogenous HETEs, produced by leukocytes during stimulation, are likely to play a role in the expression of chemotactic responsiveness to the exogenous stimulating factors. Inhibitors of lipoxygenase reactions suppress both random migration and chemotactic migration in response to several stimuli. These inhibitors also suppress neutrophil aggregation and degranulation reactions. In contrast, cyclooxygenase inhibitors are not effective and in fact can stimulate random and chemotactic migration (32-34). It is interesting that the inhibition of random and chemotactic motility achieved with the lipoxygenase inhibitors can be fully restored with purified 5-HETE at concentrations of 1-20 ng per 2 X 10^6 cells (32,35), but the same quantities of 5-HETE do not fully reverse the inhibition of enzyme release. This suggests that, in fact, more than a single lipoxygenase metabolite may be involved as an intermediate in these reactions.

While it is now known that the generation of lipoxygenase metabolites and the maintenance of intracellular pools of these metabolites are critical to the expression of chemotactic responses, the mechanism by which they function is not yet understood. Two possibilities have been explored. It was shown by Stenson and Parker (36) that human PMNs reincorporate a certain portion of the generated 5-HETE into their phospholipids. Perhaps the reincorporated 5-HETE alters the general biophysical properties of the membranes after it has been incorporated or affects very specific membrane functions. Alternatively, it has been shown that highly reactive intermediates such as the 5-hydroperoxyeicosatetraenoic acid can covalently derivatize intracellular proteins and other constituents which may be critical to cell function (37). Whether either of these potential mechanisms of action are significant will have to await the outcome of additional studies.

There is very little data on the involvement of lipoxygenase metabolites in the chemotaxis of cells other than leukocytes. In a very recent report by Mensing and Czarnetski (38), it was shown that leukotriene B_4 is chemotactic for both normal human fibroblasts and embryonic rat fibroblasts. As with other chemotactic factors, the effects were shown to be dose-responsive and migration was inhibited at high concentrations. Interestingly, the response was only observed when the attractant was presented in the form of a gradient. The lack of chemokinetic effect seen with the fibroblasts is in contrast to what has been previously reported in leukocytes. Never-the-less, this study strongly suggests that the lipid mediators of chemotaxis in leukocytes are also involved in the chemotactic responses of non-leukocytic cells as well.

Studies carried out in our laboratory has provided additional evidence for the involvement of lipoxygenase metabolites in the response of tumor cells to chemotactic factors. Our studies showed that lipoxygenase inhibitors including nordihydroguaiaretic acid (NDGA), 5,8,11,14-eicosatetraynoic acid (ETYA) and nafasatrom all dramatically inhibited the Walker 256 carcinosarcoma cells , a chemotactically-responsive cell line derived from a rat mammary tumor, from responding to the peptide chemotactic factor, N-formyl-methionyl-leucyl-phenylalanine (FMLP) and the phorbol ester chemotactic factor, 12-O-tetradecanoyl-phorbol acetate (TPA) in the adherence assay (39,40). Inhibition was obtained at concentrations of 10^{-6} - 10^{-4} M. Inhibition was also obtained with p-bromophenacyl bromide, a PLA_2 inhibitor, at the same concentrations. In contrast, indomethacin and aspirin were not effective inhibitors. Indomethacin, in fact, potentiated the adherence response to the chemotactic factors when the cells were pretreated for 18 hours.

Direct evidence for the involvement of lipoxygenase products was obtained using the nylon fiber adherence assay (Figure 1). Leukotrienes B_4

induced a response which was similar in magnitude to that of FMLP but less than that obtained with TPA. The response to the leukotriene B_4 was similar to that of FMLP in other respects as well. As with FMLP, the maximum effects were seen when the cells were tested immediately after stimulation. In contrast, the maximum response to TPA was obtained if the cells were examined 15-30 minutes after stimulation. Why there should be a lag phase with TPA but not with the other agents is unclear, but this same kinetic difference is seen in leukocytes.

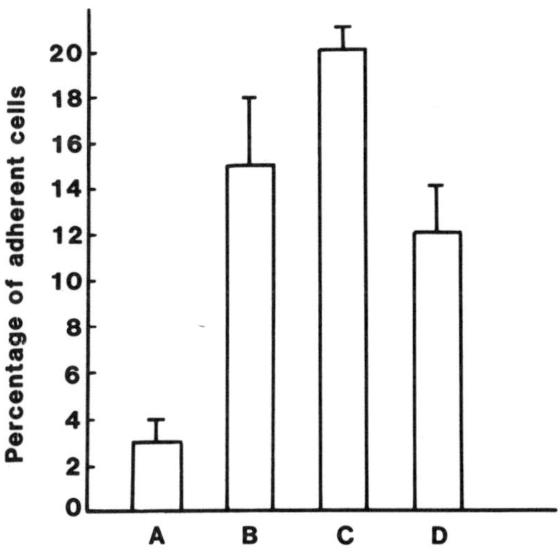

Figure 1. Adherence of Walker 256 carcinosarcoma cells in the nylon fiber adherence assay. The assay was carried out as described in our previous report (39). The factors used to stimulate the response included: A, buffer alone; B, 10^{-6} M FMLP; C, 1.6 X 10^{-8} M TPA and D, 2 X 10^{-8} M 5,12-di-HETE. With the buffer alone, FMLP and 5,12-di-HETE, the cells were stimulated and immediately applied to the column. With TPA, the cells were stimulated and applied to the column 15 minutes later. The values shown are averages ± standard errors of 4 separate columns in two different experiments.

If endogenous lipoxygenase products are intermediates in the response of the Walker cells to factors such as FMLP and TPA, the production of these metabolites should be detectable in the stimulated cells. To examine the Walker cells for the production of these metabolites we used a procedure identical to that described by Chensue et al (41). The cells were labeled with [^{3}H]-arachidonic acid and then either stimulated with optimal amounts of TPA or left unstimulated. One hour later the control and stimulated cultures were harvested, and analyzed by high performance liquid chromatography (HPLC). In the control cells, a single major peak of activity, corresponding to our 15-HETE standard was observed. A more complex pattern was seen in the TPA-stimulated cells. In addition to the peak corresponding to the 15-HETE there was also a second peak of activity which co-migrated with the 5-HETE standard. These chromatographic patterns are shown in Figure 2. Activity in the leukotriene region of the profile was also observed in some trials although this was not consistent from experiment to experiment. Obviously, much more work will have to be done before the significance of these findings are known. In spite of this, however, these results suggest that the Walker cells have the ability to metabolize arachidonic acid through the lipoxygenase pathway to produce metabolites which can mediate chemotactic responses.

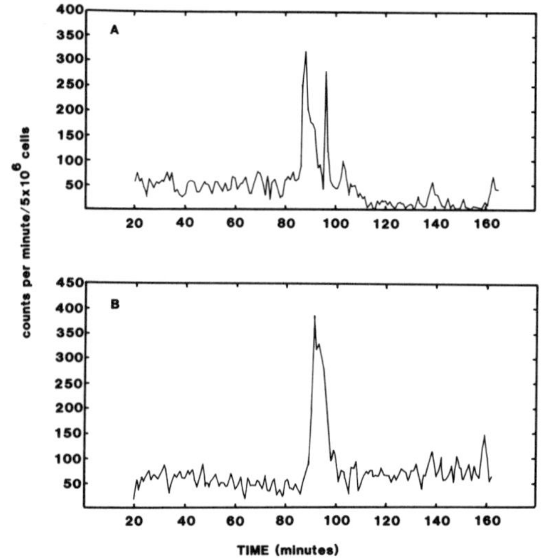

Figure 2. High performance liquid chromatography analysis of arachidonic acid metabolites from Walker 256 carcinosarcoma cells. The analysis was carried out as described by Chensue et al (41). The lower panel (control cells) shows a peak of activity comigrating with the 15-HETE standard. The upper panel (TPA-treated cells) shows a more complex pattern of mono-HETE production. The second peak of activity comigrated with the 5-HETE standard.

2.2.2 Cyclooxygenase metabolites and chemotactic responses. The cyclooxygenase metabolites of arachidonic acid, particularly the E-series prostaglandins and prostacyclin (PGI_2), have activities which inhibit chemotactic responses. *In vitro* treatment of leukocytes with prostaglandin E_1 (PGE_1) inhibits both the directional motility response and the cell to cell aggregation response induced by chemotactic factors (42,43). The structural specificity of this activity is indicated by the fact that prostaglandin $F_{2\alpha}$ ($PGF_{2\alpha}$) does not duplicate these effects. The ability of the E-series prostaglandins to inhibit leukocyte function may have *in vivo* relevance since these same agents can suppress a number of different inflammatory conditions in experimental animals (44,45) and suppression is associated with alterations in leukocyte function.

The biological basis for the suppression of leukocyte functions by the prostaglandins is unknown. An early study by Rivkin, Rosenblatt and Becker (42) showed that suppression of chemotaxis by prostaglandins was associated with an increase in intracellular cyclic AMP levels. Other agents which raised cyclic AMP levels also inhibited the chemotactic response in the same cells. Recent studies by Fantone et al (46) have also shown that PGE_1 reduces the binding affinity of the neutrophil receptor for the chemotactic peptide, FMLP. This may represent, therefore, another mechanism by which leukocyte chemotactic responses are suppressed.

Other cells which respond to chemotactic factors are also susceptible to the anti-chemotactic activity of prostanoid metabolites. Studies in our laboratory have shown that E-series prostaglandins and PGI_2 inhibit chemotactic factor-induced adherence and motility in a number of rat and mouse tumor cell lines (47-49). In the Walker 256 carcinosarcoma line, the inhibition of these responses is associated with a 4-5X increase in cyclic AMP levels (47,48). Other agents which elevate cyclic AMP including epinephrin, cholera toxin and dibutyril cyclic AMP also inhibit the response. In contrast, the pro-aggregatory thromboxane B_2 (TXB_2) does not inhibit the response or raise the cyclic AMP levels. It appears, therefore, that in susceptibility to prostaglandin-mediated inhibition, the tumor cells are very similar to leukocytes.

The use of cell lines that can be maintained in culture has certain advantages over the use of end-stage cell such as PMNs. In culture-adapted cells, for example, it is possible to obtain large numbers of cells (synchronized, if necessary) at a given time. This is important in studies aimed at identifying short-lived intermediates such as the metabolites of arachidonic acid which may be present in very small amounts. Another advantage to the use of cultured cells in chemotaxis studies is that they can be stimulated with appropriate factors and then followed indefinitely. Using this, it has been possible to examine the events which occur following treatment of the cells with a desensitizing concentration of chemotactic factor - to determine how soon afterwards the cells regain ability to respond and to identify some of the metabolic events which occur during this period (50). With end-stage cells such as leukocytes, of course, this cannot be done. Perhaps the biggest advantage of using cells in culture is that clonal populations with differing chemotactic characteristics can be isolated from a common "stem" line (51) or can be maintained under conditions which allow for a differential expression of chemotactic responsiveness. These lines can then be compared to identify the cellular and molecular characteristics that accompany the differences in responsiveness. Using this approach we have found a very interesting relationship between the production of cyclooxygenase metabolites and chemotactic responsiveness in the Walker 256 carcinosarcoma cells.

When the Walker cells are maintained as a suspension culture or as a monolayer culture, they respond to stimulation with a variety of factors including the C5-derived tumor cell chemotactic factor, the synthetic peptide, FMLP and the phorbol ester, TPA (39,40,50). TPA induces a very rapid adherence response in these cells. The response can be detected as early as 15 minutes after exposure and peaks by 30-45 minutes. At the period of maximal response, many of the attached cells have also spread. The response in transient, however, and by two hours after exposure to the stimulating agent, the cells have detached from the substratum. The adherence response extends over a dose range from 1.6×10^{-10} - 1.6×10^{-7} M. Other phorbol esters including phorbol dibenzoate, phorbol dibutyrate and phorbol diacetate also induce the adherence response in the Walker cells although they are not as effective as TPA on a molar basis. In contrast, the parent phorbol alcohol is completely without effect. The adherence-inducing effect of the phorbol esters can be detected on a variety of substrates including basement membrane collagen and intact endothelial cells as well as on substrates such as glass or plastic.

TPA also induces a chemotactic response in the Walker cells. When examined using the Boyden chamber assay, the maximum response is seen at 1.6×10^{-8} M, the same concentration which stimulate maximum adhesiveness. However, the motility response is seen only if TPA is added to the lower well of the Boyden chamber (i.e., provided to the cells in a gradient). If the agent is added to the top well of the Boyden chamber (along with the cells), motility is inhibited. Thus the response in these cells appears to be chemotactic but not chemokinetic. This is similar to what has been previously reported for murine macrophages (52).

We also examined the effects of TPA on the proliferative response of the Walker cells. This was done because previous studies by others showed that TPA stimulated growth in some cells but inhibited growth in other cells (49). Our studies showed that TPA caused a very slight ($\sim$ 5%) inhibition of Walker cell growth. This was seen at the earliest time points (1-2 days) after treatment but not at longer time points.

As well as growing in culture, the Walker cells readily proliferate as an ascites tumor when injected intraperitoneally into Sprague Dawley rats. Very interestingly, however, the ascites tumor cells show very little response to TPA when challenged immediately upon removal from the animal. Although they do not respond immediately upon removal from the animal, they regain their responsiveness as they become readapted to culture. Interestingly, treatment of the cells with indomethacin immediately upon removal from the animal dramatically increases their response.

The cells grown in culture were compared with the ascites tumor cells for the production of cyclooxygenase metabolites by radioimmunoassay procedures (53-55). The cells grown in culture produced very small amounts of these agents. In contrast, the ascites tumor cells produced much higher levels - particularly of the PGE_2 and prostacyclin metabolite, 6-keto $PGF_{1\alpha}$. Thus, there was an inverse correlation between biological responsiveness to TPA and the production of cyclooxygenase metabolites which are known to inhibit these biological responses. A comparison of the cells maintained in culture and as ascites tumors is shown in Table 1. The Walker cells are not the only chemotactically-responsive cells that produce low levels of cyclooxygenase metabolites. In a recent study (49) we compared the production of these metabolites by highly responsive murine fibrosarcoma cells and chemotactically nonresponsive Swiss 3T3 mouse fibroblasts. The 3T3 fibroblasts produced much higher levels of both PGE_2 and 6-keto $PGF_{1\alpha}$ than did the murine fibrosarcoma cells. Based on these data, we have speculated that the cyclooxygenase metabolites may play a role as endogenous regulators of chemotactic responsiveness in these cells. If this turns out to be the case, it may explain why some cells are highly-responsive to chemotactic factors and other cells not.

3. CONCLUSIONS

It is becoming increasingly clear that chemotactic responses are an

Table 1. Response in the Adherence Assay and Production of Cyclooxygenase Metabolites by Walker 256 carcinosarcoma cells

Condition of Maintenance	% response in the adherence assay ($\bar{y}$ ± SEM)[a]	Production of cyclooxygenase metabolites[b] pg 2 X 10^6 cells/4 hr[b]	
		PGE_2	6 keto $PGF_{1\alpha}$
Continuously in culture	67 ± 7	162 ± 51	<30
As ascites tumors: then in culture			
0 days	9 ± 5	2460 ± 141	2340 ± 77
5 days	58 ± 6	602 ± 41	<16

[a] The adherence assay was carried out as described previously by us (39).
[b] Cyclooxygenase metabolites were analyzed by radioimmunoassay (53-55).

integral part of many biological processes. Not only do they occur under physiologically normal conditions but also occur in many pathological states. The accumulation of leukocytes at locations of acute or chronic inflammation and the secondary localization of circulating tumor cells at metastatic sites are two conditions in which chemotactic responses contribute to effects which are deleterious to the host. The formation of fibrotic lesions and the inappropriate formation of new blood vessels are other conditions in which chemotactic responses may contribute to the pathophysiology of the lesion. The need for effective strategies to interfere with chemotactic response is clear. The identification of arachidonic acid metabolites as an important feature in these responses suggest avenues to pursue. Two broad approaches are suggested. On the one hand, it may be possible to develop agents which interfere with the generation or function of the pro-chemotactic lipoxygenase metabolites. On the other hand, it may be possible to use directly the anti-chemotactic cyclooxygenase metabolites or synthetic analogues of these agents. The anti-inflammatory (44,45) and anti-metastatic (56) activities which these agents are known to possess may be due, at least in part, to their anti-chemotactic properties.

References

1. Snyderman R, Goetzl EJ: Molecular and cellular mechanisms of leukocyte chemotaxis. Science (213):830-836, 1981.
2. Keller HU, Till GO: Leukocyte Locomotion and Chemotaxis. Birkhauser Verlag, Basel 1983.
3. Wilkinson PC: Chemotaxis and Inflammation. 2nd edition. Churchill Livingstone New York, 1982.
4. Varani J, Orr FW: Chemotaxis in non-leukocytic cells. In: Ward PA (ed). Handbook of Inflammation Vol. 4 Elsevier, Amsterdam, 1983, pp 211-244.
5. Kreutzer DL, O'Flaherty JT, Orr W, Showell HJ, Ward PA, Becker EL: Quantitative comparisons of various biological responses of neutrophils to different active and inactive chemotactic factors. Immunopharmacology (1):39-47, 1978.
6. Simchowitz L, Mehta J, Spilberg I: Chemotactic factor-induced superoxide radical generation by human neutrophils: Requirement for protease (esterase) activity. J. Lab. Clin. Med. (94):403-407, 1979.
7. Dunn GA: Mechanisms of fibroblast locomotion. In: Gurtis ASJ and Pitts JD (eds) Cell Adhesion and Motility. Cambridge University Press, New York, 1980. pp 409-424.
8. Stossel TP: The mechanism of leukocyte locomotion. In: Gallin JI and Quie PG (eds). Leukocyte Chemotaxis: Methods, Physiology and Clinical Implications. Raven Press, New York. pp 143-157.
9. Vitkauskas GG, Showell HJ, Becker EL: Specific binding of synthetic chemotactic peptides to rabbit peritoneal neutrophils: Effects of dissociability of bound peptide, receptor activity and subsequent biological responsiveness (deactivation) Mol. Immunol. (17):171-179, 1980.
10. Chiang TM, Postlethwaite AE, Beachey EH, Seyer JM and Kang AH: Binding of chemotactic collagen-derived peptides to fibroblasts: The relationship to fibroblast chemotaxis. J. Clin. Invest. (62):916-922, 1978.
11. Clarke PRH, Varani J: Phorbol ester binding to chemotacticallyresponsive and non-responsive Walker 256 carcinosarcoma cells. Cancer Res. (in press).
12. Rao KMK, Varani J: Actin polymerization induced by chemotactic peptide and concanavalin A in rat neutrophils. J. Immunol. (129):1605-1607, 1982.
13. Malech HL, Root RK, Gallin JI: Structural analysis of human neutrophil migration: Centriole, microtubule and microfilament orientation and function during chemotaxis. J. Cell Biol. (75):666-670, 1977.
14. Zigmond SH: Ability of polymorphonuclear leukocytes to orient in gradients of chemotactic factors. J. Cell Biol. (75):606-616, 1977.
15. Marasco WA, Becker EL, Oliver JM: The ionic basis of chemotaxis: Separate cation requirements for neutrophil orientation and locomotion in a gradient of chemotactic peptide. Am. J. Path. (98):749-768, 1980.
16. Kuehl FA, Egan RW: Prostaglandins, arachidonic acid and inflammation. Science (210):978-984, 1980.
17. O'Flaherty JT: Lipid mediators of inflammation and allergy. Lab. Invest. (47):314-329, 1982.
18. Hirata F, Corcoran BA, Venkatasubramanain K, Schiffmann E, Axelrod J: Chemoattractants stimulate degradation of methylated phospholipids and release of arachidonic acid in rabbit leukocytes. Proc. Nat. Acad. Sci. USA (76):2640-2643, 1979.
19. Hirata F, Axelrod J: Phospholipid methylation and biological signal transmission. Science (209):1082-1090, 1980.

20. Michell RH: Inositol phospholipids and cell surface receptor function. Biochim. Biophys. Acta. (415):81-147, 1975.

21. Lapetina EG, Billah MM, Cuatrecasas P: Rapid acylation and deacylation of arachidonic acid into phosphatidic acid of horse neutrophils. J. Biol. Chem. (225):10966-10970.

22. Kroner EE, Peskar BA, Fischer H, Ferber E: Control of arachidonic acid accumulation in bone marrow-derived macrophages by acyltransferases. J. Biol. Chem. (256):3690-3697, 1981.

23. Samuelsson B: Oxidative products of arachidonate: Leukotrienes, a new group of compounds including SRS-A. In: Becker EL, Simon AS and Austen KF (eds) Biochemistry of the Acute Allergic Reaction, Fourth International Symposium Alan R Liss, Inc. New York, 1980. pp 1-12.

24. Samuelsson B: The leukotrienes: A new group of biologically-active compounds including SRS-A. Trends Pharmacol. Sci. (1):227-230, 1980.

25. Morley J, Bray MA, Jones RW, Nugteren DH, Van Dorp DA: Prostaglandin and thromboxane production by human and guinea pig macrophages and leukocytes. Prostaglandins (17):729-736, 1979.

26. Smith MJH: Prostaglandins and the polymorphonuclear leukocyte. In: Ford-Hutchinson AW and Rainsford KD (eds) Prostaglandins and Inflammation Birkhauser Verlag, Basel, 1980, pp 91-103.

27. Goetzl EJ, Woods JM, Gorman RR: Stimulation of human eosinophil and neutrophil polymorphonuclear leukocyte chemotaxis and random migration by 12-L-hydroxy-5,8,10,14-eicosatetraenoic acid. J. Clin. Invest. (59):179-183, 1977.

28. Goetzl EJ, Pickett WC: The human PMN leukocyte chemotactic activity of complex hydroxy-eicosatetraenoic acids (HETEs). J. Immunol. (125):1789-1791, 1980.

29. Goetzl EJ, Brash AR, Tauber AI, Oates JA, Hubbard WC: Modulation of human neutrophil function by monohydroxy-eicosatetraenoic acids. Immunology (39):491-501, 1980.

30. Ford-Hutchinson AW, Bray MA, Doig MV, Shipley ME, Smith MJH: Leukotriene B, a potent chemokinetic and aggregating substance released from polymorphonuclear leukocytes. Nature (286):264-265, 1980.

31. O'Flaherty JT, Thomas MJ, Lees CJ, McCall CE: Neutrophil-aggregating activity of monohydroxyeicosatetraenoic acids. Am. J. Path. (104):55-62, 1981.

32. Goetz EJ, Weller PE, Sun FF: The regulation of eosinophil function by endogenous mono-hydroxy-eicosatetraenoic acids (HETEs). J. Immunol. (124):926-933, 1980.

33. Showell HJ, Naccache PH, Sha'afi RI, Becker EL: Inhibition of rabbit neutrophil lysozomal enzyme secretion, non-stimulated and chemotactic factor stimulated locomotion by nordihydroguaiaretic acid. Life Sciences (27):421-426, 1980.

34. Smolen JE, Weissmann G: Effects of indomethacin, 5,8,11,14-eicosatetrynoic acid and p-bromophenacyl bromide on lysozomal enzyme release and superoxide anion generation by human polymorphonuclear leukocytes. Brit. J. Pharmacol. (29):533-538, 1980.

35. Goetzl EJ, Hill HR, Gorman RR: Unique aspects of the modultion of human neutrophil function by 12-L-hydroperoxy-5,8,10,14-eicosatetraenoic acid. Prostaglandins (19):71-85, 1980.

36. Stenson WF, Parker CW: Metabolism of arachidonic acid in ionophore-stimulated neutrophils: Esterification of a hydroxylated metabolite into phospholipids. J. Clin. Invest. (64):1457-1465, 1979.

37. Goldman DW, Goetzl EJ. Novel metabolites of arachidonic acid in human neutrophils. Fed. Proc. (40):1004a, 1981.

38. Mensing H, Czarnetski BM: Leukotriene B_4 induced in vitro fibroblast chemotaxis. J. Invest. Derm. (82):9-12, 1984.
39. Varani J, Fantone JC: Phorbol myristate acetate-induced adherence of Walker 256 carcinosarcoma cells. Cancer Res. (42):190-197, 1982.
40. Varani J: Chemotaxis of metastatic tumor cells. Cancer Mets. Reviews (1):17-28, 1982.
41. Chensue SW, Kunkel SL, Higashi GI, Ward PA, Boros DL: Production of superoxide anion, prostaglandins, and hydroxyeicosatetraenoic acids by macrophages from hypersensitivity - type (Schistosoma mansoni egg) and foreign body - type granulomas. Infect. Immun. (42):1116-1125, 1983.
42. Rivkin I, Rosenblatt J, Becker EL: The role of cyclic AMP in the chemotactic responsiveness and spontaneous motility of rabbit peritoneal neutrophils. The inhibition of neutrophil movement and the elevation of cyclic AMP levels by catecholamines, prostaglandins, theophilline and cholera toxin. J. Immunol. (115):1126-1134, 1975.
43. O'Flaherty JT, Kreutzer DL, Ward PA: Effect of prostaglandin E_1, E_2 and $F_{2\alpha}$ on neutrophil aggregation. Prostaglandins (17):201-210, 1979.
44. Kunkel SL, Thrall RS, Kunkel RG, McCormick JR, Ward PA, Zurier RB: Suppression of immune complex vasculitis in rats by prostaglandins. J. Clin. Invest. (64):1525-1529, 1979.
45. Kunkel SL, Chensue SW, Plewa M, Higashi GI: Macrophage function in the Schistosoma mansoni egg-induced pulmonary granuloma: Role of arachidonic acid metabolites in macrophage Ia antigen expression. Am. J. Path. (114):240-249, 1984.
46. Fantone JC, Marasco WA, Elgas LJ, Ward PA: Anti-inflammatory effects of prostaglandin E_1: In vivo modulation of the formyl peptide chemotactic receptor on the rat neutrophil. J. Immunol. (130):1495-1497, 1983.
47. Fantone JC, Elgas LJ, Weinberger L, Varani J: Modulation of tumor cell adherence by prostaglandins. Oncology (40):421-426, 1983.
48. Fantone JC, Kunkel SL, Varani J: Inhibition of tumor cell adherence by prostaglandins. In: Mundy GR (ed) Prostaglandins and Cancer: First International Conference. Alan R. Liss, Inc. New York 1982, pp 673-677.
49. Lee EC, Situ R, Fantone JC, Varani J: Functional responses of tumor cells to phorbol esters: Role for prostaglandins. Oncology (in press).
50. Grimstad IA, Wass JA, Spirnak J, Fantone JC, Varani J: Chemotactic factor-induced adherence of tumor cells. Invasion Metastasis (2):274288, 1982.
51. Orr FW, Varani J, Delikatny J, Jain N, Ward PA: Comparison of the chemotactic responsiveness of fibrosarcoma subpopulations of differing malignancy. Am. J. Path. (102):160-167, 1981.
52. Laskin D, Laskin JD, Weinstein B, Carchman RA: Induction of chemotaxis in mouse peritoneal macrophages by phorbol ester tumor promoters. Cancer Res. (41):1923-1928, 1981.
53. Fitzpatrick FA, Bundy GL: Hapten mimic elicits antibodies recognizing prostaglandin E_2. Proc. Nat. Acad. Sci. USA (75):2689-2693, 1978.
54. Fitzpatrick FA, Wynalda MA: A rapid, solid-phase radioimmunoassay for prostaglandin $F_{2\alpha}$ and its main circulating metabolites. Analyt. Biochem. (73):198-208, 1976.
55. Fitzpatrick FA, Gorman RR, McGuire JC, Kelly RC, Wynalda MA, Sun FF: A radioimmunoassay for thromboxane B_2. Analyt. Biochem. (82):1-7, 1977.
56. Honn KV, Cicone B, Skoff A: Prostacyclin: A potent antimetastatic agent. Science (212):1270-1272, 1981.

15

PROSTANOID RECEPTORS

D. EUAN MACINTYRE

1. INTRODUCTION

PROSTAGLANDINS (PG) and THROMBOXANES (Tx) are icosanoids and can be considered as oxidised derivatives of the hypothetical compounds PROSTANOIC ACID and THROMBANOIC ACID, which form the basis of their nomenclature, Figure 1. PG and Tx can be termed collectively as PROSTANOIDS because Tx are generated from PG endoperoxides. Although mono-enoic (PG_1, Tx_1) and tri-enoic (PG_3, Tx_3) prostanoids may occur naturally, this paper will concentrate mainly upon the bis-enoic (PG_2, Tx_2) prostanoids that are derived from arachidonic acid, the major icosanoid precursor esterified in, and subsequently liberated by acyl hydrolase action on, the phospholipids of mammalian cell membranes. Although biological effects have been attributed to PGA, PGB and PGC, the occurrence in nature of these compounds is the subject of much conjecture and thus their actions must be interpreted with caution (1,2).

PG and Tx are not stored in tissues, but are produced when tissues or cells are exposed to an appropriate chemical or mechanical stimulus. Prostanoid biosynthesis is an event common to almost all mammalian cells, and the products exert a plethora of biological effects. As these biological effects are exerted at or near the site of prostanoid production, PG and Tx conform to the class of pharmacologically active agents known as AUTACOIDS.

2. BIOLOGICAL EFFECTS OF PROSTANOIDS

The vasodilator and smooth muscle stimulating activities of prostanoids were recognised at the time of their discovery (3).

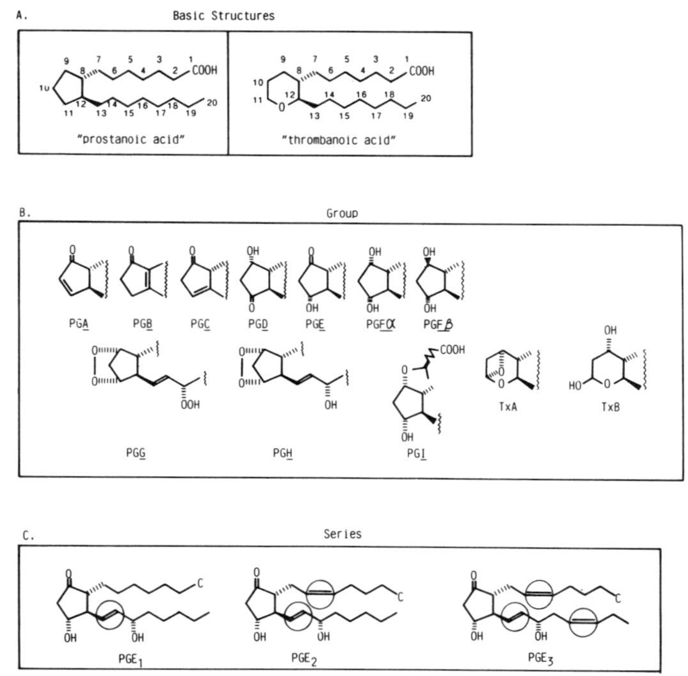

FIGURE 1. PROSTANOID NOMENCLATURE.
A. Basic structures common to all PG or Tx.
B. Groups are given letter designations according to the substituents on the pentane or tetrahydropyran ring and in the case of PGG on C_{15}. $PGF_{2\alpha}$ and $PGF_{2\beta}$ differ in the orientation of the OH group on C_9.
C. Series are given a numerical subscript following the group name. These subscripts correspond to the total number of double bonds in the alpha (C_1-C_8) and omega (C_{12}-C_{20}) side chains.

Subsequently, the ubiquity and biological versatility of the prostanoids became apparent. They are known to affect the functioning of the respiratory, digestive, renal, reproductive, nervous, endocrine and cardiovascular systems. Such a spectrum of activity has led to much speculation concerning the patho-physiological importance of prostanoids (4,5,6). For obvious reasons, it is beyond the scope of this article to present a systematic account of prostanoid actions. For such information, the interested reader is directed to one or more of the authoritative reviews and treatises on the subject (4-9). Rather, to illustrate the versatility of prostanoids, it is my intention to discuss but a few of their biological actions that may contribute to their involvement in the spectrum of diseases referred to as "cancer".

2.1 Cell Proliferation and Differentiation

Addition of prostanoids to cell cultures derived from normal or malignant cell lines resulted in inhibition (PGE, PGF, PGD) (10,11,12) or stimulation (PGE, PGF) of cell replication (13,14). Some of this variability may be attributed to the concentrations of prostanoid used: there is evidence that low concentrations of PGE_2 stimulate, whereas higher concentrations inhibit, the proliferation of lymphoid cells. Alternatively, there may exist cell to cell variations in responsiveness to prostanoids (15). Mechanistic studies of prostanoid actions on cell replication indicate that PGE_2 and $PGF_{2\alpha}$ stimulate mitogenesis of 3T3 cells, but by different mechanisms (13,14). Moreover, there is evidence that endogenous PGE_2 may mediate, at least in part, the mitogenic effects of platelet-derived growth factors (PDGF) on fibroblastic cells (15).

The best characterised effects of prostanoids on the proliferation and differentiation of cells concern the haematopoietic cells. Prostanoids appear to be physiologically important regulators of committed stem cell proliferation and differentiation. Erythropoiesis is enhanced by PGA. Thymocyte differentiation and proliferation are stimulated by PGE > PGF, but these same prostanoids inhibit T cell proliferation and the production of lymphokines by T cells. PGE also influences the differentiation of B cells into immunoglobulin-producing plasma cells. These observations suggest that prostanoids may play an important immunoregulatory role. (17)

2.2 Bone Resorption

Hypercalcaemia associated with solid tumours is mediated by a humoral factor. Tashjian et al (18) showed that PGE_2 induces osteolysis and release of calcium from bone in vitro. This effect of PGE_2 is shared by PGI_2 whereas $PGF_{2\alpha}$ is much less active (19). Further studies showed that, in animal models, hypercalcaemia was induced by tumours that secrete large amounts of PGE_2 (20). In clinical conditions, it remains to be seen whether prostanoid production is the cause of hypercalcaemia or is merely a coincident event. Moreover, the inter-relationships, if any, among prostanoid formation, hypercalcaemia and the development of osseous metastases remain to be evaluated.

2.3 Vascular Homeostasis

The elegant studies of Vane and colleagues (6) showed that vascular homeostasis is regulated, in part, by the opposing actions of the major platelet and vascular arachidonate metabolites: TxA_2, produced by activated platelets, is a vasoconstrictor and potent inducer of platelet aggregation whereas PGI_2, produced by vascular tissue, is a vasodilator and potent inhibitor of platelet activation.

There is evidence that circulating tumour cells can associate with blood platelets, and may induce platelet activation (21). This platelet-tumour cell interaction may be of importance in the spread of tumours. The mechanisms whereby platelets promote metastasis include:-

a. the adhesion of platelet-tumour cell aggregates to the vascular endothelium.
b. release by platelets of certain tumour growth factors e.g. PDGF.
c. shielding by the platelet of the tumour cells from immune surveillance. (22)

Honn and coworkers (23) proposed that metastasis results when tumour cells perturb the delicate balance between TxA_2 formation and PGI_2 formation towards the former. They showed that administration of PGI_2 but not PGE_2 or $PGF_{2\alpha}$ inhibited pulmonary tumour colony formation resulting from injection of B16 amelanotic melanoma cells into mice. These findings suggest that metastasis can be regulated by prostanoids and that endogenous inhibitors of platelet activation e.g. PGI_2 or PGD_2 may be natural deterrents to metastasis.

3. STIMULUS-RESPONSE COUPLING: GENERAL

Although prostanoids are formed and may act intracellularly, they are generally considered to be extracellular mediators - newly synthesized prostanoids are liberated into the extracellular milieu whence they elicit their biological effects. In order for cells to respond to external chemical stimuli (e.g. prostanoids) there must exist some information transfer system whereby events at the cell surface influence the rates of the key biochemical reactions that actually mediate the cellular response. These key biochemical reactions are controlled by stimulus-induced changes in the intracellular concentrations of biochemical intermediaries (or second messenger molecules) including cAMP, cGMP, Ca^{2+} and Diacylglycerol (24). These second messengers activate specific protein kinases, respectively cAMP-dependent protein kinase, cGMP-dependent protein kinase, Ca^{2+}-calmodulin-dependent protein kinase and protein kinase C.

Chemical stimuli can alter the concentration of second messenger(s) by two distinct mechanisms.

3.1 Ionophoric

The chemical stimulus per se forms a lipid-soluble dynamically reversible complex with ions (e.g. Ca^{2+}) and acts as a vesicle for transporting Ca^{2+} across biological membranes.

3.2 Receptor-mediated

The chemical stimulus combines with a specific recognition site, or RECEPTOR on the cell surface and the stimulus-receptor complex initiates the change in concentration of the second messenger molecule(s).

The receptor-occupancy theory of drug action suggests that the magnitude of the response is proportional to the concentration of the drug-receptor complex, and that the formation of the drug-receptor complex is dictated by the law of mass action. A receptor must recognise the chemical features of the drug (agonist or antagonist) which are appropriate for the formation of the drug-receptor complex. In addition, the receptor must be capable of being activated by certain drugs (agonists) to initiate a biological response. The effectiveness of the drug-receptor complex at eliciting a biological response is termed EFFICACY whereas the property of a drug that determines the magnitude of the pharmacological response produced following receptor occupation is termed INTRINSIC ACTIVITY. The avidity with which drugs combine with a receptor is termed AFFINITY. Thus agonists and antagonists have affinity for a receptor: agonists have intrinsic activity and efficacy whereas pure antagonists have zero intrinsic activity and zero efficacy.

3.3 Receptor characterization

Receptors may be characterised by several techniques.

3.3.1 Structure-activity relationships and the effects of antagonists. The specificity of a receptor is not defined by a single agonist but rather by the range of chemical structures that interact with it - for example, alpha and beta adrenoceptors were characterised by the rank order of agonist potency on different tissues. An alternative approach is to use a specific antagonist to define a receptor. Such an approach would discriminate between receptors for different agonists and may distinguish receptor sub-types activated by a single agonist - for example, histamine H_1 and H_2 receptors were indicated by the lack of effect of Mepyramine (an H_1 antagonist) on histamine-induced gastric acid secretion.

3.3.2 Radioligand binding studies. In recent years, the increased availability of radiolabelled agonists and antagonists of high specific activity has facilitated the identification of numerous receptors by direct ligand binding techniques. A binding site may be identified as a receptor only after a careful comparison of the properties of the binding site with the properties expected fo the receptor. The following criteria should be satisfied:- a. The distribution of binding sites should accord with that of the receptor. b. The affinity, specificity, reversibility and rate of binding should be consistent with those observed between the agonist and the cell. c. There should be a functional correlation between binding the putative receptor and the physiological consequences of receptor activation.

3.3.3 Transduction mechanisms. As the effects of agonists ultimately depend upon the regulation of intracellular biochemical reactions by second messenger molecules, receptors for different agonists and receptor sub-types for a single agonist may be differentiated on the basis of the information transduction system to which the agonist-receptor complex is coupled - for example, muscarinic cholinoceptors, α_1 adrenoceptors, H_1 histamine receptors and V_1 vasopressin receptors are coupled to inositol phospholipid hydrolysis, Ca^{2+} flux, diacylglycerol formation and cGMP formation whereas β adrenoceptors, H_2 histamine receptors and V_2 vasopressin receptors are coupled to activation of adenylate cyclase and consequent elevation in the concentration of cAMP (24).

4. STIMULUS-RESPONSE COUPLING: PROSTANOIDS

4.1 Ionophoric effects

Kirtland and Baum (25) first suggested that PGE_1 could act as a mitochondrial Ca^{2+} ionophore. Subsequently, several groups have reported that natural prostanoids and PG analogues possess ionophoric activity in certain organelles, subcellular fractions and in artificial in vitro systems. The relevance of such observations to physiological (or pathological) responses in an intact cell remains to be established. Lipophilic acids would be expected to ion pair with cations and render them lipid soluble, but prostanoids lack the backbone rigidity characteristic of true ionophores and cannot form well-defined complexation cages nor do they flip-flop between limited conformational options in the manner required to facilitate translocation of cations across membranes. As prostanoids are quantitatively much less efficacious Ca^{2+} ionophores than, for example, A23187, it is unlikely that ionophoric activity is a major mechanism whereby extracellular prostanoids induce cell activation (26). However, this limited Ca^{2+} ionophoric activity may be relevant to the effects, if any, of prostanoids that act intracellularly. For example, the site of platelet TxA_2 biosynthesis is contiguous with the major intracellular Ca^{2+} storage organelle, the dense tubular system, and liberation of stored Ca^{2+} by endogenous, intracellular TxA_2 may contribute towards platelet activation (27).

4.2 Receptor-mediated effects

In view of the ubiquity, potency and biological specificity of the different prostanoids, on teleological grounds alone it could be inferred that their various effects would best be explained by combination with distinct receptors. That this is so, and the effects of the natural prostanoids are mediated by combination with distinct receptors is indicated by:-

1. Radioligand binding analyses.
2. A comparison of the structure-activity relationships of agonists.
3. The use of selective antagonists.
4. Analyses of transduction mechanisms.

In most aspects of pharmacology, biochemical approaches to receptor characterization (e.g. radioligand binding, analyses of transduction mechanisms) lagged behind physiological investigations. This was not the case for prostanoid pharmacology where physiological and biochemical studies proceeded in parallel. This was due, in large part, to the relatively recent discovery purification and structural elucidation of prostanoids, especially PGD_2 (1974), TxA_2 (1975) and PGI_2 (1976) - by which time biochemical techniques for examining receptors had been established for steroid hormones, peptide hormones and neurotransmitters.

4.3 Characterization of Prostanoid Receptors

4.3.1 Radioligand Binding. Kuehl and Humes (28) identified, on rat adipocytes, a saturable, displaceable binding site for $[^3H]$-PGE_1 that possessed the pharmacological characteristics of a PGE_1 receptor. Shortly thereafter, PGE receptors in corpora lutea (29), liver (30) and platelets (31) were identified. Powell et al (32) first demonstrated receptors for $[^3H]$-$PGF_{2\alpha}$ on corpora lutea, and these have been characterized in detail by Rao and coworkers (33). By the end of 1979, specific, high-affinity binding sites for $[^3H]$-PGD_2 and $[^3H]$-PGI_2 had been detected on human platelets (34,35). The detection of TxA_2 receptors by radioligand binding analysis proved somewhat more difficult. This was due mainly to the instability of TxA_2 and its rapid hydrolysis to the relatively inert TxB_2. Mechanistic studies of TxA_2 became feasible with the synthesis of stable analogues such as 9α,11α-epoxymethano-PGH_2 (U44069) and 11α,9α-epoxymethano-PGH_2 (U46619) which act as TxA_2 mimetics. Armstrong et al (36) reported that $[^3H]$-U44069 bound to a saturable, high affinity binding site on human platelets and could be displaced by unlabelled U44069, by TxA_2 and by PGH_2.

4.3.2 Structure-Activity Relationships. Andersen and Ramwell (37) made the first systematic attempt to characterise the actions, and hence receptors, for prostanoids by comparing the rank order of agonist potency of PGA, PGE and PGF on several pharmacological preparations. Four classes of prostanoid actions/receptors were distinguished:-

Class I: non-polar prostaglandin action (A > E > F) e.g. gastric acid secretion.

Class II: E-specific action (E > A > F). e.g. inhibition of platelet aggregation.

Class III: mixed E and F action (E > F > A) e.g. induction of labour.

Class IV: polar prostaglandin action (F > E > A) e.g. luteolysis.

They reasoned that non-specific effects should display a rank order of potency related to the polarity sequence, and speculated that Class I and Class IV effects may fall into this category. The same group proceeded to examine the effects of systematic structure variations including:- a. stereochemical changes; b. group substitutions or deletions; c. changes in hydrocarbon backbone - on the effects of prostaglandins on several preparations. They concluded that the 15(S)-hydroxyl group, the constituents on carbon atoms 9 and 11 of the cyclopentane ring, and the close alignment of the alpha and omega side chains were important for the interaction of prostanoids with their cellular receptors. This 'hairpin conformation hypothesis' implied a multi-point contact of a prostaglandin molecule with its cellular receptor, regulated in part by the configuration of the side chains. This concept was later supported by physiological evidence of the existence of such hairpin conformations of PGE and PGF in the crystalline state and in solution (38,39).

With the discovery and conformational analysis of TxA_2, TxB_2 and PGI_2, it soon became obvious that all natural prostanoids did not exist in hairpin conformation (40). To test further the significance of prostanoid 'hairpin conformation' in bio-recognition phenomena, Andersen et al (41) synthesised a series of analogues of 2 basic types: the first were geminally methylated at C16, C17 or C18 so as to enhance or limit the tendency for side chain alignment; the second were macrolide structures in which the side chains are maintained in alignment or non-alignment by covalent bonding. These analogues were tested in preparations sensitive to PGE_2, $PGF_{2\alpha}$ and PGI_2. It was concluded that hairpin alignment of prostanoids improved potency only at $PGF_{2\alpha}$ receptors and decreased potency at PGE_2 and PGI_2 receptors.

More recently, Kennedy and colleagues (42,43,44) compared the rank order of agonist potency of PGD_2, PGE_2, $PGF_{2\alpha}$, PGI_2 and U46619 (a TxA_2-mimetic) on a range of smooth muscle preparations. They separated the preparations examined into 3 distinct groups according to the rank order of agonist potency of each preparation:

Group I: PGE_2-sensitive preparations, e.g. guinea-pig ileum, where the rank order of potency was

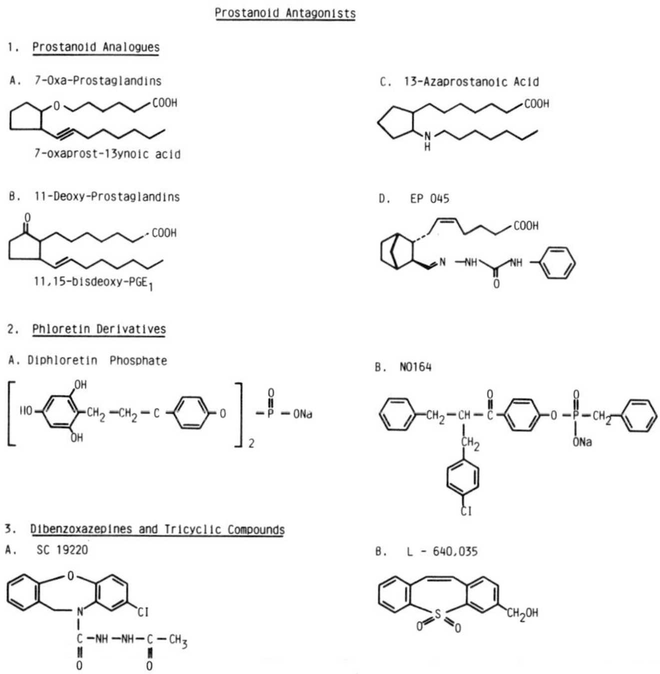

FIGURE 2. PROSTANOID ANTAGONISTS.

$PGE_2 > PGI_2 = PGF_{2\alpha} > PGD_2 = U46619 = 0$.

Group II: $PGF_{2\alpha}$-sensitive preparations, e.g. cat iris sphincter muscle, where the rank order of potency was $PGF_{2\alpha} > PGD_2 > U46619 > PGE_2 > PGI_2$.

Group III: TxA_2-sensitive preparations, e.g. rat aorta, where the rank order of agonist potency was $(TxA_2) > U46619 >> PGF_{2\alpha} = PGE_2 = PGD_2 = PGI_2$.

Because of the striking similarities within groups and differences between groups, they postulated the existence of 3 distinct classes of prostanoid receptors, for PGE_2, $PGF_{2\alpha}$ and TxA_2. On the basis of rank order of agonist potency it was impossible to preclude the existence of receptor sub-types for any of the prostanoids examined. Such a differentiation is best made using antagonists, or by comparing the structure-activity relationships of a series of analogues of an individual prostanoid (see section 4.4). In the studies of Kennedy et al (44), PGD_2 and PGI_2 behaved as weak but selective agonists at PGF_2 receptors and at PGE_2 receptors respectively. Earlier studies using platelets of different species indicated that there also exist distinct receptors for PGD_2 and for PGI_2 (45).

4.3.3 Use of Antagonists. Progress in the characterization of prostanoid receptors has been hampered by the lack of potent, selective antagonists of prostanoid action(s), especially as different prostanoids can act on a single tissue and exert the same or opposite effects. There are three major classes of compound that have been reported to inhibit the actions, rather than the synthesis, of prostanoids.

i. Prostaglandin analogues

ii. Dibenzoxazepines

iii. Phloretin derivatives

In addition, other miscellaneous compounds including tetrahydroisoquinolines, morphine and fenamates have been reported to act as prostanoid antagonists. Figure 2.

i. Prostaglandin analogues

A rational method of devising antagonists is to make chemical analogues of the active substance that may have affinity for its receptor, but will not produce a response and thus will act as a competitive antagonist. When the structure-activity relationships of prostanoid action were assessed, it was observed that the structural modifications of prostanoids that maintained receptor affinity but resulted in loss of intrinsic activity were (a) removal of the 15(S)-hydroxyl substituent which is essential for agonist activity, (b) deletion or replacement of substituents or carbon atoms 9 and 11, which are important for receptor combination, (c) insertion of an oxygen atom in the alkyl side chain to reduce its lipophilicity (37).

Fried and colleagues (46) synthesised racemic 7-oxaprostaglandin analogues and showed that derivatives with acetylenic linkages in the 13,14 position e.g. B7-oxa-prost-13-ynoic acid (PYA), were the most specific prostaglandin inhibitors. PYA inhibited the actions of PGE and PGF on gastro-intestinal smooth muscle, although the selectivity of the compound for prostanoids was disputed (47). MacIntyre and Salzman (48) showed that PYA inhibited the action of TxA_2-mimetics but not of PGD_2 or PGI_2 on human platelets. This lack of selectivity (or broad specificity) of the 7-oxa-prostaglandin analogues is not evident in other classes of prostaglandin analogues that act as TxA_2 antagonists. 13-Aza prostanoic acid, EP 045, pinane thromboxane A_2 (PTA_2) and AH 19437 all behave as potent, selective antagonists of the actions of TxA_2 and TxA_2 mimetics on platelets (49,50,51,52).

<u>ii</u>. <u>Dibenzoxazepines</u>

In view of their broad spectrum of antagonist activity against seemingly unrelated autacoids and neurotransmitters, it was not unexpected that tricyclic compounds might also inhibit the actions of prostanoids. Sanner (53) first reported that SC 19220 inhibited PGE_2-induced contractions of the guinea-pig ileum. Others confirmed these findings and showed that SC 19220 was equally effective against PGE and PGF on the guinea-pig ileum (47,54), a PGE-sensitive preparation according to Kennedy et al (44). MacIntyre and Salzman (48) showed that SC 19220 inhibited the effects of TxA_2-mimetics on platelets, whereas higher concentrations inhibited the effects of PGD_2 and PGI_2 also.

Carrier et al (55) have shown that L-640,035 acts as an antagonist of the contractile effects of prostanoids, thought to be mediated via TxA_2 receptors and PGE_2 receptors (56) on guinea-pig trachea. The compound also blocks the effects of TxA_2-mimetics on platelets.

iii. Phloretin Derivatives

Polyphloretin phosphate (PPP) and related polymers were found to inhibit a variety of enzymes, e.g. hyaluronidase, and to display some anti-inflammatory activity (57). Beitch and Eakins (58) showed that PPP antagonised the increase in intra-ocular pressure induced in rabbit eyes by PGE_2. This stimulated an investigation of the anti-prostaglandin action of PPP, and Eakins and Karim (59) concluded that it was more effective against PGF than PGE on various smooth muscle preparations. Upon fractionation of PPP, the prostaglandin antagonist activity was localised to the lower molecular weight residues e.g. Di-4-phloretin phosphate (DPP). Ohter phloretin derivatives e.g. N0164 and related compounds are more potent than DPP and display some selectivity for PGD_2 over PGI_2 when examined on human platelets (48). However, N0164 and related compounds also act as antagonists of TxA_2 mimetics on platelets (60).

4.4 Heterogeneity of Prostanoid Receptors

By analogy with other autacoids, it is possible, indeed likely, that receptors for the same ligand differ in different tissues and that there exist sub-types of prostanoid receptors. The pharmacological differentiation of prostanoid receptor sub-types is most readily achieved by comparing the effects of antagonists on different tissues. Where no selective antagonists are available, a comparison of the structure-activity relation-ships of a series of analogues on different tissues may also yield information on receptor heterogeneity.

4.4.1 TxA_2 Receptors. Fitzpatrick and colleagues (61) showed that 9,11-epoxyimiprostadienoic acid antagonized the actions of TxA_2 and TxA_2-mimetics on human platelets, yet directly induced contraction of vascular smooth muscle. Lefer et al (62) showed that carbocyclic thromboxane A_2 (CTA_2) exerted a similar spectrum of action and that the contractile effects of CTA_2 were suppressed by the TxA_2-antagonist, pinane thromboxane A_2. Collectively, these results suggest that the receptor sites for TxA_2 on platelets and on vascular smooth muscle may be different.

4.4.2 <u>PGE_2 Receptors</u>. Kennedy and colleagues (44) examined the effects of the dibenzoxazepine SC 19220 on five different preparations (guinea-pig ileum, dog fundus, guinea-pig fundus, chick ileum, cat trachea) that they had characterized as PGE_2-sensitive on the basis of rank orders of agonist potency. They showed that SC 19220 behaved as a competitive receptor antagonist of the effects of PGE_2 on the guinea-pig ileum and on the guinea-pig and dog fundus, but that it had no effect on contraction of the chick ileum or relaxation of the cat trachea induced by PGE_2. These findings indicate that PGE_2 receptors are heterogeneous.

4.4.3 <u>PGI_2 receptors</u>. The biological effects and specificity of the only reported PGI_2 receptor antagonist, 5(Z,E)-11-deoxy-20-methylcarbo-PGI_2 (63) have not been examined in detail. Consequently, characterisation of PGI_2 receptor sub-types depends on the comparison of the structure-activity relationships of PGI_2 analogues on different tissues. The therapeutic potential of PGI_2 is restricted by its instability and by its profound vasodilator and anti-aggregatory effects. In an attempt to obtain stable compounds with platelet-specific or vascular-specific actions large numbers of PGI_2 analogues have been synthesised. Some of these analogues display enhanced vasodilator relative to anti-aggregatory activity and vice versa. For example, carbacyclin is 7 fold more active on platelets than on vascular tissue (64) whereas 2'a-9-didehydro-PGI_2 is at least 10 fold more active on vascular tissue than on platelets (65). Such a dissociation of vasodilator and anti-aggregatory activity of PGI_2 analogues implies that there exist distinct sub-types of PGI_2 receptors on platelets and on blood vessels.

4.4.4 <u>PGD_2 receptors</u>. There exist no selective PGD_2 receptor antagonists. Although N0164 selectively inhibits the anti-aggregatory effects of PGD_2, the compound cannot be considered a selective PGD_2 receptor antagonist (section 4.3.3.iii). Therefore, PGD_2 receptor sub-types must be delineated by comparing the structure-activity relationships of PGD_2 analogues on different tissues. Unfortunately, due to the paucity of analogues synthesised (or documented) little information is available concerning the structural requirements for PGD_2-like activity on different tissues. However, Jones (66) showed that the structure-activity relationships for PGD_2-induced anti-aggregatory activity differed from those for vasoconstriction. Moreover, Fukushima et al (67) showed that the growth inhibitory (antineoplastic) activity of PGD_2 could be dissociated from the anti-aggregatory activity. Collectively, these results suggest that there may exist distinct sub-types of PGD_2 receptor on different tissues.

4.4.5 $PGF_{2\alpha}$ receptors. As for PGI_2 and PGD_2, there exist no selective $PGF_{2\alpha}$-receptor antagonists. Although it has been suggested that 20-methyl-13,14-didehydro-$PGF_{2\alpha}$ has such activity (68), this remains to be confirmed. Thus, heterogeneity of $PGF_{2\alpha}$ receptors must also be assessed by comparing the potencies of $PGF_{2\alpha}$ analogues on different tissues. As $PGF_{2\alpha}$ analogues offer therapeutic potential as anti-fertility agents, numerous derivatives have been synthesised and tested for luteolytic and smooth muscle activity. McCracken and colleagues (69) showed that 13-dehydro-$PGF_{2\alpha}$ analogues retained luteolytic activity but had diminished smooth muscle activity. This dissociation between the luteolytic and smooth muscle actions of $PGF_{2\alpha}$ analogues was confirmed by Andersen et al (41) who observed that ω chain modifications of $PGF_{2\alpha}$, such as 15-epimerization or methylation conferred luteal selectivity. These findings strongly suggest that there exist two distinct sub-types of $PGF_{2\alpha}$ receptors.

4.5 Transduction mechanisms of prostanoids

As the available evidence suggests that prostanoids exert their effects by combining with distinct receptors, the involvement of some information transduction process is necessary to explain the intracellular sequelae of agonist-receptor interaction at the surface of cells.

4.5.1 PGE. That the effects of PGE may relate to changes in cAMP was first observed by Steinberg et al (70) who showed that PGE_1 inhibited the elevation in cAMP induced by lipolytic agents acting on fat cells. In contrast, Butcher and Baird (71) showed that PGE actually stimulated cAMP formation in several tissues. At first these observations seemed mutually irreconcileable. However, as the molecular events involved in the regulation of cAMP formation were investigated, it became clear that adenylate cyclase may be under stimulatory or inhibitory control by agonists. The agonist-receptor complex communicates with guanine nucleotide-binding regulatory proteins (G), one of which mediates stimulation (Gs) of adenylate cyclase activity, while the other (Gi) is responsible for inhibition (72).

There is now much evidence in numerous systems that PGE receptors are coupled to adenylate cyclase. In most cases, the PGE-receptor complex interacts with the Gs regulatory protein leading to stimulation of adenylate cyclase and consequent elevation of the intracellular concentration of cAMP. In these tissues, the biological effects depend upon enhanced intracellular phosphorylation as a consequence of the action of cAMP-dependent protein kinases. In some instances, the PGE-receptor complex interacts with the Gi regulatory protein leading to inhibition of adenylate cyclase and presumably to a reduction in the intracellular cAMP concentration. The biochemical sequelae of these inhibitory effects are not fully understood.

4.5.2 PGF. The failure of PGF, except at very high concentrations, to mimic the effects of PGE on cAMP formation led to the suggestion that PGF receptors activated a distinct transduction mechanism. Kuehl et al (73) showed that $PGF_{2\alpha}$ stimulated the formation of cGMP by uterine tissue. The mechanisms underlying cGMP formation in intact cells are not fully understood, although it is alleged to be Ca^{2+}-dependent and frequently is associated with agonist-induced inositol phospholipid hydrolysis (24). Polyunsaturated fatty acids or their metabolites activate guanylate cyclase (74). It is thought that the elevation in cytosolic free Ca^{2+} concentration, as a consequence of agonist-induced inositol phospholipid hydrolysis, leads to acyl hydrolase activation and liberation of poly-unsaturated fatty acid.

Using cultured luteal cells (75) and 3T3 cells (76) it has been shown that $PGF_{2\alpha}$ stimulates inositol phospholipid turnover and diacylglycerol formation. Thus $PGF_{2\alpha}$-receptor combination most probably activates a phospholipase C type enzyme that hydrolyses membrane (poly) phosphoinositides. This, in turn, yields diacylglycerol and inositol phosphate(s) and leads to an elevated intracellular free Ca^{2+} concentration, and to cGMP formation. Consequently, the biochemical effects of PGF receptor activation may depend upon enhanced intracellular phosphorylation resulting from the actions of Ca^{2+}-calmodulin-dependent enzymes, protein kinase C and cGMP-dependent protein kinases.

4.5.3 <u>PGD</u>. Receptors for PGD_2 have been best characterized on the human platelet. There, PGD_2-receptor combination leads to activation of adenylate cyclase and elevation of the intracellular cAMP concentration (77). Thus the cellular effects of PGD_2 are mediated via cAMP-dependent protein kinases.

4.5.4 <u>PGI</u>. Receptors for PGI_2, as for PGD_2, have been best characterized on the human platelet. PGI_2 is a more potent stimulant of platelet adenylate cyclase than is PGD_2 : cAMP formation is stimulated to a greater extent and remains elevated for much longer (77). The effects of PGI_2 on tissues other than platelets are also coupled to cAMP formation (79). Thus it is reasonable to conclude that the cellular effects of PGI_2 are mediated via cAMP dependent protein kinases.

4.5.5 <u>TxA_2</u>. Stemming from its discovery as a potent inducer of platelet activation, the vast majority of data on receptors for TxA_2 have been obtained using platelets and mostly using stable TxA_2-mimetics in place of the evanescent TxA_2. That the platelet TxA_2 receptor is also acted upon by the endoperoxides PGG_2 and PGH_2, which also stimulate platelet activation, is indicated by radioligand binding studies and by the use of selective antagonists. It has been suggested that the platelet TxA_2 receptor is coupled to inhibition of adenylate cyclase (80), although this is disputed (81). We have shown that TxA_2 receptor occupancy is coupled to inositol phospholipid hydrolysis (and presumably DAG formation) and elevation in the concentration of Ca^{2+} in the cytosol (82). cGMP formation by platelet agonists occurs too late to be involved in the activation process. Thus the cellular effects of TxA_2 receptor occupancy are most probably mediated by Ca^{2+}-calmodulin-dependent enzymes and via protein kinase C.

Whether or not prostanoid receptor sub-types share the same transduction process (as, for example, do β_1 and β_2 adrenoceptors) or utilise different transduction processes (as, for example, do α_1 and α_2 adrenoceptors) remains to be determined.

5. CONCLUSIONS

The available evidence indicates that there are at least 5 distinct receptors for prostanoids - for PGD_2, PGE_2, $PGF_{2\alpha}$, PGI_2 and TxA_2 and it is probable that there are sub-classes of each of these receptors. Kennedy and colleagues (44) have proposed a nomenclature for prostanoid receptors in which each receptor is designated by the letter P preceeded by a letter signifying the most potent natural prostanoid agonist at that receptor. thus, receptors sensitive to PGD_2, PGE_2, $PGF_{2\alpha}$, PGI_2 and TxA_2 become DP-, EP-, FP-, IP- and TP-receptors respectively. Where sub-division is required within a receptor group, e.g. EP receptors, subscript numerals can be used to indicate a particular sub-type, e.g. EP_1, EP_2 receptors.

How can we exploit this knowledge of prostanoid receptors? Where endogenous prostanoids are implicated in pathological conditions, the major therapeutic stratagem is to suppress the biosynthesis of the putative prostanoid disease mediator, using inhibitors of cyclo-oxygenase. As the different prostanoids may exert mutually antagonistic biological effects on certain systems, e.g. TxA_2 and PGI_2 on vascular homeostasis, inhibition of cyclo-oxygenase is prostanoid non-specific and thus may result in the removal of a protective as well as a noxious mediator. In conditions where exogenous prostanoid administration offers therapeutic potential, the desired beneficial actions are often compromised by (side) effects on other tissues. 'Selective toxicity' is the essence of drug therapy. Selectivity in suppressing the effects of endogenous prostanoids will best be obtained using specific receptor antagonists. Selectivity in mediating the desireable effects of exogenous prostanoids on a particular target tissue will best be achieved using analogues that display specificity for a receptor sub-type and thus display a bias of activity.

The existing TxA_2 (or TP-) receptor antagonists or analogues that selectively activate the platelet PGI_2 (or IP-) receptor may impair the process of metastasis by suppressing the platelet-tumour cell interaction or the process of platelet activation. The development of specific PGE_2 (or EP-) or $PGF_{2\alpha}$ (or FP-) receptor antagonists may be of benefit in the prevention of osteolysis and in reducing the effects of PGE_2 and $PGF_{2\alpha}$ on cell proliferation and differentiation. PGD_2 (or DP-) receptor agonists that have been modified to retain growth inhibitory activity but lack other biological effects of PGD_2 may also have antineoplastic activity.

In view of the diversity of actions of prostanoids, antagonists or agonists that have affinity for a particular prostanoid receptor or receptor sub-type may be of therapeutic benefit in a number of disease processes. The challenges that remain are to categorize prostanoid receptor sub-types and to synthesise and characterize such agents.

Acknowledgements

The author wishes to thank Dr. Roma Armstrong for her constructive criticism and Mrs. M. Wright for her careful typing.

REFERENCES

1. Jonsson HT, Powers RE: Endogenous prostaglandin As : fact or artifact in biological systems. Prog Lipid Res (20): 787-790, 1981.
2. Willis AL, Stone KJ: Properties o& prostaglandins, thromboxanes, their precursors, intermediates, metabolites and analogs: a compendium. In :Fasman GD (ed) Handbook of biochemistry and molecular biology, pp312-395, 1976.
3. von Euler US: A depressor substance in the vesicular gland. J. Physiol (84) 21-22P, 1935.
4. Horton EW: Hypothesis on physiological roles of prostaglandins. Physiol Rev (49): 122-161, 1969.
5. Horton EW: Prostaglandins and smooth muscle. Brit Med Bull (35) : 295-300, 1979.
6. Moncada S, Vane JR: Pharmacology and endogenous roles of prostaglandin endoperoxides, thromboxane A_2 and prostacyclin. Pharmacol Rev (30) : 293-331, 1979.
7. Karim SMM: Prostaglandins: chemical and biochemical aspects. MTP Press, Lancaster, 1976.
8. Vane JR, Bergstrom S: Prostacyclin, Raven Press, New York, 1979.
9. Moncada S (ed): Prostacyclin, thromboxane and leukotrienes. Brit Med Bull (39) : 1983.
10. Santoro MG, Phillpot GW, Jaffe BM: Inhibition of tumour growth in vivo and in vitro by prostaglandin E. Nature (263) : 777-779, 1976.
11. Sakai T, Yamaguchi N, Shiroko Y, Sekiguchi M, Fujii G, Nishino H: Prostaglandin D_2 inhibits the proliferation of human malignant tumour cells. Prostaglandins (27) : 17-26, 1984.
12. Fukushima M, Kato T, Ueda R, Ota K, Narumiya S, Hayaishi O: Prostagland). D_2: a potential anti-neoplastic agent. Biochem Biophys Res Commun (105) : 956-964, 1982.
13. de Asua LJ, Otto AM, Ulrich MO, Martin-Perez J, Thomas G: The regulation of DNA replication in animal cells by $PGF_{2\alpha}$. In: Prostaglandins and cancer: first international conference. Alan R Liss Inc, New York, pp309-331, 1982.
14. Rozengurt E, Collins MKL, Keehan M: Mitogenic effect of prostaglandin E_1 in Swiss 3T3 cells: role of cyclic AMP. J Cell Physiol (116): 379-384, 1983.
15. Rozengrut E, Stroobant P, Waterfield M, Deuel TF, Keehan M: Platelet-derived growth factor elicits cyclic AMP accumulation in Swiss 3T3 cells: role of prostaglandin production. Cell (34) : 265-272, 1983.
16. Honn KV, Bockman RS, Marnett LJ: Prostaglandins and cancer: a review of tumor initiation through tumor metastasis. Prostaglandins (21) : 833-866, 1981.
17. Goldyne E, Stobo JD: Immunoregulatory roles of prostaglandins and related lipids. CRC Crit Rev Immunol (2) : 189-224, 1981.
18. Tashjian A, Voelkel E, Levine L, Goldhaver P: Evidence that the bone resorbing-stimulating factor produced by mouse fibrosarcoma cells is prostaglandin E_2. J Exp Med (136) : 1329-1343, 1972.

19. Bennett A, Edwards D, Ali NN, Auger D, Harris M: Prostacyclin potently resorbs bone in vitro. Adv PG Tx Res (6) : 547-558, 1980.
20. Tashjian AH, Voelkel EF, Goldhaver P, Levine L: Prostaglandins, calcium metabolism and cancer. Fed Proc (33) : 81-86, 1974.
21. Gasic GJ, Gasic TB, Galanti N, Johnson T, Murphy S: Platelet-tumor cell interactions in mice. The role of platelets in the spread of malignant disease. Int J Cancer (11) : 704-718, 1973
22. Mehta P: Potential role of platelets in the pathogenesis of tumor metastasis. Blood (63) : 55-63, 1984.
23. Honn KV, Cicone B, Skoff A, Rominie M, Gossett D, Thompson K, Hodge D: Thromboxane synthetase inhibitors and prostacyclin can control tumor cell metastasis. J Cell Biol (87) : 64-71, 1980.
24. Berridge MJ: Phosphatidylinositol hydrolysis: a multifunctional transducing mechanism. Mol Cell Endocrinol (24) : 115-141, 1981.
25. Kirtland SJ, Baum H: Prostaglandin E_1 may act as a calcium ionophore. Nature New Biol (236) : 47-49, 1972.
26. Pressman BC, Fahim M: Pharmacology and toxicology of the monovalent carboxylic ionophores. Ann Rev Pharmacol Toxicol (22) : 465-490, 1982.
27. Gerrard JM, Peterson DA, White JG: Calcium mobilization. In: Gordon JL (ed) Platelets in biology and pathology 2. North Holland, Amsterdam, 1981, pp407-436.
28. Kuehl FA, Humes JL: Direct evidence for a prostaglandin receptor and its application to prostaglandin measurements. Proc Natl Acad Sci (69) : 480-484, 1972.
29. Rao CV: Steroid hormone modulation of [^{3}H]-prostaglandin E_1 binding to bovine corpus luteum cell membranes. Prostaglandins (9) : 569-578, 1975.
30. Smigel M, Fleischer S: Characterization of prostaglandin E_1 receptors in rat liver plasma membranes. Biochim Biophys Acta (332) : 358-371, 1974.
31. Gorman RR: Specific PGE_1 and PGE_2 binding sites on platelet membranes. Prostaglandins (6) : 542, 1974.
32. Powell WS, Hammarstrom S, Samuelsson B: Prostaglandin $F_{2\alpha}$ receptors in ovine corpora lutea. Eur J Biochem (41): 103-107, 1974.
33. Rao CV, Mitra SB: Distribution of PGE and $PGF_{2\alpha}$ receptor proteins in the intracellular organelles of bovine corpora lutea. Meth Enzymol (86) : 192-202, 1982.
34. Cooper B: Agonist regulation of the human platelet prostaglandin D_2 receptor. Life Sci (25) : 1361-1368, 1979.
35. Siegl AM, Smith JB, Silver MJ: Selective binding site for [^{3}H]-prostacyclin on platelets. J Clin Invest (63) : 215-220, 1979.
36. Armstrong RA, Jones RL, Wilson NH: Ligand binding to thromboxane receptors on human platelets; correlation with biological activity. Br J Pharmac (79) : 953-964, 1983.
37. Andersen NH, Ramwell PW: Biological aspects of prostaglandins. Archvs Int Med (133) : 30-50, 1974.
38. Andersen NH, Ramwell PW, Leovey EMK, Johnson M: Biological consequences of prostaglandin molecular conformations. Adv PG Tx Res (1) : 271-289, 1976.
39. Langs DA, Erman M, De Titta GT: Conformations of prostaglandin $F_{2\alpha}$ and recognition of prostaglandins by their receptors. Science (197) : 1003-1005, 1977.

40. Langs DA, De Titta GT, Erman M, Fortier S: Thromboxane molecules are not hairpin conformers. Adv PG Tx Res (6) : 477-480, 1980.
41. Andersen NH, Imamoto S, Subramanian N, Picker DH, Ladner DW, De B, Tynan SS, Eggerman TL, Harker LA, Robertson RP, Oien HG, Rao CV: Molecular basis for prostaglandin potency: tests of the significance of the hairpin conformation in bio-recognition phenomena. Prostaglandins (22) : 841-856, 1981.
42. Kennedy I, Coleman RA, Humphrey PPA, Lumley P: Studies on the characterization of prostanoid receptors. Adv PG Tx LT Res (11) : 327-332, 1983.
43. Coleman RA, Humphrey PPA, Kennedy I, Levy GP, Lumley P: Preliminary characterization of three types of prostanoid receptor mediating smooth muscle contraction. Br J Pharmac (69) : 265P, 1980.
44. Kennedy I, Coleman RA, Humphrey PPA, Lumley P: Studies on the characterisation of prostanoid receptors: a proposed classification. Prostaglandins (24) : 667-688, 1982.
45. MacIntyre DE: Platelet prostaglandin receptors. In: Gordon JL (ed) Platelets in biology and pathology 2. North Holland, Amsterdam, 1981, pp211-247.
46. Fried J, Santhanakrishnan T, Himizu J, Lin CH, Ford SH, Rubin B, Grigas EO: Prostaglandin antagonists: synthesis and smooth muscle activity. Nature (223) : 208-210, 1969.
47. Bennett A, Posner J: Studies on prostaglandin antagonists. Br J Pharmac (42) : 584-594, 1971.
48. MacIntyre DE, Salzman EW: Antagonism of human platelet responses to stimulatory and inhibitory prostaglandins. Prog Lipid Res (20): 453-459, 1980.
49. Le Breton GC, Venton DL, Enke SE, Halushka PV: 13-Aza-prostanoic acid: a specific antagonist of the human blood platelet thromboxane/endoperoxide receptor. Proc Natl Acad Sci (76) : 4097-4101, 1979.
50. Jones RL, Wilson NH: Thromboxane receptor antagonism shown by a prostanoid with a bicyclo 2.2.1 heptane ring. Br J Pharmac (73) : 220-221P, 1981.
51. Nicolau KC, Magolda RL, Smith JB, Aharony D, Smith EF, Lefer AM: Synthesis and biological activity of pinane TxA_2, a selective inhibitor of coronary artery constriction, platelet aggregation and thromboxane formation. Proc Natl Acad Sci (76) : 2566-2570, 1979.
52. Geisow HP, Hornby EJ, McCabe PJ: Inhibition of platelet aggregation by AH 19437, a thromboxane receptor blocking drug. Br J Pharmac (73) : 219P, 1981.
53. Sanner JH: Antagonism of prostaglandin E_2 by SC19220. Archvs Int Pharmacodyn Ther (180) : 46-56, 1969.
54. Coleman RA, Kennedy I, Levy GP: SC19220, a selective prostanoid receptor antagonist. Br J Pharmac (69) : 266-267P, 1980.
55. Carrier R, Cragoe EJ, Ethier D, Ford Hutchinson AW, Girard Y, Hall RA, Hamel P, Rokach J, Share NN, Stone CA, Yusko P: Studies on L-640,035: a novel antagonist of contractile prostanoids in the lung. Br J Pharmac in press, 1984.

56. Jones RL, Peesapati V, Wilson NH: Antagonism of the thromboxane sensitive contractile systems of the rabbit aorta, dog saphenous vein and guinea-pig trachea. Br J Pharmac (76) : 423-428, 1982.
57. Fries B: The edema-inhibiting action of polyphloretin phosphate in some types of capillary damage: an experimental investigation. Acta Chir Scand (119) : 1-7, 1960.
58. Beitch BR, Eakins KE: The effects of prostaglandins on the intraocular pressure of the rabbit. Br J Pharmac (37) : 158-167, 1969.
59. Eakins KE, Karim SMM, Miller JD: Antagonism of some smooth muscle actions of prostaglandins by polyphloretin phosphate. Br J Pharmac (39) : 556-563, 1970.
60. MacIntyre DE, Gordon JL: Discrimination between platelet prostaglandin receptors with a specific antagonist of bis-enoic prostaglandins. Thromb Res (11) : 705-713, 1977.
61. Fitzpatrick FA, Bundy GL, Gorman RR, Honohan T: 9,11-Epoxyiminoprosta 5,13 dienoic acid is a thromboxane antagonist in human platelets. Nature (275) : 764-766, 1978.
62. Lefer AM, Smith EF, Araki H, Smith JB, Aharony D, Claremon DA, Magolda RL, Nicolau KC: Dissociation of vasoconstrictor and platelet aggregatory activities of thromboxane by carbocyclic thromboxane A_2, a stable analog of thromboxane A_2. Proc Natl Acad Sci (77) : 1706-1710, 1980.
63. Marcelli N, Magnio O, Gandolfi C, Grossoni M, Ceseroni R: 11- and 15- Deoxy-carboprostacyclins: hypothetic prostacyclin antagonists. Adv PG Tx LT Res (11) : 281-285, 1983.
64. Whittle BJR, Moncada S, Whiting F, Vane JR: Carbacyclin - a potent, stable prostacyclin analogue for the inhibition of platelet aggregation. Prostaglandins (19) : 605-627, 1980.
65. Aristoff PA, Harrison AW, Aiken JW, Gorman RR, Pike JE: Synthesis and structure-activity relationships of novel stable prostacyclin analogs. Adv PG Tx LT Res (11) : 267-274, 1983.
66. Jones RL: Definition of prostaglandin-sensitive arterial constrictor systems. Acta Biol Med Germ (37) : 837-844, 1978.
67. Fufushima M, Kato T, Ota K, Arai Y, Narumiya S, Hayaishi H: 9-Deoxy 9- prostaglandin D_2, a prostaglandin D_2 derivative with potent anti-neoplastic and weak smooth muscle contracting activities. Biochem Biophys Res Commun (109) : 626-633, 1982.
68. Ceserani R, Gandolfi C, Longiave D, Mandelli K: A new selective antagonist of prostaglandins: 20-methyl-13,14-didehydro-$PGF_{2\alpha}$. PG Med (2) : 455-457, 1979.
69. McCracken JA, Glew ME, Hull SS, Boraird L, Underwood L, Fried J: Comparative luteolytic effects of $PGF_{2\alpha}$ and its 13-dehydro analogs in vivo. Adv PG Tx Res (6) : 365-380, 1980.
70. Steinberg D, Vaughan M, Nestel PJ, Bergstrom S: Effects of prostaglandins on hormone-induced mobilization of free fatty acids. J Clin Invest (43) : 1533-1540, 1963.
71. Butcher RW, Baird CE: Effects of prostaglandins on adenosine 3'5'-monophosphate levels in fat and other tissues. J Biol Chem (243) : 1713-1717, 1968.
72. Gilman AG: Guanine nucleotide-binding regulatory proteins and dual control of adenylate cyclase. J Clin Invest (73) : 1-4, 1984.

73. Kuehl FA: Prostaglandins, cyclic nucleotides and cell function. Prostaglandins (5) : 325-340, 1974.
74. Asano T, Hidaka H: Purification of guanylate cyclase from human platelets and effect of arachidonic acid peroxide. Biochem Biophys Res Commun (78) : 910-918, 1977.
75. Raymond V, Leung PCK, Labrie F: Stimulation by $PGF_{2\alpha}$ of phosphatidic acid-phosphatidylinositol turnover in rat luteal cells. Biochem Biophys Res Commun (116) : 39-46, 1983.
76. Macphee CH, Drummond AH, Otto AM, de Asua LJ: Prostaglandin $F_{2\alpha}$ stimulates phosphatidylinositol turnover and increases the cellular content of 1,2-diacylglycerol in confluent resting Swiss 3T3 cells. J Cell Physiol (119) : 35-40, 1984.
77. Mills DCB, Macfarlane DE: Stimulation of human platelet adenylate cyclase by prostaglandin D_2. Thromb Res (5) : 406-412, 1974.
78. Miller OV, Gorman RR: Evidence for distinct prostaglandin I_2 and D_2 receptors in human platelets. J Pharmac Exp Ther (210) : 134-140, 1979.
79. Vesin MF, Khac LD, Harbon S: Prostacyclin as an endogenous modulator of adenosine cyclic 3',5'-monophosphate levels in rat myometrium and endometrium. Mol Pharmacol (16) : 823-840, 1979.
80. Miller OV, Johnson RA, Gorman RR: Inhibition of PGE_1-stimulated cAMP accumulation in human platelets by thromboxane A . Prostaglandins (13) : 599-609, 1977.
81. Best LC, McGuire MB, Martin TJ, Preston FE, Russell RGG: Effects of epoxymethano analogues of prostaglandin endoperoxides on aggregation, on release of 5HT and on the metabolism of cyclic AMP and cyclic GMP in human platelets. Biochim Biophys Acta (583) : 344-351, 1979.
82. Pollock WK, Armstrong RA, Brydon LJ, Jones RL, MacIntyre DE: Thromboxane-induced phosphatidate formation in human platelets: relationship to receptor occupancy and to changes in cytosolic free calcium. Biochem J in press, 1984.

16

LEUKOTRIENES: BIOLOGICAL PROPERTIES, EVIDENCE FOR SPECIFIC RECEPTOR SITES AND EVIDENCE FOR THE INVOLVEMENT OF LEUKOTRIENES IN PATHOLOGICAL SITUATIONS

A.W. FORD-HUTCHINSON

1. INTRODUCTION

The leukotrienes are a novel group of mediators derived from the metabolism of arachidonic acid through the action of the 5-lipoxygenase enzyme pathway (1,2). The initial step in this pathway involves insertion of molecular 0_2 at position C5 to produce 5-hydroperoxyeicosatetraenoic acid. This is followed by a dehydration step to produce an unstable epoxide intermediate (leukotriene A_4). There are two enzymic routes of metabolism from leukotriene A_4. The first involves a stereospecific hydrolysis by an epoxide hydrolase (LTA_4 hydrolase) to produce a dihydroxyeicosatetraenoic acid (Leukotriene B_4) with a precise stereochemistry (5[S],12[R]-dihydroxy-6,14-cis-8,10-trans eicosatetraenoic) (3,4). Metabolism of leukotriene B_4 involves ω-oxidation to produce 20 OH-leukotriene B_4 and 20 COOH-leukotriene B_4 (5). This metabolism results in loss of biological activity (6,7). The second route of metabolism of leukotriene A_4 involves the enzymic addition of glutathione to produce the peptidolipid conjugate, leukotriene C_4. Subsequent metabolism of leukotriene C_4 involves loss of glutamic acid to produce leukotriene D_4, loss of glycine to produce leukotriene E_4 and readdition of glutamic acid to produce leukotriene F_4 (1,2). Leukotriene C_4 and D_4 and to a lesser extent leukotriene E_4 collectively account for the biological activity known as slow reacting substance of anaphylaxis. Detailed, stereoselective routes of synthesis have been described for all the leukotrienes (8,9,10).

2. BIOLOGICAL PROPERTIES OF LEUKOTRIENE B_4

Leukotriene B_4 is a potent stimulator of leukocyte function in vitro (11). Thus it has been shown to stimulate the chemotaxis, chemokinesis and aggregation of polymorphonuclear leukocytes (12,13,14). Associated with these events was a stimulation of Ca^{2+} and Na^+ fluxes (15), enhancement of C3b receptor site expression (16), elevation of intracellular cAMP levels (17) and enhancement of the release of lysosomal enzymes and the production of superoxide anions (14,18). These effects on neutrophil function occured at very low concentrations and were similar to those seen with other potent chemotactic factors such as the complement-derived peptide, C5a, and the synthetic peptide, formyl-methionyl-leucyl-phenylalanine. In addition to effects on polymorphonuclear leukocytes, leukotriene B_4 also was a potent chemotactic and chemokinetic agent for eosinophils and enhanced C3b receptor site expression in these cells, thus influencing complement-dependent killing of Schistosoma mansoni (16,19). Finally leukotriene B_4 has been reported to stimulate the chemokinesis of macrophages and monocytes (13) and the chemotaxis of fibroblasts (20).

There are a number of pieces of evidence which have indicated that the potent effects of leukotriene B_4 on leukocyte function are mediated through interactions with specific, high affinity receptor sites. First, the effects of leukotriene B_4 on polymorphonuclear leukocyte function were highly structurally specific and any changes in either the triene structure or the configuration of the hydroxyl groups resulted in dramatic losses of biological activity (7,11). Secondly, it has been suggested that 5[S],12[S], dihydroxy-5-trans,8-cis,10-trans,14-cis-eicosatetraenoic acid and leukotriene B_4-dimethylamide antagonized the biological actions of leukotriene B_4. Finally, direct radioligand binding studies have been carried out using [^{3}H]-leukotriene B_4. Binding analysis indicated the expression of two classes of receptor sites for leukotriene B_4 on human polymorphonuclear leukocytes with respective mean affinities of

3.9×10^{-10}M and 6.1×10^{-8}M and mean densities of 4.4×10^{3} and 2.7×10^{5} sites per cell. The presence of high affinity receptor sites was also directly correlated with the chemotactic response (23,24). Thus deactivation of polymorphonuclear leukocytes by a prior exposure to leukotriene B_4 resulted in a loss of both high affinity binding sites and a reduction in the chemotactic response. Also, structural isomers of LTB_4 inhibited the binding of [^{3}H]-leukotriene B_4 to polymorphonuclear leukocytes at concentrations similar to those required to elicit chemotaxis. Other chemotactic factors did not interfere with the binding. Thus leukotriene B_4 has receptor sites which are similar to, but distinct from those for other chemotactic factors, such as formyl-methionyl-leucyl-phenylalanine. For this synthetic peptide, the high affinity receptor sites have been related to chemotaxis and the low affinity receptor sites to induction of enzyme release.

In addition to the effects of leukotriene B_4 on neutrophil function, evidence has now accumulating that this leukotriene may have effects on lymphocyte function which also may be mediated through interaction with receptor sites (25). Thus, fluorescence activated cell sorting studies have been carried out using as a probe, leukotriene B_4 coupled to fluorescein-labelled bovine serum albumin (25). Approximately 11% of human blood T-lymphocytes were shown to bind leukotriene B_4 using this probe. The leukotriene B_4-reactive T-lymphocytes were distributed between the helper-inducer (8%) subsets and the suppressor-cytotoxic (14%) subsets. Functional consequences of these interactions were also demonstrated and leukotriene B_4 was shown to enhance significantly the proliferation of mitogen-stimulated suppressor-cytotoxic T-lymphocytes (10^{-12} - 10^{-6}M) and to inhibit significantly the proliferation of mitogen-stimulated helper-inducer T-lymphocytes (10^{-8} - 10^{-6}M) (25). The net effect of leukotriene B_4 on blood T-lymphocytes was an immunosuppression. In addition, other studies have been carried out and leukotriene B_4 has been reported to induce one class of human suppressor cells through a cyclooxygenase-dependent mechanism (26), to stimulate the activity of human natural cytotoxic cells in mixed human lymphocyte preparations (27) and to replace either the interleukin-2 or helper cell requirement for γ-interferon production by murine Lyt-1^-2^+ cells (28). These preliminary results obtained in vitro suggest that leukotriene B_4, in addition to its potent effects on neutrophil function, may also have important actions upon lymphocyte activation and function.

Many of the in vitro effects of leukotriene B_4 were also observed in vivo. Thus leukocyte accumulation was induced following injection of leukotriene B_4 into rabbit skin (29), the rabbit eye (30), the guinea-pig peritoneal cavity (13) and human skin (29,31). In addition leukotriene B_4 induced vascular permeability changes in the skin of the rabbit and the pig, these responses being potentiated by co-administration of a vasodilator prostaglandin (32, Chan and Ford-Hutchinson, unpublished observations). In these experiments dermal blood flow immediately following intradermal injection of leukotriene B_4 was significantly enhanced in the pig but was unchanged in the rabbit. In the rabbit it has been shown that the vascular permeability changes induced by leukotriene B_4 were dependent of the presence of neutrophils (33). It has been proposed that neutrophils adhere to the vascular endothelium in the presence of leukotriene B_4 or other chemotactic factors, inducing changes in vascular permeability through a secondary mechanism, probably involving the release of lysosomal enzymes (33). In the rabbit skin this increase in vascular permeability is only expressed in the presence of a vasodilator such as prostaglandin E_1. Leukotriene B_4 may also have effects on pain responses and following injection into the paw of the rat, leukotriene B_4 induced a slight hyperalgesia similar to that observed with prostaglandin E_1 (34). Both leukotriene B_4 and prostaglandin E_1 were shown to prevent the initial hypoalgesia and significantly reduce the latency of development of yeast-induced hyperalgesia in the rat paw (34). There was no synergy between the two agents suggesting a common mechanism of action.

3. BIOLOGICAL ACTIVITIES OF LEUKOTRIENE C_4 AND D_4

Leukotriene C_4 and D_4, and to a lesser extent the other peptidolipid leukotrienes, account for the biological activity known as slow reacting substance of anaphylaxis (1,2,35). These peptidolipid leukotienes have potent actions on a variety of smooth muscle preparations such as those derived from the respiratory tract (36,37,38), the vasculature (35), the gastrointestinal tract (38) and the gall bladder (40). There are considerable species variations in the responses of smooth muscle preparations to leukotrienes. Thus leukotrienes were potent contractile agents upon human (41) and guinea-pig trachea but had little or no action upon the trachea of the rat, cat or dog (42). However, with regard to the human respiratory tract leukotriene C_4 and D_4 are potent contractile agents on human bronchus (36), human parenchymal strips and human trachea (41) in vitro and when given by aerosol in vivo to normal volunteers induce prolonged bronchoconstriction at low doses (43,44). Leukotriene C_4 and D_4 also have potent effects in the cardiovascular system and were potent vasoconstrictors in a number of vascular beds. As with respiratory smooth muscle there are profound species variations (39,45). In addition to these direct effects on smooth muscle, leukotriene C_4 and D_4 when administered to man or to animal preparations may exert indirect pharmacological effects mediated through the production of other mediators. Of particular interest are effects on the secretion of mucous in the respiratory tract (46), cardiovascular effects, such as negative inotropic (47) and vascular dilator actions and increases in dermal blood flow which have been shown after intradermal injections in human skin (48).

The structural specificity and the high potency of peptidolipid leukotrienes on smooth muscle preparations are indicative of interactions with specific receptor sites. Studies using the leukotriene antagonist, FPL-55712, have suggested that there are multiple receptor sites for leukotrienes on smooth muscle preparations (49,50,51). In particular, it has been proposed that there are separate leukotriene C_4 and D_4 receptor sites. Binding studies with [^{3}H]-leukotrienes have confirmed the presence of multiple receptor sites. In guinea-pig lung, separate receptor sites for leukotriene D_4 and C_4 have been defined and affinity constants obtained (52). Studies with the leukotriene antagonist, FPL-55712, have shown that it has a high affinity for the leukotriene D_4 receptor site which is consistent with its more effective antagonism of leukotriene D_4-induced responses on guinea-pig pulmonary smooth muscle preparations. The binding of leukotriene D_4 to its receptor site is inhibited by Na^+ and GTP suggesting that the action of leukotriene D_4 as associated with a depression in intracellular cyclic AMP levels. In addition to the studies on guinea-pig lung (53), a leukotriene C_4 receptor site has been described in rat lung (54), multiple leukotriene C_4 receptor sites described in guinea-pig ileal longtitudinal smooth muscle and separate leukotriene C_4 and D_4 receptor sites defined in guinea-pig brain homogenates (55).

4. EVIDENCE FOR LEUKOTRIENE INVOLVEMENT IN PATHOLOGICAL PROCESSES

There have been two major problems that have prevented an evaluation of the role of leukotrienes in pathological processes. These were; first the lack of any specific antagonists of the actions or inhibitors of the synthesis of leukotrienes which could be used in vivo in man. Secondly, there were problems in developing specific and sensitive assays for determining the levels of leukotrienes in biological fluids. Recently, specific and sensitive radioimmunoassays for peptidolipid leukotrienes and leukotriene B_4 have been developed and these can be used for measuring leukotrienes within biological fluids (56,57). Unfortunately, it appeared that leukotrienes were rapidly broken down within the circulatory system through extensive β-oxidation or else were rapidly excreted in the bile. Therefore there are no stable urinary or plasma metabolites of leukotrienes that can be measured to give an overall assessment of leukotriene release in vivo. However, evidence for leukotriene involvement in pathological processes has been obtained from the measurement of these compounds in other biological fluids and these studies are summarized below.

4.1 Leukotrienes and inflammation

In animal models inflammatory exudates may be readily produced in the rat by the implantation of polyester sponges. Using this technique, high levels of leukotriene B_4 have been detected in sponges previously soaked in various irritants prior to implantation in the rat (58,59). In particular, high levels of leukotriene B_4, were detected in exudates from sponges pretreated with zymosan, the presence of leukotriene B_4 being confirmed by high pressure liquid chromatographic analysis (59). In time course studies, using zymosan-impregnated sponges, the maximal rate of leukocyte infiltration was found to coincide with the maximal levels of leukotriene B_4-immunoreactivity. In these inflammatory exudates there was no evidence for significant levels of peptidolipid leukotrienes (59). As a result of a number of studies with various irritants and different drugs it was concluded that leukotriene B_4 can induce leukocyte migration into sponge exudates in the rat but that measurements of leukotriene B_4 in such exudates could not be correlated with the degree of leukocyte infiltration (59). In man, leukotriene B_4 has been reported to be present within synovial fluids from patients with inflammatory diseases (60). In particular, increased levels were observed in patients with gouty arthritis (61) and in this context uric acid crystals have been shown in vitro to be a good stimuli for the production of leukotriene B_4 by human polymorphonuclear leukocytes (62).

4.2 Leukotriene release following antigen challenge

In vitro studies carried out with lung slices obtained from actively or passively sensitized guinea pigs or rats challenged with antigen have demonstrated the release of both leukotrienes C_4, D_4, and B_4. The release is related to the dose of IgE and in rats an increase in production of leukotrienes was observed in rats with non-specific bronchial hyperreactivity (Brunet and Ford-Hutchinson, unpublished observations). Leukotrienes have been implicated in the bronchial contraction produced in human asthmatic bronchi following exposure to specific allergen (63). Purified human lung mast cells have been shown to release leukotrienes following allergen challenge (64).

Preliminary studies in which leukotriene C_4 levels were determined in plasma from either human asthmatic patients following antigen challenge (E. Hayes, unpublished observations) or rat plasma following antigen challenge (Hamel and Ford-Hutchinson, unpublished observations) have failed to show significant increases in the levels of leukotriene C_4-immunoreactivity. However, significant increases in the levels of leukotriene C_4-immunoreactivity have been determined by the radioimmunoassay in tear fluid obtained from allergic subjects challenged with specific allergen directly in the eye (65). The levels of leukotriene C_4-immunoreactivity in allergen-challenged tear fluid were significantly higher than those in control tear fluid. These effects were not due to non-specific irritation because subjects challenged with ammonia showed no increase in leukotriene C_4-immunoreactivity in tear fluid compared to normal control tear fluid (Bisgaard and Ford-Hutchinson, unpublished observations). The levels of leukotriene C_4-immunoreactivity in tear fluid were significantly reduced by pretreatment with topical sodium cromoglycate demonstrating that it is possible to reduce the levels of leukotrienes by therapeutic intervention in man. In human skin evidence for the release of leukotrienes following allergen challenge has also been obtained either in atopic dermatitis or in the Praüsnitz-Küstner reaction in normal volunteers (Bisgaard and Ford-Hutchinson, unpublished observations).

4.3 Leukotrienes and inflammatory bowel disease

Sharon and Stenson (66) have studied colonic mucosa from patients with inflammatory bowel disease. Diseased colonic mucosa when incubated with ionophore A23187 converted 2.2% of added arachidonic acid to leukotriene B_4 as opposed to 0.4% in normal mucosa. Direct extraction of inflammatory bowel disease mucosa indicated the presence of leukotriene B_4 whereas analysis of normal mucosa showed that leukotriene B_4 levels in this tissue were below the detection limit of the assay. The authors also suggested that sulfasalazine, a drug known to be effective in this disease, could inhibit the production of leukotriene B_4 by inflammatory bowel disease mucosa. A problem with these studies was the presence of large amounts of neutrophils within diseased mucosa. It is therefore difficult to ascertain whether the changes simply represented influx of neutrophils or were related in some way to the etiology of the disease. However, the studies suggested that leukotriene B_4 may be an important mediator of inflammation in inflammatory bowel disease.

4.4 Leukotrienes and psoriasis

There is now a considerable body of evidence to suggest that leukotrienes and other lipoxygenase products may have a role in skin diseases such as psoriasis. The evidence is as follows: 1) topical application of leukotriene B_4 to normal human skin resulted in the induction of intraepidermal micro-abscesses (67). These abscesses were similar to lesions observed in pustular psoriasis. 2) neutrophil accumulation is an important aspect of the pathology of this skin disease and has been reported to be one of the earliest pathological events occurring prior to the development of a prepapillilary lesions (68). 3) leukotriene B_4 can be produced by human keratinocytes following stimulation with the calcium ionophore A23187 (69). 4) human keratinocytes were stimulated to proliferate when exposed to leukotrienes B_4, C_4 and D_4 in vitro (Kragballe and Vorhees, results to be published). 5) leukotrienes C_4 and D_4 were potent vasodilators in human skin with a prolonged duration of action (48). Enhanced blood flow is another manifestation of psoriasis (70). 6) there are a number of pieces of evidence suggesting abnormalities in arachidonic acid metabolism in psoriasis. In particular, there were elevated levels of arachidonic acid as well as the lipoxygenase product 12-hydroxyeicosatetraenoic acid in involved psoriatic skin (71). With regard to leukotrienes, elevated levels of both leukotriene B_4 and C_4 have been recovered from lesional skin in patients with psoriasis (72). In addition, leukotrienes B_4 and C_4 have been detected in psoriatic scales (results to be published).

These observations, together with suggestions that drugs that are effective in psoriasis may have some effects on the production of leukotrienes, have been put forward to suggest that leukotrienes may be an important mediators of this human skin disease (73,74).

5. CONCLUSIONS

There are specific, high affinity receptor sites for leukotrienes on a number of tissues. Interaction with these receptor sites leads to the induction of a number of important biological actions. Because of the potency of these actions it has been suggested that leukotrienes may be important pathological mediators in human disease. There is now some evidence for leukotriene release during pathological processes in man. However, conclusive proof of the involvement of leukotrienes in human disease will be dependent on the development and use of specific inhibitors of the production of leukotrienes or antagonists of their action in human disease.

References

1. Samuelsson B: The leukotrienes: an introduction. In: Samuelsson B, Paoletti, R (eds) Adv in Prostaglandin, Thromboxane Leukotriene Res. Raven Press, New York, 1982, vol. 9, pp 1-17.

2. Samuelsson B: Leukotrienes: a new class of mediators of immediate hypersensitivity reactions and inflammation. Adv Prostaglandin Thromboxane Leukotriene Res. Raven Press, New York, 1983, vol. 11, pp 1-13.

3. Borgeat P, Samuelsson B: Arachidonic acid metabolism in polymorphuclear leukocytes: Effects of ionophore A23187. Proc Natl Acad Sci USA (76): 2148-2152, 1979.

4. Borgeat P, Samuelsson B: Transformations of arachidonic acid by rabbit polymorphonuclear leukocytes. Formation of novel dihydroxyeicosatetraenoic acids. J Biol Chem (254): 2643-2646, 1979.

5. Powell WS: Properties of leukotriene B_4 20-hydroxylase from polymorphonuclear leukocytes. J Biol Chem (259): 3082-3089, 1984.

6. Camp RDR, Wollard PM, Mallet AJ, Fincham NJ, Ford-Hutchinson AW, Bray MA: Neutrophil aggregating and chemokinetic properties of 5,12,20-trihydroxy-6,8,10,14-eicosatetraenoic acid isolated from human leucocytes. Prostaglandins (23): 631-641, 1982.

7. Ford-Hutchinson AW, Rackham A, Zamboni R, Rokach J, Roy S: Comparative biological activities of synthetic leukotriene B_4 and its ω-oxidation products. Prostaglandins (25): 29-37, 1983.

8. Rokach J, Girard Y, Guindon Y, Atkinson JG, Larue M, Young RN, Masson P and Holme G: The synthesis of leukotrienes. In: Progress in lipid research. 1981, (20), pp 905-907.

9. Guindon Y, Zamboni, R, Lau CK, Rokach J: Stereospecific synthesis of leukotriene B_4 (LTB_4). Tet Letts (23): 739-742, 1982.

10. Green RH, Lambeth PG: Leukotrienes. Tet Letts (39): 1687-1721, 1983.

11. Ford-Hutchinson AW: The role of leukotriene B_4 as a mediator of leukocyte function. In: Keller HU, Till GO (eds) Leukocyte locomotion and chemotaxis. Basel, Birkhäuser Verlag, 1983, pp 154-164.

12. Ford-Hutchinson AW, Bray MA, Doig MV, Shipley MF, Smith MJH: Leukotriene B: A potent chemokinetic and aggregating substance released from polymorphonuclear leukocytes. Nature (286): 264-265, 1980.

13. Smith MJH, Ford-Hutchinson AW, Bray MA: Leukotriene B: a potential mediator of inflammation. J Pharm Pharmacol (32): 517-518, 1980.

14. Goetzl EJ, Pickett, WC: The human PMN leukocyte chemotactic activity of complex hydroxy-eicosatetraenoic acids (HETEs). J Immunol (125): 1789-1791, 1980.

15. Molski TFP, Naccache PH, Borgeat P and Shaafi RI: Similarities in the mechanisms by which formyl-methionyl-leucyl-phenylalanine, arachidonic acid and leukotriene B_4 increase calcium and sodium influxes in rabbit neutrophils. Biochem Biophys Res Commun (103): 227-232, 1981.

16. Nagy L, Lee TH, Goetzl EJ, Pickett WC, Kay AB: Complement receptor enchancement and chemotaxis of human neutrophils and eosinophils by leukotrienes and other lipoxygenase products. Clin Exp Immunol (47): 541-547, 1982.

17. Claesson H: Leukotrienes A_4 and B_4 stimulate the formation of cyclic AMP in human leukocytes. FEBS letter (139): 305-308, 1982.

18. Rae SA, Smith MJH: The stimulation of lysosomal enzyme secretion from human polymorphonuclear leucocytes by leukotriene B_4. J Pharm Pharmac (33): 616-617, 1981.

19. Moqbel R., Sass-Kuhn SP, Goetzl EJ, Kay AB: Enhancement of neutrophil- and eosinophil-mediated complement-dependent killing of Schistosomula of Schistosoma Mansoni in vitro by leukotriene B_4. Clin Exp Immunol (52): 519-527, 1983.

20. Mensing H, Czarnetzki BM: Leukotriene B_4 induces in vitro fibroblast chemotaxis. J Invest Dermatol (82): 9-12, 1984.

21. Feinmark SJ, Lindgren JA, Claesson HE, Malmsten C, Samuelsson B: Stimulation of leukocyte degranulation by leukotriene B_4 and its ω-oxidized metabolites. Febs Lets (136): 141-144, 1981.

22. Showell HJ, Otterness IG, Marfat A, Corey EJ: Inhibition of leukotriene B_4-induced neutrophil degranulation by leukotriene B_4-dimethylamide. Biochem Biophys Res Commun (106): 741-747, 1982.

23. Goldman DW, Goetzl EJ: Specific binding of leukotriene B_4 to receptors on human polymorphonuclear leukocytes. J Exp Med (129): 1600-1604, 1982.

24. Goldman DW, Goetzl EJ: Heterogeneity of human polymorphonuclear leukocyte receptors for leukotriene B_4. Identification of a subset of high affinity receptors that transduce the chemotactic response. J Exp Med (159): 1027-1041, 1984.

25. Payan DG, Missirian-Basteian A, Goetzl EJ: Human T-lymphocyte subset specificity of the regulatory effects of leukotriene B_4. Proc Natl Acad Sci USA, In press.

26. Rola Pleszczynski M, Borgeat P, Sirois P: Leukotriene B_4 induces human suppressor lymphocytes. Biochem Biophys Res Commun (108): 1531-1537, 1982.

27. Rola-Pleszczynski M, Gagnon L, Sirois P: Leukotriene B_4 augments human natural cytotoxic cell activity. Biochem Biophys Res Commun (113): 531-537.

28. Johnson HM, Torres BA: Leukotrienes: positive signals for regulation of γ-interferon production. J Immunol (132): 413-416, 1984.

29. Bray, MA, Ford-Hutchinson AW, Smith MJH: Leukotriene B_4: an inflammatory mediator _in vivo_. Prostaglandins (22): 213-222, 1981.

30. Bhattacherjee P, Hammond B, Salmon JA, Stepney R , Eakins KE: Chemotactic response to some arachidonic acid lipoxygenase products in the rabbit eye. Eur J Pharmacol (73): 21-28, 1981.

31. Camp RDR, Coutts AA, Greaves MW, Kay AB, Walport MJ: Responses of human skin to intradermal injection of leukotrienes C_4, D_4 and B_4. Br J Pharmacol (80): 497-502, 1983.

32. Bray MA, Cunningham FM, Ford-Hutchinson AW, Smith MJH: Leukotriene B_4: a mediator of vascular permeability. Br J Pharmacol (72): 483-486, 1981.

33. Wedmore CV, Williams TJ: The control of vascular permeability by polymorphonuclear leukocytes in inflammation. Nature (284): 646-650, 1980.

34. Rackham A, Ford-Hutchinson AW: Inflammation and pain sensitivity: effects of leukotrienes D_4, B_4 and prostaglandin E_1 in the rat paw. Prostaglandins (25): 193-203, 1983.

35. Piper PJ: Formation and action of leukotrienes. Physiol Rev. (64): 744-761, 1984.

36. Dahlen SE, Hedqvist P, Hammarström S, Samuelsson B: Leukotrienes are potent constrictors of human bronchi. Nature (228): 484-486, 1980.

37. Lewis RA, Austen KF, Drazen JM, Clark DA, Marfat A, Corey EJ: Slow reacting substances of anaphylaxis: identification of leukotrienes C-1 and D from human and rat sources. Proc Natl Acad Sci USA (77): 3710-3714, 1980.

38. Piper PJ, Samhoun MN, Tippins JR, Williams TJ, Palmer MA, Peck MJ: Pharmacological studies on pure SRS-A, and synthetic leukotrienes C_4 and D_4. In: Piper PJ (ed) SRS-A and Leukotrienes. New York: Wiley, 1981, pp 81-99.

39. Letts LG, Piper PJ, Newman DL: Leukotrienes and their action in the coronary circulation. In: Piper PJ (ed) Leukotrienes and other lipoxygenase products. New York: Wiley, 1983, pp 94-107.

40. Yusko P, Hall RA, Ford-Hutchinson AW: Contraction of guinea-pig bladder strips by leukotrienes and other agonists. Prostaglandins (25): 397-403, 1983.

41. Jones TR, Davis C, Daniel EE: Pharmacological study of the contractile activity of leukotriene C_4 and D_4 on isolated human airway smooth muscle. Can J Physiol Pharmacol (60): 638-643, 1982.

42. Krell RD, Osborn R, Vickery L, Falcone K, O'Donnell M, Gleason J, Kinzig C, Bryan D: Contraction of isolated airway smooth muscle by synthetic leukotrienes C_4 and D_4. Prostaglandins (22): 387-409, 1981.

43. Weiss JW, Drazen JM, Coles N, McFadden ER, Weller PW, Corey EJ, Lewis RA, Austen KF: Bronchoconstrictor effects of leukotriene C in humans. Science (216): 196-198, 1982.

44. Barnes N, Costello JF, Piper PJ: The comparative bronchoconstrictor actions of leukotrienes C_4, D_4 and histamine in normal human volunteers. Thorax, in press.

45. Burke JA, Levi R, Guo Z, Corey EJ: Leukotrienes C_4, D_4 and E_4: effects on human and guinea pig cardiac preparations in vitro. J Pharm Exp Therap (221): 235-241, 1982.

46. Richardson PS, Peatfield AC, Jackson DM, Piper PJ: The effect of leukotrienes on the output of mucine from the cat trachea. In: Piper PJ (ed) Leukotrienes and Other Lipoxygenase Products. New York: Wiley, 1983, pp 178-187.

47. Michelassi F, Landa L, Hill RD, Lowenstein E, Watkins WD, Petkau AJ, Zapol WM: Leukotriene D_4: a potent coronary artery vasoconstrictor associated with imparied ventricular contraction. Science (217): 841-843, 1982.

48. Bisgaard H, Kristensen J, Sondergaard JE: The effect of leukotriene C_4 and D_4 on cutaneous blood flow in humans. Prostaglandins (23): 797-801, 1982.

49. Fleisch JH, Rinkema LE, Baker SR: Evidence for multiple leukotriene D_4 receptors in smooth muscle. Life Sci (31): 577-581, 1982.

50. Krell RD, Tsai BS, Berdoulay A, Barone M, Giles RE: Heterogeneity of leukotriene receptors in guinea pig trachea. Prostaglandins (25): 171-178, 1983.

51. Jones T, Denis D, Hall R, Ethier D: Pharmacological study of the effects of leukotrienes C_4, D_4, E_4 and F_4 on guinea pig trachealis: interaction with FPL-55712. Prostaglandins (26): 833-843, 1983.

52. Pong S, DeHaven RN: Characterization of a leukotriene D_4 receptor in guinea pig lung. Proc Natl Acad Sci USA (80): 7415-7419, 1983.

53. Pong S, DeHaven RN, Kuehl FA, Egan RW: Leukotriene C_4 binding to rat lung membranes. J Biol Chem (258): 9816-9819, 1983.

54. Nicosia S, Crowley HJ, Oliva D, Welton AF: Binding sites for ^{3}H-LTC$_4$ in membranes from guinea pig ileal longtitudinal muscle. Prostaglandins (3): 483-494, 1984.

55. Cheng, JB, Townley RG: Effect of the serine-borate complex on the relative ability of leukotriene C_4, D_4 and E_4 to inhibit lung and brain [^{3}H] leukotriene D_4 and [^{3}H] leukotriene C_4 binding: demonstration of the agonists potency order for leukotriene D_4 and leukotriene C_4 receptors. Biochem Biophys Res Commun, (119): 612-617, 1984.

56. Hayes EC, Lombardo DL, Girard Y, Maycock AL, Rokach J, Rosenthal AS, Young RN, Egan RW and Zweerink HJ: Measuring leukotrienes of slow-reacting substance of anaphylaxis: development of a specific radioimmunoassay. J Immunol (131): 429-433, 1983.

57. Rokach J, Hayes EC, Girard Y, Lombardo DL, Maycock AL, Rosenthal AS, Young RN, Zamboni R, Zweerink HJ: The development of sensitive and specific radioimmunoassays for leukotrienes. Prostaglandins Leukotrienes and Med (13): 21-25, 1984.

58. Simmons PA, Salmon JA, Moncada S: The release of leukotriene B_4 during experimental inflammation. Biochem Pharmacol (32): 1353-1359, 1983.

59. Ford-Hutchinson AW, Brunet G, Savard P, Charleson S: Leukotriene B_4, polymorphonuclear leukocytes and inflammatory exudates in the rat. Prostaglandins, in press.

60. Klickstein LB, Shapleigh C, Goetzl EJ: Lipoxygenation of arachidonic acid as a source of polymorphonuclear leukocyte chemotactic factors in synovial fluid and tissue in rheumatoid arthritis and spondyloarthritis. J Clin Invest (66): 1166-1170, 1980.

61. Rae SA, Davidson EM, Smith MJH: Leukotriene B_4, an inflammatory mediator in gout. Lancet (11): 1122-1123, 1983.

62. Serhan CN, Lundberg U, Weissman G, Samuelsson B: Formation of leukotrienes and hydroxy acids by human neutrophils and platelets exposed to monosodium urate. Prostaglandins (27): 563-581, 1984.

63. Dahlén S, Hansson G, Hedqvist P, Björk, T, Granström E, Dahlén B: Allergen challenge of lung tissues from asthmatics elicits bronchial contraction that correlates with the release of leukotrienes C_4, D_4 and E_4. Proc Natl Acad Sci USA (80): 1712-1716, 1983.

64. MacGlashan DW, Schleimer RP, Peters SP, Schulman ES, Adams GK, Newball HH, Lichtenstein LM: Generation of leukotrienes by purified human lung mast cells. J Clin Invest (70): 747-751, 1982.

65. Bisgaard H, Ford-Hutchinson AW, Charleson S, Taudorf E: Detection of leukotriene C_4-like immunoreactivity in tear fluid from subjects challenged with specific allergen. Prostaglandins (27): 369-375, 1984.

66. Sharon P, Stenson WF: Production of leukotrienes by colonic mucosa from patients with inflammatory bowel disease. Gastroenterology (86): 453-460, 1984.

67. Camp R, Jones RR, Brian S, Wollard P, Greaves M: Production of intraepidermal micro-abscessess by topical application of leukotriene B_4 - a potential experimental model of psoriasis. J Invest Dermatol (82): 202-204, 1984.

68. Chowaniec O, Jablonska S, Beutner EH, Proniewska M, Jarzabek-Cherzelska M, Rzesa GE: Earliest clinical and histological changes in psoriasis. Dermatologica (163): 42-51, 1981.

69. Brain SD, Camp RDR, Leigh IM, Ford-Hutchinson AW: Synthesis of leukotriene B_4-like material by cultured human keratinoctes. J Invest Dermatol (78): 328, 1982.

70. Klemp P, Staberg B: Cutaneous blood flow in psoriasis. J Invest Dermatol (81): 503-506, 1983.

71. Hammarström S, Hamberg M, Samuelsson B, Duell EA, Stamski M, Voorhees JJ: Increased concentrations of free arachidonic acid, prostaglandins, E_2, $F_{2\alpha}$ and of 12 L-hydroxy-5,8,10,14-eicosatetraenoic acid (HETE) in epidemis of psoriasis: evidence for perturbed regulation of arachidonic acid levels in psoriasis. Proc Natl Acad Sci USA (725): 5130-5134, 1975.

72. Brain SD, Camp RDR, Dowd RM, Kobza-Black A, Wollard PM, Mallet AI, Greaves MW: Psoriasis and leukotriene B_4. Lancet ii: 762-763, 1982.

73. Kragballe K, Voorhees, JJ: Arachidonic acid and leukotrienes in dermatology. J Invest Dermatol (81): 293-296, 1983.

74. Greaves MW: Neutrophil polymorphonuclears, mediators and the pathogenesis of psoriasis. Brit J Dermatol (109): 115-118, 1983.

17

CYCLIC NUCLEOTIDES

DOUGLAS L. ARCHER AND JOANN A. WESS

The advent of improved methods for detecting cyclic nucleotides led to associations of increased or decreased levels of cyclic adenosine monophosphate (cAMP) and cyclic guanosine monophosphate (cGMP) with particular disease states (reviewed by Coffey (1)). Likewise, the concept of the two cyclic nucleotides acting as opposing signals in cell regulation (2) led to further disease associations. Inevitably, then, research was carried out on the role of cyclic nucleotides in cancer processes. Clearly cGMP levels are elevated in some tumor types with or without a decrease in cAMP; the underlying mechanism for the shifted cGMP:cAMP ratio is unknown.

A growing body of evidence suggests a role for cGMP as a positive signal in the carcinogenic process. The studies of Kimura and Murad (3) indicated that the increased levels of cGMP observed in regenerating liver, fetal liver, and hepatoma cells were due to an increase in the ratio of particulate (membrane-bound) to soluble GTP pyrophosphate lyase (cyclizing) (guanylate cyclase). Thus, not only the increase in cGMP but also the intracellular location of its synthesis may be important in proliferative events.

Many of the studies on the role of cyclic nucleotides as second messengers in cell proliferation have been conducted in lymphocytes. Recently, Ananthakrishnan et al. (4) demonstrated (a) the ability of phytohaemagglutinin (PHA) to increase cGMP levels in resting lymphocytes and (b) the ability of isolated resting-lymphocyte nuclei to respond to cGMP (and/or calcium) with increased RNA synthesis rates through a modification of chain initiation and/or elongation. The ability of cGMP to stimulate incorporation of phosphate into specific

W.E.M. Lands (ed.), BIOCHEMISTRY OF ARACHIDONIC ACID METABOLISM.

nuclear acidic proteins of lymphocytes had been established by Johnson and Hadden (5). Studies on the T-lymphocyte-related gamma interferon system by Johnson et al. (6) revealed second messenger functions of cGMP unrelated to proliferation. The murine T-cell subpopulation Lyt 1^-2^+ is responsible for gamma interferon production in response to T-cell mitogens, interleukin 2 (IL2) being the membrane-level messenger molecule (product of Lyt 1^+2^- helper cell) required for interferon production. Cyclic GMP, or inducers of cGMP such as arginine vasopressin and the tumor promoter 12-O-tetradecanoylphorbol-13-acetate (TPA), were able to mimic IL2's activity, suggesting that cGMP may function as the second messenger for IL2 (6). The effect of cGMP or its inducers could be antagonized by cAMP or its inducers (6). More recently, Johnson and Torres (7) showed that leukotriene B_4, C_4, and D_4 were also able to mimic IL2's function in this well-defined system. More will be said about leukotrienes later in this chapter.

The carcinogenic process has been divided into multiple steps termed initiation and promotion. Initiation is an irreversible genetic alteration of the cell's DNA by a dose of carcinogen which alone would not lead to tumor formation. The promotion phase is carried out by repetitive application of chemicals termed promoters, which are generally noncarcinogenic in themselves; the promotion process is partially reversible. The critical matter in initiation/promotion is the sequence of application: a single dose of initiator, followed by multiple doses of promoter. This process is defined separately from cocarcinogenesis, in which a chemical (cocarcinogen) may enhance the tumorigenicity of a carcinogen when applied as a single dose, before, at the same time as, or after the carcinogen (but not multiple, long-term application) to express its effect. The most thoroughly studied model of initiation/promotion is mouse skin, recently reviewed by Boutwell et al. (8), although the process has been shown to occur in other organ systems as well. The study of the initiation/promotion process, as well as specific inhibitors of the process, will be the focus of the remainder of this chapter with regard to the role of cyclic nucleotides in cancer.

In actuality, initiation/promotion in mouse skin is more complex than the description given above. Promotion has been further divided into at least two stages (reviewed by Slaga et al. (9)). The best characterized promoter is TPA, which is the standard by which most other compounds are judged. Mezerein, a diterpene similar to TPA, causes many of the same morphological and biochemical changes caused by TPA, but is a very weak promoter. However, when the initiator is followed first with one to four applications of TPA (which would result in no papillomas or carcinomas), and then with multiple applications of mezerein, the net result is the same as would have been obtained with TPA alone for the same time period (9). Various inhibitors, relatively specific for each stage of the promotion process, have been identified, such as protease inhibitors for stage I and retinoic acid for stage II. The induction of the enzyme ornithine decarboxylase (ODC) with resultant accumulation of the polyamines putrescine and spermidine is a biochemical process associated with stage II promotion (9). Cell proliferation is also association with stage II.

Several lines of evidence indicate a role for cyclic nucleotides in the promotion process, although the data are at times confusing and conflicting. Perchellet and Boutwell (10), using the mouse skin model, first suggested that cyclic nucleotides may regulate ODC activity and may be part of the molecular mechanism by which TPA induces ODC. In a follow-up paper, Perchellet and Boutwell (11) reported that papilloma formation in initiated mice was inhibited by applying cAMP or cGMP, or isobutylmethylxanthine (IBMX), a phosphodiesterase inhibitor which raises cyclic nucleotide levels, before applying promoter (TPA). This finding would not be unexpected for cAMP, which inhibits proliferation effectively (stage II promotion), but generally cGMP is associated with potentiating a variety of cellular activities. What is often overlooked is that addition of a single cyclic nucleotide (e.g., cGMP) to a system may lead to a rise in the counterpart cyclic nucleotide (e.g., cAMP) as the cell attempts to reestablish the normal ratio of cGMP to cAMP and maintain homeostasis. Data derived from investigations on the effects of TPA in both lymphocytes and mouse epithelium are fortunately more clear-cut. Using human peripheral blood lymphocytes (PBL), Coffey and Hadden (12) established that TPA elevates cGMP, guanylate cyclase activity, and cGMP phosphodiesterase activity without increasing cAMP or cAMP phosphodiesterase. Further work by Coffey and Hadden (13) demonstrated that (a) human PBL guanylate cyclase activities are increased in a dose-dependent fashion by TPA, and (b) the ability of a given member of the phorbol ester series to activate guanylate cyclase correlates well with its relative potency as a promoter and its ability to function as a PBL mitogen. At this point, it should be noted that calcium and other divalent cations play a critical role in the TPA-guanylate cyclase-cGMP story, but this role cannot possibly be addressed adequately in a chapter of this length. Those interested in this aspect of cyclic nucleotides will find ample material on calcium's role in the bibliography (12, 13, 14).

Like lymphocytes, increases in cGMP or at least increases in the cGMP:cAMP ratio have been reported in mouse skin treated with TPA. Belman and Troll (15) reported that treatment of mouse skin with TPA decreases cAMP and/or raises cGMP, and that other promoters such as phorbol dibenzoate and anthralin have the same effect. Furthermore, Belman and Troll (15) pointed out the vagaries of direct skin application of dibutyryl cGMP (DBcGMP) in that (a) cGMP is normally present intracellularly in concentrations one-tenth the amount of cAMP, and (b) at concentrations of DBcGMP (or cGMP) of 10^{-4}M, cAMP phosphodiesterase is inhibited, with an increase in cAMP as the net result. This may partly explain some of the discrepancies in the literature. Belman and Garte (16), extending these observations, reported that TPA causes a decreased β-hormone responsiveness of adenylcyclase in mouse epidermis, that increased cGMP levels are due to products of arachidonic acid metabolism, and that protease inhibitors suppress TPA-induced promotion by preventing activation of phospholipase A_2 and thus the formation of arachidonic acid. The increases in cGMP caused by promoters have also been reported for carcinogens. N-methyl-N-nitro-N-nitrosoguanidine, N-methylnitrosourea, hydrazine and 4-nitroquinoline-1-oxide increased guanylate cyclase activity and caused cGMP accumulation in liver and colonic mucosal tissues (17). Retinol and the antioxidants butylated hydroxyanisole (BHA) and butylated hydroxytoluene (BHT), all putative anticarcinogens, significantly suppressed the carcinogen-mediated increase in cGMP (17). The mechanism by which this occurs was discussed by Craven and DeRubertis (17) and several possibilities were suggested: (a) by anti-oxidant activity affecting generation of reactive intermediates; (b) by acting as free radical scavengers; (c) by blocking oxidation of critical tissue sites by reactive oxides of nitrogen generated from the parent compounds; (d) by blocking sulfhydryl groups required for guanylate cyclase activation; or (e) by direct suppression of guanylate cyclase.

The suppression of cGMP accumulation and carcinogenesis by antioxidants becomes important in light of work carried out with lymphocytes and mouse epidermis.

Studies conducted in our laboratory over a number of years showed that a variety of antioxidants inhibited several in vitro immune responses (18-22), but at that time, the underlying mechanism was uncertain. The same immunosuppressive antioxidants were studied by Blalock et al. (23) and shown to mimic the ability of murine α/β interferon to prevent clone formation of malignant cells in soft agar; human malignant cells were likewise affected in a non-cytotoxic fashion. Further parallels between the activities of phenolic antioxidants and virus-type interferon were detailed by Archer et al. (24), and the involvement of sulfhydryl groups on lymphocyte membranes was suggested. In 1981, Wess and Archer (25) reported that the immunosuppressive effects of BHA on the murine in vitro antibody response could be reversed by exogenous cGMP but not cAMP. Further studies by Wess and Archer (26) showed that the in vitro suppressive effects of methyl paraben, gallic acid, and BHT, like those of BHA, could also be reversed by exogenous cGMP, as could the B-cell subset-specific suppression caused by nordihydroguaiaretic acid (NDGA) (27), a reported inhibitor of lipoxygenase. These findings, coupled with the findings of Coffey and Hadden (13) relative to promoter strength being linked to cGMP-inducing ability of the phorbol ester series, led Wess and Archer (26) to conclude that the anti-carcinogenic activities of the phenolic antioxidants may lie in part in their ability to function as antipromoters owing to their interference with cGMP metabolism. Support for anti-oxidants functioning as antipromoters has subsequently been published. Slaga et al. (9) list BHA and BHT (also vitamin A derivatives as addressed by Craven and DeRubertis (17)) as inhibitors of TPA promotion in mouse skin. Nakadate et al. (28) reported that NDGA inhibited TPA-induced ODC activity (stage II-associated) in mouse epidermis. Kozumbo et al. (29) reported that BHA, BHT, and propyl gallate also prevented TPA-induced ODC activity in mouse skin. Likewise, Fischer et al. (30) concluded from their studies that lipoxygenase products of arachidonic acid metabolism appear to be important in tumor promotion.

The specific sites of antioxidant interference with arachidonic acid metabolism or the precise mechanism by which they interfere are unknown, but probably vary with the antioxidant, based on our experiences with in vitro systems (25, 26, 27). Our work also makes clear a role for sulfhydryl groups (19, 22, 24, 26, 27) and calcium (25, 26, 27), and a role for membrane sulfhydryl groups and calcium in guanylate cyclase activation is supported by a great number of literature reports.

One brief note about calcium: Verma and Boutwell (14) showed that calcium was required for ODC induction by TPA. Although the role of ODC inductions in promotion has been controversial, the inhibition of TPA-induced mouse skin tumor promotion by alpha-difluoromethylornithine, an irreversible inhibitor of ODC, would argue that induction of this enzyme and subsequent accumulation of polyamines must play an important role in the promotion process. A role for cGMP in ODC induction was discussed earlier in this chapter. Recently,

Levine (31) reported that the calcium ionophore A23187, a reported stage I tumor promoter and known cGMP activator, caused biosynthesis and release of leukotrienes and prostaglandins from rat basophil leukemia cells and that biosynthesis could be inhibited by anti-inflammatory drugs, antioxidants (BHA, NDGA), and calcium channel blockers. Leukotrienes are recognized as being mediators of immediate hypersensitivity and inflammation. Another mediator of immediate hypersensitivity, histamine, has recently been given an implied role in promotion and tumor survival (32). The antigen-induced, IgE-dependent release of mediators from human lung tissue was shown to result in simultaneous rises in both cGMP (14-fold) and cAMP (3-fold); exogenous histamine likewise produced a prompt increase in cGMP and cAMP (33). In both cases (antigen-induced and histamine-induced release of mediators) H-1 antagonists prevented cGMP accumulation, and H-2 antagonists prevented cAMP accumulation. A possible role in lung tumor promotion for these mediators, both histamine and the leukotrienes, should not be overlooked.

In summary, a role for cGMP in the carcinogenic process is strongly suggested by the literature. The basis for this may be loss of control of cGMP, translocation of guanylate cyclase to the membrane-bound form leading to hyper-reactivity of the cell, loss of ability to produce adequate cAMP, or some other as yet undisclosed mechanism. Perhaps more consideration should be given to cGMP metabolism in proposing chemotherapeutic approaches to cancer treatment.

References

1. Coffey RG: Assays for cyclic nucleotides including clinical application. In: Hadden JW, Coffey RG and Spreafico F (eds) Comprehensive immunology, Vol 3, Immunopharmacology. Plenum Medical Book Co., New York, 1977, pp 389-412.

2. Goldberg ND, Haddox MK, Durham E, Lopez C, Hadden JW: Evidence for opposing influences of cyclic GMP and cyclic AMP in the regulation of cell proliferation and other biological processes. In: Clarkson B and Baserga R (eds) Control of Proliferation in Animal Cells. Cold Spring Harbor Press, New York, 1974, pp 609-626.

3. Kimura H, Murad F: Increased particulate and decreased soluble guanylate cyclase activity in regenerating liver, fetal liver, and hepatoma. Proc Nat Acad Sci USA (72): 1965-1969, 1975.

4. Ananthakrishnan R, Coffey RG, Hadden JW: Cyclic GMP and calcium in lymphocyte activation by phytohaemagglutinin. Lymphocyte Differentiation (1): 183-196, 1981.

5. Johnson EM, Hadden JW: Phosphorylation of lymphocyte nuclear acidic proteins: regulation by cyclic nucleotides. Science (187): 1198-1200, 1975.

6. Johnson HM, Archer DL, Torres BA: Cyclic GMP as the second messenger in helper cell requirement for gamma-interferon production. J Immunol (129): 2570-2572, 1982.

7. Johnson HM, Torres BA: Leukotrienes: positive signals for regulation of gamma-interferon production. J Immunol (132): 413-416, 1984.

8. Boutwell RK, Verma AK, Ashendel CL, Astrup E: Mouse skin: a useful model system for studying the mechanism of chemical carcinogenesis. In: Hecker E, Fusenig NE, Kunz W, Marks F, Thielman HW (eds) Carcinogenesis, Vol 7: Cocarcinogenesis and Biological Effects of Tumor Promoters. Raven Press, New York, 1982, pp 1-12.

9. Slaga TJ, Fischer SM, Weeks CE, Nelson K, Mamrack M, Klein-Szanto AJP: Specificity and mechanism(s) of promoter inhibitors in multistage promotion. In: Hecker, E, Fusenig NE, Kunz W, Marks F, Thielman HW (eds) Carcinogenesis, Vol 7: Cocarcinogenesis and Biological Effects of Tumor Promoters. Raven Press, New York, 1982, pp 19-34.

10. Perchellet J-P, Boutwell RK: Enhancement by 3-isobutyl-1-methylxanthine and cholera toxin of 12-O-tetradecanoyl-phorbol-13-acetate-stimulated cyclic nucleotide levels and ornithine decarboxylase activity in isolated epidermal cells. Cancer Res (40): 2653-2659, 1980.

11. Perchellet J-P, Boutwell RK: Effects of 3-isobutyl-1-methylxanthine and cyclic nucleotides on the biochemical processes linked to skin tumor promotion by 12-O-tetradecanoyl-phorbol-13-acetate. Cancer Res (41): 3927-3935, 1981.

12. Coffey RG, Hadden JW: Phorbol myristate acetate stimulation of lymphocyte guanylate cyclase and cyclic guanosine 3':5'-monophosphate phosphodiesterase and reduction of adenylate cyclase. Cancer Res (43): 150-158, 1983.

13. Coffey RG, Hadden JW: Phorbol myristate acetate

stimulation of lymphocyte guanylate cyclase. Biochem Biophys Res Commun (101): 584-590, 1981.

14. Verma AK, Boutwell RK: Intracellular calcium and skin tumor promotion: calcium regulation of the induction of epidermal ornithine decarboxylase activity by the tumor promoter 12-O-tetradecanoylphorbol-13-acetate. Biochem Biophys Res Commun (101): 375-383, 1981.

15. Belman S, Troll W: Phorbol-12-myristate-13-acetate effect on cyclic adenosine 3':5'-monophosphate levels in mouse skin and inhibition of phorbol-myristate-acetate-promoted tumorigenesis by theophylline. Cancer Res (34): 3446-3455, 1974.

16. Belman S, Garte SJ: Phorbol myristate acetate uncouples the relationship between beta-adrenergic receptors and adenyl cyclase in mouse epidermis: a phenotypic trait in papillomas. In: Hecker E, Fusenig NE, Kunz W, Marks F, Thielman HW (eds) Carcinogenesis, Vol 7: Cocarcinogenesis and Biological Effects of Tumor Promoters. Raven Press, New York, 1982, pp 561-563.

17. Craven PA, DeRubertis FR: Inhibition by retinol and butylated hydroxyanisole of carcinogen-mediated increases in guanylate cyclase activity and guanosine 3':5'-monophosphate accumulation. Cancer Res (37): 4088-4097, 1977.

18. Archer DL, Bukovic-Wess JA, Smith BG: Inhibitory effect of an antioxidant, butylated hydroxyanisole, on the primary in vitro immune response. Proc Soc Exp Biol Med (154): 289-294, 1977.

19. Archer DL, Bukovic-Wess JA, Smith BG: Suppression of macrophage-dependent T-lymphocyte functions by gallic acid, a food additive metabolite. Proc Soc Exp Biol Med (156): 465-469, 1977.

20. Archer DL, Smith BG, Bukovic-Wess JA: Use of an in vitro antibody-producing system for recognizing potentially immunosuppressive compounds. Int Arch Allergy Appl Immunol (56): 90-93, 1978.

21. Archer DL, Johnson HM: Blockade of mitogen induction of the interferon lymphokine by a phenolic food additive metabolite. Proc Soc Exp Biol Med (157): 684-687, 1978.

22. Archer DL, Wess JA: Chemical dissection of the primary and secondary in vitro antibody response with gallic acid and butylated hydroxyanisole. Drug Chem Toxicol (2): 155-166, 1979.

23. Blalock JE, Archer DL, Johnson HM: Anticellular and immunosuppressive activities of foodborne phenolic compounds. Proc Soc Exp Biol Med (167): 391-393, 1981.

24. Archer DL, Smith BG, Johnson HM: Direct and induced suppression of in vitro antibody production by selected phenols: a comparison to virus-type interferon. In: Dean J, Padarathsingh M (eds) Biological Relevance of Immune Suppression. Van Nostrand, New York, 1981, pp 199-212.

25. Wess JA, Archer DL: Restoration by cyclic guanosine monophosphate and extracellular calcium of butylated hydroxyanisole-suppressed primary murine thymus-dependent antibody response. Immunopharmacology (3): 361-366, 1981.

26. Wess JA, Archer DL: Evidence from in vitro murine immunologic assays that some phenolic food additives may function as antipromoters by lowering intracellular cyclic GMP levels. Proc Soc Exp Biol Med (170): 427-430, 1982.

27. Wess JA, Archer DL: Differential effects of nordihydroguaiaretic acid (NDGA) on B-cell subsets: reversal of NDGA-induced antibody suppression by cyclic GMP is subset specific. Int J Immunopharmacol (6): 27-34, 1984.

28. Nakadate T, Yamamoto S, Ishii M, Kato R: Inhibition of 12-O-tetradecanoylphorbol-13-acetate-induced epidermal ornithine decarboxylase activity by lipoxygenase inhibitors: possible role of product(s) of lipoxygenase pathway. Carcinogenesis (3): 1411-1414, 1982.

29. Kozumbo WJ, Seed JL, Kensler TW: Inhibition by 2(3)-tert-butyl-4-hydroxyanisole and other antioxidants of epidermal ornithine decarboxylase activity induced by 12-O-tetradecanoylphorbol-13-acetate. Cancer Res (43): 2555-2559, 1983.

30. Fischer SM, Mills GD, Slaga TJ: Inhibition of mouse skin tumor promotion by several inhibitors of arachidonic acid metabolism. Carcinogenesis (3): 1243-1245, 1982.

31. Levine L: Inhibition of the A-23187-stimulated leukotriene and prostaglandin biosynthesis of rat basophil leukemia (RBL-1) cells by nonsteroidal anti-inflammatory drugs, antioxidants, and calcium channel blockers. Biochem Pharmacol (32): 3023-3026, 1983.

32. Bartholeyns J, Bouclier M: Involvement of histamine in growth of mouse and rat tumors: antitumoral properties of monofluoromethylhistidine, and enzyme-activated irreversible inhibitor of histidine decarboxylase. Cancer Res (44): 639-645, 1984.

33. Platshon LF, Kaliner M: The effects of the immunologic release of histamine upon human lung cyclic nucleotide levels and prostaglandin generation. J Clin Invest (62): 1113-1121, 1978.

18

CELL-CELL SIGNALLING

JACQUES MACLOUF

1. INTRODUCTION

The concept of cellular cooperation has long been recognized as being fundamental in several physio-pathological systems such as: 1) In immunology, B-cells require interaction with T-cells to induce antibody responses. 2) The development of atherosclerotic processes involves interactions between blood cells and vascular cells. 3) Inflammatory reactions necessitate a cooperation among various cell types. 4) Tumor cell metastases are governed by an inter-relation of malignant cells with blood platelets, leukocytes, red blood cells ... followed by interactions with vascular endothelium. Although it seems reasonable to assume that cellular cooperation needs cell-cell signalling in the initiation and/or amplification of such processes, little is known about the biochemical nature of these events. For example, cell-cell recognition can be explained by membrane interactions such as between glycoproteins (or glycoproteins and structural macromolecules) followed by metabolic events of one cell that can turn on or off an activation of neighbouring cells. In this respect, autacoids (e.g., icosanoids) may provide valuable clues for cell-cell signalling. Since the understanding of the role of icosanoids in *in vivo* situations is poor, a reappraisal of their function in *in vitro* cellular interactions (either cause or consequence) may turn out to be a valuable complementary approach. At least, the vast variety of their pharmacological effects could suggest their implication in some of the processes involved in the metastatic cascade.

The role of icosanoids in cell-cell signalling has been suggested by considering these molecules as amplifiers (or modulators) of the cell reactivity from which they originate. Such an approach is largely exemplified by in vitro studies of platelet reactivity. The pro-aggregatory thromboxane A_2 produced by activated platelets 1) serves as an intracellular messenger that can translocate internal Ca^{2+}, 2) binds on the membrane of the cell that has produced it and 3) recruits other platelets to amplify the formation of a thrombus. A similar approach has so far been unsuccessful to unravel the function of the platelet 12-lipoxygenase pathway. Thromboxane A_2 is also known to be a potent vasoconstrictor and has been suggested to play a role in cell adhesion (1). Reports that involve its role in the latter situations, i.e., messenger of one cell type to others, have been outnumbered by studies of its effects on the functions of the cell from which it is synthesized (e.g., platelet-platelet interactions).

In order to assess the role of icosanoids most of the studies have been carried out using exogenous metabolites. All these studies imply that under certain conditions involving cellular interactions, one cell type could produce such metabolites that could influence the metabolism and/or the function of the other cell type. Unfortunately, in a large number of studies, unrealistic concentrations have been used. Until now, few studies have been designed to provide direct biochemical evidence of cell-cell cooperation mediated by icosanoids.

I have put together situations in which various aspects of cellular cooperation involving arachidonic acid metabolites have been studied. The selection and the number of examples that follow are quite arbitrary. They only represent examples of a new line of investigation of arachidonic acid metabolism in cellular interactions. Such works can be the basis for a reappraisal of the role of icosanoids in the various situations of cellular interactions involved in the cancer process.

2. EICOSANOIDS AND CELLULAR COOPERATION

2.1 One cell synthesizes an eicosanoid that can influence the function of other cells (homologous or heterologous)

Such a system can be represented by platelet-derived thromboxane A_2 that can serve to recruit other platelets or to constrict the vessel wall. Alternatively, vascular cells (e.g., endothelial cells or smooth muscle cells) can produce prostacyclin that can either dilate the vessel wall or prevent platelet aggregation. Such a process, frequently described, represents signalling because the transduction of the signal to homologous or heterologous cells is accompanied by other biochemical reactions (i.e., Ca^{2+} fluxes, cAMP variation ...). Platelet aggregation mediated by thromboxane A_2 has been perceived as one "pathway" of the cell reactivity that can be overcome easily when it is blocked. This has led to the view that this mediator is part of the synergism for other biochemical pathways to induce platelet aggregation. The role of thromboxane A_2 on ADP-induced fibrinogen binding has established common and independent sequences for these two mediators in platelet-platelet adhesion (2). This concept, based on the articulation of various biochemical pathways to produce cell activation, may explain why the blockade of one pathway (e.g., cyclo-oxygenase by aspirin) does not result in a dramatic bleeding. It does not imply that the pharmacologically inhibited step is unnecessary for the hemostatic process but that its role is perhaps more subtle and less dominant than originally thought.

2.2 One cell can synthesize a product that is further metabolized by other cells: Transcellular metabolism

These studies require a co-incubation of different cell types known to have a specific metabolic pattern of arachidonate metabolites. As a prerequisite an appropriate purification of each cell type and study of their individual metabolism should be done prior to the study of a potential metabolic cooperation. A limited number of examples have appeared recently.

2.2.1 <u>Platelet-derived endoperoxides are transformed into prostacyclin by adjacent endothelial cells</u>. The "steal hypothesis" (i.e., prostacyclin synthase from endothelial cells utilizes platelet-derived endoperoxides) was first postulated in 1976 (3). It was only in 1980 that Marcus *et al*. (4) provided the first biochemical evidence of such a cooperation. In a very elegant study these authors presented firm evidence that co-incubation of [^{3}H] C20:4-labeled platelets with aspirin-treated endothelial cells stimulated with the calcium ionophore A23187, thrombin or collagen resulted in the formation of [^{3}H] 6-keto-$PGF_{1\alpha}$ and [^{3}H] TXB_2. Attention was drawn to the fact that an appropriate ratio of endothelial cells per platelet had to be respected in order to achieve the optimal demonstration. The proximity of platelets and endothelial cells was also critical. Although at first, strictly confined to *in vitro* work, the occurrence of such a situation has recently been evoked in *in vivo* conditions where similar cooperation could possibly exist: 1) The administration of thromboxane synthase inhibitors to human volunteers was followed by a significant reduction of the platelet-derived urinary metabolite of thromboxane A_2 (i.e., 2,3 dinor-thromboxane B_2) accompanied by an increase of the vascular-endothelial cell-derived urinary metabolite of prostacyclin (i.e., 2,3 dinor-6-keto-prostaglandin $F_{1\alpha}$; 5). 2) The occurrence of elevated levels of 2,3 dinor-6-keto-prostaglandin $F_{1\alpha}$ in the urine of patients with atherosclerotic complications could be interpreted as an example of a hyperactivation of platelets leading to increased production of endoperoxides converted into prostacyclin (6). The exact role of these findings in the homeostatic balance is still obscure but they provide the first evidence of a connection between the theoretical prediction, biochemical *in vitro* demonstration and pharmacological or pathological *in vivo* situations in which such cellular cooperation mediated by arachidonic acid metabolites may take place. A still fundamental unanswered question remains: what is the exact role of such cooperative metabolism in the homeostatic balance?

2.2.2 Arachidonic acid liberated from lymphocytes contributes to an increase in the synthesis of prostaglandin E_2 and thromboxane B_2 by monocytes (7). T-lymphocytes in the presence or in the absence of phytohemagglutinin failed to generate TXB_2 or PGE_2. This finding contrasted with the stimulation of monocytes by the same inducer which resulted in synthesis of the two cyclooxygenase compounds. However, co-incubation of unlabeled monocytes with [^{14}C] arachidonic acid prelabeled T-lymphocytes resulted in the synthesis of [^{14}C] thromboxane B_2 as the major monocyte-derived icosanoid. The authors concluded that T-lymphocyte derived arachidonic acid was used by monocytes for the synthesis of icosanoids. Such a finding could have relevance in inflammatory situations where there is proximity of these cells. The *in vivo* existence of such metabolic cooperation is still unknown.

2.2.3 Platelet-derived 12-lipoxygenase products can be converted by the leukocyte enzymes into 5 S, 12 S-diHETE or the 12, 20-diHETE (8,9,10). A metabolite of arachidonic acid from leukocytes was recently isolated, namely the (5 S, 12 S)-5,12-dihydroxy-(E,Z,E,Z)-6,8,10,14-icosatetraenoic acid (5 S, 12 S-diHETE) a stereoisomer of leukotriene (LT) B_4 (8). This compound was formed upon addition of 12-HETE to leukocytes stimulated with the calcium ionophore A23187. It could also be formed after incubation of 5-HETE with platelets. These experiments suggested the occurrence of biochemical interactions between platelets and leukocytes.

Additional *in vitro* work has provided examples for such an hypothesis. It was shown that in contrast to the stimulation of human peripheral leukocytes by arachidonic acid, that produce little LT, the co-incubation of those cells with platelets in the presence of C20:4 resulted in the formation of C-5 lipoxygenase derived metabolites as well as of 5 S, 12 S-diHETE (9). The finding that direct addition of 12-HPETE but not of 12-HETE could activate the C-5 lipoxygenase of leukocytes incubated with arachidonic acid alone indicated the nature of the activator. In further work it was also shown that 12-HETE could even inhibit to some extent the activity of C-5 lipoxygenase. This inhibition was also accompanied by the synthesis of 5 S, 12 S-diHETE

(10). These studies indicated that although the biological significance of the appearance of the mixed compound is unclear, its occurrence results from biochemical interactions (positive or negative) between the C-5 and C-12 lipoxygenases suggesting a potential cooperation between the two cells. Further results confirmed these findings (11). It was found that in addition to the 5 S, 12 S-diHETE the co-incubation of [^{3}H] C20:4-labeled platelets with unlabeled leukocytes resulted in the formation of [^{3}H] 5-HETE and of [^{3}H]-LTB_4 suggesting that platelet C20:4 can also serve as substrate for icosanoid formation by neutrophils. More recently another metabolite, 12 S,20-diHETE was identified in combined suspensions of thrombin- or collagen-stimulated platelets and unstimulated neutrophils (12,13).

The results suggest that platelet-neutrophil interactions could act in synergy for the production of lipoxygenase products that could be implied in the development of inflammatory and allergic reactions.

2.3 Regulation of arachidonic acid oxidative metabolism by cellular cooperations: Glomeruli-macrophage interactions (14).

The oxidation of arachidonic acid by rat isolated glomeruli is mainly directed to 12-HETE, PGE_2 and $PGF_{2\alpha}$. Resident macrophages convert C20:4 to equivalent amounts of 12-HETE and 6-keto-$PGF_{1\alpha}$. When both cells were co-incubated, the production of 6-keto-$PGF_{1\alpha}$ was increased and considerable amounts of TXB_2, PGD_2 and HHT were also produced whereas 12-HETE synthesis remained unchanged. The addition of a lipid extract of glomeruli that contained oxygenated metabolites of C20:4 to macrophages was followed by an increase of 12-HETE production. On the contrary, when a lipid extract from macrophages was added to glomeruli, 12-HETE production by glomeruli was nearly completely abolished. The authors have suggested that interaction between glomeruli and macrophages could possibly result in the activation of C20:4 metabolism by macrophages followed by inhibition of this metabolism by glomeruli. If such interactions occurred *in vivo*, they could play an important role in the pathophysiology of certain forms of glomerulonephritis.

2.4 One cell can release a constituent that can influence the arachidonic acid metabolism of another cell (15).

Preliminary work has been presented in which protease-rich granule fractions of polymorphonuclear leukocytes could alter prostacyclin production by endothelial cells as a reflection of injury. Cathepsin G consistently depressed thrombin-induced PGI_2 synthesis in a time and dose related manner. Exposure of cathepsin G-treated cells to sodium arachidonate did not impair prostacyclin production suggesting an inactivation of a membrane-bound surface receptor of thrombin by the protease. The receptor could be synthesized *de novo* after some time since the cells responded again to thrombin. The synthesis of prostacyclin that follows the injury of endothelial cells has been assumed to decrease the extent of cellular damage, e.g., subsequent formation of a thrombus. Neutrophil protease digestion of endothelial cell surface proteins could therefore modify vessel wall integrity by altering cell-substrate interaction and indirectly affect prostacyclin production. Such an effect could have dramatic consequences on vessel permeability and thrombogenicity. In contrast, purified elastase produced rapid detachment of endothelial cells from the culture plate without causing an increase in 6-keto-$PGF_{1\alpha}$. This finding indicates that cell injury may occur in the absence of cell activation, i.e., without any release of the beneficial prostacyclin. Such events that can indirectly affect the arachidonic acid metabolism seem very important in regard to the recent hypothesis ascribing to prostacyclin a major beneficial role in reducing tumor cell metastasis (16).

2.5 One cell depleted of arachidonic acid oxidative metabolism can interact with another cell and induce the synthesis of an oxygenated metabolite (17).

Increased adhesion of red blood cells to normal endothelial cells has been reported in sickle cell anemia and diabetes mellitus. The adhesion was correlated with the severity of vascular disease in both disorders. The amount of 6-keto-$PGF_{1\alpha}$ released under those circumstances was statistically higher with the abnormal red cells. A rank correlation could be established between 6-keto-$PGF_{1\alpha}$ and the adhesion ratio. These results suggest that abnormal cells can induce endothelial damage and induce prostacyclin release; alternatively, stimulation of the arachidonic acid metabolism results from cell-cell contact after membrane interactions. This increased synthesis could prevent the extension of complex thrombotic events.

3. TOWARDS A REAPPRAISAL OF THE SIGNIFICANCE OF EICOSANOID FORMATION: SOME CONSIDERATIONS FOR THEIR APPRECIATION IN TUMOR CELL-METASTASIS

Both malignant cells and host cells involve a wide variety of cells with a broad spectrum of compounds. Until now too little attention has been paid to the self-evident fact that not all the cells present in malignant tumors are cancer cells. In many tumors, macrophages and leukocytes may outnumber the cancer cells. Very few data are available concerning the exact contribution of one cell type (i.e., host or tumoral) to the overall production of arachidonate metabolites. It is very likely that most studies have already measured the metabolic production of cellular cooperations without unraveling the specifics: the progressive changes in the icosanoid content observed on tumors undergoing therapy may reflect a change in the cell composition (i.e., in cellular cooperation) rather than direct causal relationship with the tumor regression. Further one should not forget that two distinct metabolites can have synergistic effects (LTB_4 and PGE_2 for the potentiation of inflammatory reactions; 18). Experiments should be designed in order to demonstrate the existence of a (or various) biochemical signalling between different cell types. One possible approach to monitor cellular cooperation could be the development of assay methods capable of detecting products resulting from a mixed metabolism. Such estimation includes measurement of 5 S, 12 S-diHETE or 12 S, 20-diHETE as good criteria to estimate interaction between cells containing C-5 lipoxygenase activity and those containing C-12 lipoxygenase. However, it may only represent special situations.

Interspecies metabolic differences should also be considered. In contrast to man, pig platelets are nearly devoid of C-12 lipoxygenase (19) whereas their leukocytes contain both a C-5 and C-12 lipoxygenase resulting in the formation of 5 S, 12 S-diHETE by the white cells alone. If the formation of the transcellular metabolite is part of the cell-cell signalling, distinct features of this cooperation (e.g., platelet-leukocyte interaction) or of the polymorphonuclear cell properties (e.g., activation, adhesion ...) have to be looked for in man as compared to pig.

In order to demonstrate the involvement of icosanoids in various situations drugs have been used to inhibit the cyclooxygenase and/or the lipoxygenase pathway (20). These studies have not usually totally ruled out a non-specific effect of the inhibitors (as opposed to a putative unique target effect on oxidative enzymes). Although many agents that affect tumor promotion <u>in vivo</u> affect arachidonic acid transformation in cell culture, no unequivocal proof has been brought that agents that inhibit tumor growth and inhibit arachidonate metabolism in cell culture - or conversely - do it on the basis of related events. Few reports have hypothesized about the possible involvement of a drug acting to suppress a cell messenger that could have an indirect role in cell differentiation or transformation. The recent finding that arachidonic acid, 12- and 15-HETE induce starfish oocyte maturation suggest a close relationship of these metabolites with mitotic or meiotic processes (21). Such a role for HETE's in cell differentiation (i.e., synthesized by one cell type and acting on another one) has also been suggested by the induction of suppressor cells from human peripheral blood T cells by 15-HPETE a product of mouse macrophages and human T-lymphocytes (22). The fact that 15-HPETE or 15-HETE administered intraperitonally to mice induced the generation of suppressor cells from these animals offers a tempting role for considering these compounds as being messengers of cell function and/or cellular cooperation by a membrane-derived mechanism (23). In other studies, a portion of 12-HETE was incorporated to the membrane phospholipids of polymorphonuclear cells. These findings suggest that membrane-associated-HETE could serve in the expression of critical leukocyte (or other cell) functions.

The rate of unstable active metabolites (e.g., HPETE, PGG_2, PGH_2, LTA_4, PGI_2, TXA_2) is an important point to consider. In close cellular interactions, the vicinity of the cells may facilitate transcellular reactions. Further the usual presence of proteins may turn out to deeply affect the stability of the intermediate compounds (e.g., $t\frac{1}{2}$ = 30 s for TXA_2 in buffer); albumin has been shown to stablize TXA_2 ($t\frac{1}{2}$ = 8 min; 24), PGI_2 (25), LTA_4 (26) whereas it catalyzed the destruction of PGD_2 (27). Plasma proteins have recently been shown to convert LTA_4 to LTB_4 (28). Such an active role of proteins on the fate of cellular messengers can have dramatic unpredictable effects that have to be looked for.

The variety of oxygenated metabolites of arachidonic acid (from cyclooxygenase and lipoxygenase) produced by one cell should be re-evaluated when cellular cooperation takes place. Some modifications of the oxidative pathways may result in the synthesis of positive or negative regulators for cellular interactions that can also affect further cell differentiation. In addition, apparently unrelated events (e.g., degranulation of the lysosomal content) can modify indirectly the arachidonic acid metabolism resulting in unpredictable consequences. Among the various biochemical responses to tumor induction the modified synthesis of icosanoids is accompanied by variations in the level of cyclic nucleotides, which inhibit phospholipase and thereby reduce phospholipid turnover ... also known to be part of a response to cellular signalling. These responses may correlate with tumor formation but the essential nature of some of these effects in this formation is yet to be proved.

The fact that the exact role of icosanoids could not be determined in *in vitro* or *in vivo* situations does not imply that their function is artifactual or limited but rather that the appropriate experiments still remain to be designed. Viewing their function as part of cell-cell signalling may turn out to be a new promising approach. Such an approach may also help in understanding the mode of action of already recognized active drugs as well as in designing new active molecules.

REFERENCES

1. Spagnuolo PJ, Ellner JJ, Hassid A, Dunn MJ: Thromboxane A_2 mediated augmented polymorphonuclear leukocytes adhesiveness. J. Clin. Invest. (66): 406-414, 1980.
2. Bennett JS, Vilaire G, Burch JW: A role for prostaglandins and thromboxanes in the exposure of platelet fibrinogen receptor. J. Clin. Invest. (68): 981-987, 1981.
3. Bunting S, Gryglewski R, Moncada S, Vane JR: Arterial walls generate from prostaglandin endoperoxides a substance (prostaglandin X) which relaxes strips of mesenteric and coeliac arteries and inhibits platelet aggregation. Prostaglandins (12): 897-913, 1976.
4. Marcus AJ, Weksler BB, Jaffe EA, Broekman MJ: Synthesis of prostacyclin from platelet-derived endoperoxides by cultured human endothelial cells. J. Clin. Invest. (66): 979-986, 1980.
5. Fitzgerald GA, Baash AR, Oates JA, Pederson AK: Endogenous prostacyclin biosynthesis and platelet function during selective inhibition of thromboxane synthase in man. J. Clin. Invest. (72): 1336-1343, 1983.
6. Fitzgerald GA, Smith B, Pederson AK, Brash AR: Increased prostacyclin biosynthesis in patients with severe atherosclerosis and platelet activation. N. Eng. J. Med. (310): 1065-1068, 1984.
7. Goldyne ME and Stobo JD: Human monocytes synthesize eicosanoids from T lymphocyte-derived arachidonic acid. Prostaglandins (24): 623-630, 1982.
8. Borgeat P, Picard S, Vallerand P, Sirois P: Transformation of arachidonic acid in leukocytes. Isolation and structural analysis of a novel dihydroxy derivative. Prostaglandins Med. (6): 557-570, 1980.
9. Maclouf J, Fruteau de Laclos B, Borgeat P: Stimultion of leukotriene biosynthesis in human blood leukocytes by platelet-derived 12-hydroperoxy-icosatetraenoic acid. Proc. Natl. Acad. Sci. USA (79): 6042-6046, 1982.
10. Borgeat P, Fruteau de Laclos B, Maclouf J: New concepts in the modulation of leukotriene biosynthesis. Biochem. Pharmacol. (32): 381-387, 1983.

11. Marcus A, Broekman MJ, Safier LB, Ullman HL, Islam N, Serhan CN, Rutherford LE, Morchan HM, Weissmann G: Formation of leukotriene and other hydroxy acids during platelet-neutrophil interactions in vitro. Biochem. Biophys. Res. Commun. (109): 130-137, 1982.
12. Marcus AJ, Safier LB, Ullman HL, Broekman MJ, Islam N, Oglesby TD, Gorman RR: 12S, 20-dihydroxyicosatetraenoic acid: A new icosanoid synthesized by neutrophils from 12S-hydroxyicosatetraenoic acid produced by thrombin- or collagen-stimulated platelets. Proc. Natl. Acad. Sci. USA (81): 903-907, 1984.
13. Wong PY-K, Westlund P, Hamberg M, Granstrom E, Chao PH-W, Samuelsson B: Hydroxylation of 12L-hydroxy-5,8,10,14-eicosatetraenoic acid in human polymorphonuclear leukocytes. J. Biol. Chem. (259):2683-2686, 1984.
14. Sfaer J, Baud L, Bens M, Podjarny E, Schlondorff D, Ardaillou R, Sraer JD: Glomeruli cooperate with macrophages in converting arachidonic acid to prosstaglandins and hydroxyeicosatetraenoic acids. Prostaglandins Leukotrienes Med.
15. Weksler BB, Brower M, de Roque R, Jaffe EA, Tack-Goldman K: Alterations in endothelial cell function induced by proteases from human polymorphonuclear leukocytes. Blood (60):225a (abstract), 1982.
16. Honn KV, Cicone B, Skoff A: Prostacyclin: A potent antimetastatic agent. Science (212):1270-1272, 1981.
17. Wautier JL, Pintigny D, Wautier MP, Maclouf J, Corvazier E, Caen JP: Release of prostacyclin after erythrocyte adhesion to endothelial cells. Thromb. Haemostas. (50):241 (abstract) 1983.
18. Smith MJH: Biological activities of leukotriene B_4. Adv. Prostaglandin, Thromboxane, Leukotriene Res. (9):283-292, 1982.
19. Yoshimoto T, Miyamoto Y, Ochi K, Yamamoto S: Arachidonate 12-lipoxygenase of porcine leukocyte with activity for 5-hydroxyeicosatetraenoic acid. Biochim. Biophys. Acta (713):638-646, 1982.
20. Honn KV, Busse WD, Sloane BF: Prostacyclin and thromboxanes. Implications for their role in tumor cell metastasis. Biochem. Pharmacol. (32):1-11, 1983.
21. Meijer L, Guerrier P, Maclouf J: Arachidonic acid, 12- and 15-hydroxyeicosatetraenoic acids, eicosapentaenoic acid and phospholipase A_2 induce starfish oocyte maturation. Dev. Biol. (In press).
22. Gualde N, Rabinovitch H, Fredon D, Rigaud M: Effects of 15-hydroperoxy-eicosatetraenoic acid on human lymphocyte sheep erythrocyte rosette formation and response to concanavalin A associated with HLA system. Eur. J. Immunol. (12):773-777, 1982.
23. Mexmain S, Gualde N, Aldigier JC, Motta C, Chable-Rabinovitch H, Rigaud M: Specific binding of 15-HETE to lymphocytes. Effects on the fluidity of the plasmatic membranes. Prostaglandins, Leukotrienes Med.
24. Folco G, Grandstrom E, Kindahl H: Albumin stabilizes thromboxane A_2. FEBS Lett. (82):321-324, 1977.
25. Wynalda MA, Fitzpatrick FA: Albumins stablize prostaglandin I_2. Prostaglandins (20):853-861, 1980.
26. Fitzpatrick FA, Morton DR, Wynalda MA: Albumin stabilizes leukotriene A_4. J. Biol. Chem. (257):4680-4683, 1982.

27. Fitzpatrick FA, Wynalda MA: Albumin-lipid interactions: Prostaglandin stability as a probe for characterization binding sites on vertebrate albumins. Biochemistry (20):6219-6134, 1981.
28. Fitzpatrick FA, Haeggstrom J, Granstrom E, Samuelsson B: Metabolism of leukotriene A_4 by an enzyme in blood plasma: A possible leukotactic mechanism. Proc. Natl. Acad. Sci. USA (80):5425-5429, 1983.

19

PROTEOLYTIC ENZYMES AND ARACHIDONIC ACID METABOLITES

BONNIE F. SLOANE AND KENNETH V. HONN

1. MEDIATION OF LYSOSOMAL ENZYME RELEASE BY 20:4 METABOLITES

Borgeat and Sirois (1) have summarized the relationship between arachidonic acid (20:4) metabolites and lysosomal enzyme release thusly: agents which increase cellular levels of cAMP inhibit lysosomal enzyme release, whereas agents which increase cGMP induce release. In 3T3L1 fibroblasts the most potent stimulator of increased cAMP is PGI_2 followed by PGE_1 (2). However, the relative potency of the prostaglandins for cAMP accumulation appears to be a function of the cell type. Thromboxane A_2 prevents an increase in cellular cAMP due to an external stimulus (3). In contrast to the cyclooxygenase products (PGI_2, PGE_1, etc.), products of the lipoxygenase pathway have been shown to increase cellular cGMP (4).

The role of 20:4 metabolites in release of lysosomal enzymes has been extensively studied in neutrophils. The enzymes which have been studied are the glycosidases: lysozyme, β-glucuronidase and β-N-acetyl-glucosaminidase. Most of these studies have used inhibitors of varying specificity for the cyclooxygenase/lipoxygenase pathways and thus only provide indirect evidence that 20:4 metabolites induce lysosomal enzyme release. Indomethacin at low concentrations which inhibit the cyclo-oxygenase pathway (10^{-5} M) does not affect release of β-glucuronidase from neutrophils (5,6). Naccache et al. (7) found that lysozyme release could be inhibited by 5,8,11,14-eicosatetraynoic acid (ETYA), an inhibitor of both the cyclooxygenase and lipoxygenase pathways. Nordihydroguaiaretic acid (NDGA), a lipoxygenase pathway inhibitor, blocks both production of hydroxyeicosatetraenoic acids (HETE) and lysozyme release (8). Recent work by Smith et al. (9) has suggested that the lipoxygenase products responsible for lysosomal enzyme release in neutrophils may be leukotrienes. Using a selective inhibitor of leukotriene biosynthesis, 6,9-diepoxy-6,9-(phenylimino)-Δ6,8-prostaglandin I_1 (U-60, 257), they demonstrated parallel inhibition in response to f-Met-Leu-Phe of leukotriene biosynthesis and release of β-glucuronidase and lysozyme.

Direct evidence that lipoxygenation products of 20:4 metabolism regulate lysosomal enzyme release from neutrophils can be found in the following studies. Naccache et al. (7) demonstrated that lysozyme release from neutrophils is stimulated by addition of exogenous 20:4, thus proving that 20:4 metabolites do play a role in lysosomal enzyme release. Walenga et al. (8) confirmed an increase in lysozyme release by added 20:4 and demonstrated that under these conditions 20:4 is metabolized to HETEs. At low concentrations leukotriene B_4 (LTB_4), another lipoxygenase product, has been reported to stimulate lysozyme release from leukocytes, whereas at higher concentrations (10^{-6} to 10^{-5} M) LTB_4 stimulates release of both lysozyme and β-N-acetylglucosaminidase (10). Bokoch and Reed (10) did not find any stimulation of β-glucuronidase release by LTB_4. In contrast, studies by Samuelsson and co-workers (11), Smith et al. (9) and Showell et al. (12) found that LTB_4 stimulates release of both lysozyme and β-glucuronidase. Samuelsson and co-workers (11) demonstrated that LTB_4 stimulation of lysosomal enzyme release is stereospecific. Showell et al. (12) have extended these studies on mediation of lysosomal enzyme release by LTB_4 to demonstrate that a derivative of LTB_4 (LTB_4 dimethylamide), which is purported to act as a LTB_4 receptor antagonist, inhibits LTB_4-stimulated release of both lysozyme and β-glucuronidase from rabbit neutrophils.

LTB_4 may not be the only lipoxygenase product involved in the control of lysosomal enzyme release. O'Flaherty et al. (13) have recently demonstrated that 5-L-HETE potentiates the release from human neutrophils of both β-glucuronidase and lysozyme in response to LTB_4 and platelet activating factor. The potentiating effects of 5-L-HETE appear to be selective since degranulation induced by C5a, f-Met-Leu-Phe or A23187 is not potentiated. These studies suggest that lipoxygenase products, principally leukotriene B_4, of 20:4 metabolism may play a role in the induction of lysosomal enzyme release in neutrophils.

2. MEDIATION OF PROTEOLYTIC ENZYME SYNTHESIS AND RELEASE BY 20:4 METABOLITES

2.1 Proteolytic enzymes

There are two major classes of proteolytic enzymes or proteases: 1) the endopeptidases or proteinases which cleave in the middle of a peptide chain and 2) the exopeptidases which cleave sequentially from the carboxy or amino terminus of a peptide chain. Some proteases exhibit properties of both major classifications. There are four subclasses of proteinases: 1) serine, 2) cysteine, 3) metallo and 4) aspartic. Proteinases of the serine and cysteine subclasses have been linked to arachidonic acid metabolism.

Serine proteinases are the most abundant class of proteinases. Trypsin and the proteinases of the blood coagulation cascade and the fibrinolysis cascade are all examples of serine proteinases. These enzymes are active at neutral pH and often exist in the form of zymogens or inactive precursors. The serine proteinase which has been linked to 20:4 metabolism is an enzyme of the fibrinolysis cascade, plasminogen activator (PA). PA activates by limited proteolysis the zymogen plasminogen to the active serine proteinase, plasmin.

Cysteine proteinases require activation by thiol reagents. This group of enzymes plays a major role in intracellular protein degradation. Examples of this class include the lysosomal cysteine proteinases, cathepsins B, H and L, and the cytosolic Ca^{2+}-activated neutral proteinases (CANP). Lysosomal cysteine proteinases are active below pH 7 and inactivated above pH 7, whereas CANP are active at neutral pHs. Two cysteine proteinases have been linked to 20:4 metabolism, cathepsin H and a cathepsin B-like cysteine proteinase associated with tumor cells.

2.2 Plasminogen activator

Plasminogen activator (PA) has been implicated in cellular transformation from the normal to the malignant state, tumor cell proliferation and tumor cell invasion and metastasis (for review see 14). There does appear to be a link between 20:4 metabolism and PA induction, however, conflicting reports in the literature makethe nature of this link unclear.

Crutchley et al. (15) first examined the relationship between PA production and 20:4 metabolism in human foreskin fibroblasts, HeLa cells and MDCK cells. Changes in prostaglandin biosynthesis precede release of PA in response to 12-O-tetradecanoylphorbol-13-acetate (TPA). Production of PA after 18 hr is linearly related to PGE_2 concentration over a TPA dose range of 0.5-50 nM. However, when human foreskin fibroblasts, HeLa or MDCK cells are incubated with indomethacin at doses which completely suppress cyclooxygenase activity no inhibition of PA production is observed. These authors concluded that the TPA effects on PA are not mediated via prostaglandins.

In subsequent studies Crutchley and co-workers (16,17) directly examined the effects of exogenous prostaglandins on PA production by human foreskin fibroblasts. PGE_1, PGE_2 and PGI_2 stimulate PA biosynthesis and release (16). PGD_2 and 6-keto-$PGF_{1\alpha}$ are ineffective. These stimulatory effects are slow in onset and abolished by cycloheximide or actinomycin D. Since similar increases in PA are observed with cAMP modulating compounds, the authors concluded that the prostaglandin effects are mediated via the generation of cAMP (16). Conanan and Crutchley (17) also demonstrated that induction of PA by PGE_1 in human foreskin fibroblasts is dependent on the presence of serum.

Additional studies from the same laboratory suggest a role for products of the lipoxygenase pathway in PA production by cancer cells. In the first of two papers, Crutchley and Maynard (18) examined the role of cyclooxygenase and lipoxygenase products in the synthesis of PA in TPA-stimulated HeLa cells. Indomethacin at concentrations which inhibit prostaglandin production does not inhibit PA synthesis, whereas the lipoxygenase pathway inhibitors ETYA and NDGA block PA synthesis. In the second paper, Crutchley and Maynard (19) demonstrated a role for lipoxygenase products in PA biosynthesis and release by HeLa and rat basophilic leukemia (RBL-1) cells. Indomethacin inhibits prostaglandin production but has no effect on PA synthesis or release by A23187-stimulated RBL-1 cells or by TPA-stimulated HeLa cells. In contrast, NDGA and ETYA inhibit PA synthesis and release concomitantly with an inhibition of unidentified lipoxygenase-derived hydroxylated fatty acids. Incubation of HeLa cells with the lipoxygenase product 5-HETE (0.5-2.0 µg/ml) results in a 2.5-fold stimulation of PA synthesis.

The above studies implicate lipoxygenase products of 20:4 metabolism in PA synthesis and release. The role of prostaglandins is somewhat unclear but might be clarified if studies with cyclooxygenase pathway inhibitors and exogenous prostaglandins were performed in the same cell type. Endogeneously synthesized prostaglandins may not affect the biosynthesis and/or release of plasminogen activator by either normal or cancer cells. However, both normal and cancer cells may respond to pharmacologic levels of exogenous prostaglandins (which activate adenylate cyclase) in the same way they respond to isoproterenol or phosphodiesterase inhibitors (16).

2.3 Cathepsin B-like cysteine proteinase

A tumor cysteine proteinase which appears to be related to lysosomal cathepsin B has been implicated in the ability of tumor cells to invade into the extracellular matrix and to metastasize to secondary sites (for review see 20,21). Lysosomal cathepsin B can degrade such components of the extracellular matrix as collagen, fibronectin and proteoglycans (22-25). Activity of the tumor cathepsin B-like cysteine proteinase (CB) has been correlated with tumor malignancy in a number of tumor lines yet not in all tumor lines studied. CB activity in tumors seems to be associated with the viable tumor cells, prossibly with the plasma membrane of these tumor cells. CB activity has been measured in the sera, urine, ascites fluid and pancreatic fluid of tumor-bearing patients. CB is released from tumor explants and tumor cells in vitro as well as from normal subcutaneous tissue exposed to tumor-conditioned medium. CB in tumor cell membranes or released from tumor cells (or from host cells in response to tumor cells) may not possess proteolytic activity at neutral pH and thus may not be able to facilitate tumor cell invasion. However, CB exhibits enhanced stability at neutral or slightly alkaline pH's. There is not yet definitive proof that CB plays a role in tumor invasion and metastasis; there is, however, an increasing body of correlative evidence relating CB activity and tumor malignancy. This correlative evidence plus preliminary evidence that tumor CB can degrade components of the extracellular matrix in vitro suggests that CB may be one proteinase active in a proteolytic cascade resulting in tumor invasion and metastasis.

Lysosomal enzyme release from neutrophils seems to be mediated by lipoxygenase products of 20:4 metabolism. CB is related to lysosomal cathepsin B and has been shown to be released from tumor cells in culture. Therefore, we tested the hypothesis that 20:4 metabolites and specifically metabolites of the lipoxygenase pathway may mediate CB release by tumor cells (26). Indomethacin, a cyclooxygenase pathway inhibitor, does not affect release of CB. An inhibitor of the lipoxygenase and cyclooxygenase pathways, 5,8,11,14-eicosatetraynoic acid (ETYA), decreases release of CB from B16 amelanotic melanoma (B16a) cells.

Nafazatrom is a lipoxygenase pathway inhibitor in tumor cells (26,27; Figure 1) and in neutrophils (28). Nafazatrom inhibits the formation of an unidentified monohydroxy fatty acid from the lipoxygenase pathway by B16a cells (26,27; Figure 1; ED_{50} = 5 µm).

Synthesis of this product is not inhibited by indomethacin (50 μM), but is inhibited by ETYA (ED_{50} = 0.9 μM). Nafazatrom does not inhibit formation of cyclooxygenase products at the doses studied (27). Nafazatrom added daily at 1, 5 and 25 μg/ml also inhibits release of CB from Bl6a tumor cells after 3 da of exposure (26; Figure 2). After

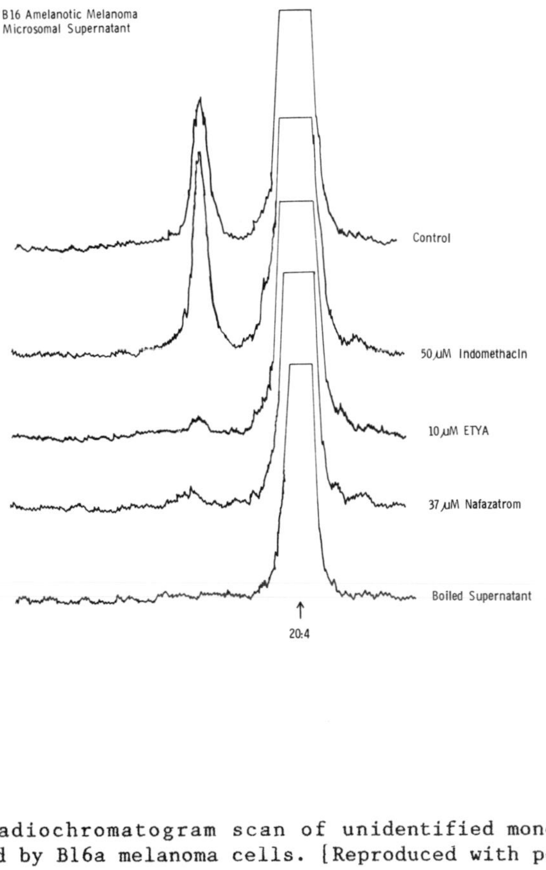

Figure 1. Radiochromatogram scan of unidentified monohydroxy fatty acid produced by Bl6a melanoma cells. [Reproduced with permission from (26).]

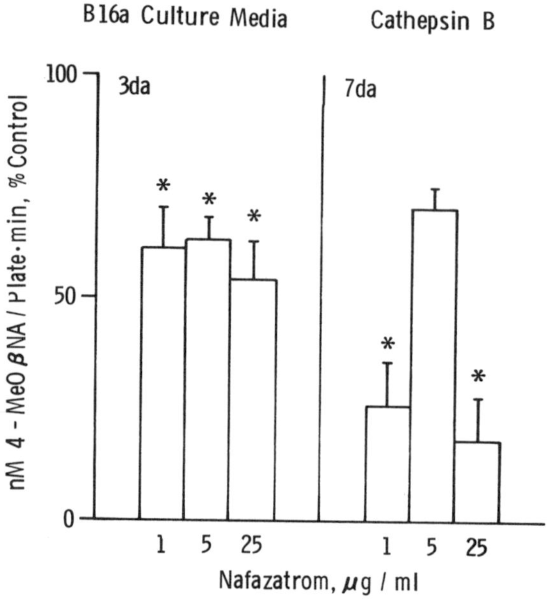

FIGURE 2. CB activity released into the media upon exposure to nafazatrom (Bay g 6575). Media and drug were changed daily so that release was measured over a 24 hr period. [Reproduced with permission from (26).]

7 da of exposure all doses inhibit CB release, but the dose response curve is bell-shaped. In addition, administration of nafazatrom in vivo at 0.3 mg significantly decreases CB activity in the primary tumors as well as increasing the life span of the tumor bearing animals (unpublished observations).

Release of CB from tumor cells can be directly stimulated by lipoxygenase products of 20:4 metabolism. CB release from B16a cells was stimulated by a 24 hr exposure to 5-hydroperoxyeicosatetraenoic acid (5-HPETE; unpublished observation). In contrast a 24 hr exposure to 15-HPETE (0.03, 0.3 and 3.0 μg/ml) did not enhance release of CB but significantly decreased (42%) CB release at 0.03 µg/ml. These results suggest that products of the 5-lipoxygenase pathway, i.e., leukotrienes, may play a role in CB release from tumor cells much as in the release of other lysosomal enzymes from neutrophils.

2.4 Cathepsin H

A recent paper by Katunuma and co-workers (29) indicates that a reciprocal relationship may exist between 20:4 metabolites and proteolytic enzymes. The lysosomal cysteine proteinase cathepsin H (purified from rat liver) transforms leukotriene D_4 to leukotriene E_4 with the release of glycine. Cathepsin H does not affect the transformation of leukotriene C_4 to leukotriene D_4. The cysteine proteinases cathepsin B and cathepsin C do not affect leukotriene biosynthesis.

3. CONCLUSION

There is suggestive evidence that lipoxygenase products of 20:4 metabolism may mediate the synthesis and release of proteinases of two classes in tumor cells or in normal cells treated with a phorbol ester. Studies in this area are clearly preliminary in nature, however, a broader role for lipoxygenase products in the regulation of proteinase release might be predicted from the studies on neutrophils.

REFERENCES

1. Borgeat P, Sirois P: Leukotrienes: A major step in the understanding of intermediate hypersensitivity reactions. J. Med. Chem. 24:121-126, 1981.
2. Hopkins NK, Gorman RR: Regulation of 3T3-L1 fibroblast differentiation by prostacyclin (prostaglandin I_2). Biochim. Biophys. Acta. 663:457-466, 1981.
3. Marcus AJ: The role of lipids in platelet function: With particular reference to the arachidonic acid pathway. J. Lipid Res. 19:793-826, 1978.
4. Hidaka H, Asano T. Stimulation of human platelet guanylate cyclase by unsaturated fatty acid peroxides. Proc. Natl. Acad. Sci. USA 74:3657-3661, 1977.

5. Northover BJ: Effect of indomethacin and related drugs on the calcium ion-dependent secretion of lysosomal and other enzymes by neutrophil polymorphonuclear leucocytes in vitro. Br. J. Pharmacol. 59:253-259, 1977.
6. Smith RJ: The guinea pig neutrophil calcium-dependent lysosomal enzyme secretory process. Inhibition by nonsteroid anti-inflammatory agents. Biochem. Pharmacol. 28:2739-2746, 1979.
7. Naccache PH, Showell HJ, Becker EL, Sha'afi RI: Arachidonic acid induced degranulation of rabbit peritoneal neutrophils. Biochem. Biophys. Res. Commun. 87:292-299, 1979.
8. Walenga RW, Showell HJ, Feinstein MB, Becker EL: Parallel inhibition of neutrophil arachidonic acid metabolism and lysosomal enzyme secretion by nordihydroguaiaretic acid. Life. Sci. 27:1047-1053, 1980.
9. Smith RJ, Sun FF, Bowman BJ, Iden SS, Smith HW, McGuire JC: Effect of 6,9-diepoxy-6,9-(phenylimino)-Δ6,8-prostaglandin I_1. (U-60-257) an inhibitor of leukotriene synthesis on human neutrophil function. Biochem. Biophys. Res. Commun. 109:943-949, 1982.
10. Bokoch GM, Reed PW: Effect of various lipoxygenase metabolites of arachidonic acid on degranulation of polymorphonuclear leukocytes. J. Biol. Chem. 256:5317-5320, 1981.
11. Hafstrom I, Pamblad J, Malsten CL, Radmark O, Samuelsson B: Leukotriene B_4 - a stereospecific stimulator for release of lysosomal enzymes from neutrophils. FEBS Lett. 130:146-148, 1981.
12. Showell HJ, Otterness IG, Marfat A, Corey EJ: Inhibition of leukotriene B_4-induced neutrophil degranulation by leukotriene B_4-dimethylamide. Biochem. Biophys. Res. Commun. 106:741-747, 1982.
13. O'Flaherty JT, Thomas MJ, McCall CE, Wykle RL: Potentiating actions of hydroxyeicosatetraenoates on human neutrophil degranulation responses to leukotriene B_4 and phorbol myristate acetate. Res. Commun. Chem. Pathol. Pharmacol. 40:475-487, 1983.
14. Quigley J, Cramer E, Fairbairn S, Gilbert R, Lacovara J, Ojakian G, Schwimmer R: Interaction of malignant cells with substrata: Adhesion, degranulation and migration. In: Honn KV, Crissman JD, Powers WE, Sloane BF (eds) Mechanisms of Metastasis: Potential Therapeutic Implications. Martinus Nijhoff, The Hague (In press).
15. Crutchley DJ, Conanan LB, Maynard JR: Induction of plasminogen activator and prostaglandin biosynthesis in HeLa cells by 12-O-tetradecanoylphorbol-13-acetate. Cancer Res. 40:849-852, 1980.
16. Crutchley DJ, Conanan LB, Maynard JR: Stimulation of fibrinolytic activity in human skin fibroblasts by prostaglandins E_1, E_2 and I_2. J. Pharmacol. Exp. Therap. 222:544-549, 1982.
17. Conanan LB, Crutchley DJ: Serum-dependent induction of plasminogen activator in human fibroblasts by catecholamines and comparison with the effects of prostaglandin E_1. Biochim. Biophys Acta 759:146-153, 1983.

18. Crutchley DJ, Maynard JR: Arachidonic acid metabolism and induction of plasminogen activator in HeLa cells by tumor-promoting phorbol ester. In: Powles TJ, Bockman RS, Honn KV, Ramwell PW (eds) Prostaglandins and Cancer. Alan R. Liss Inc., New York, pp. 281-287, 1982.
19. Crutchley DJ, Maynard JR: Induction of plasminogen activator by 12-0-tetradecanoylphorbol-13-acetate and calcium ionophore. Suppression by inhibitors of fatty acid lipoxygenase. Biochim. Biophys. Acta 762:76-85, 1983.
20. Sloane BF, Cavanaugh PG, Honn KV: Tumor cysteine proteinases, platelet aggregation and metastasis. In: Honn KV, Sloane BF (eds). Hemostatic Mechanisms and Metastasis. Martinus Nijhoff, The Hague, pp. 170-191, 1984.
21. Sloane BF, Honn KV: Cysteine proteinases and metastasis. Cancer Metastasis Rev. (In press).
22. Morrison RIG, Barrett AJ, Dingle JT: Cathepsin Bl and D action on human cartilage proteglycans. Biochim. Biophys Acta. 602:411-419, 1973.
23. Recklies AD, Poole AR: Proteolytic mechanisms of tissue destruction in tumor growth and metastasis. In: Weiss L, Gilbert HA (eds). Liver Metastasis. G.K. Hall, Boston, pp. 77-95, 1982.
24. Burleigh MC, Barrett AJ, Lazarus GS: Cathepsin Bl. A lysosomal enzyme that degrades native collagen. Biochem. J. 137:387-398, 1974.
25. Etherington DS, Evans PJ: The action of cathepsin B and collagenolytic cathepsin in the degradation of collagen. Acta Biol. Med. Germ. 36:1555-1563, 1977.
26. Sloane BF, Makim S, Dunn JR, Lacoste R, Theodorou M, Battista J, Alex R, Honn KV: Lipoxygenase products as mediators of tumor cell lysosomal enzyme release: Inhibition by nafazatrom. In: Powles TJ, Bockman RS, Honn KV, Ramwell PW (eds) Prostaglandins and Cancer. Alan R. Liss, New York, pp. 789-792, 1982.
27. Honn KV, Dunn JR: Nafazatrom (Bay g 6575) inhibition of tumor cell lipoxygenase activity and cellular proliferation. FEBS Lett. 139:65-68, 1982.
28. Busse WD, Mardin M, Grutzmann R, Dunn JR, Theodoreau M, Sloane BF, Honn KV: Nafazatrom (Bay g 6575): An inhibitor of cellular lipoxygenase. Fed. Proc. Fed. Am. Soc. Exp. Biol. 41:333, 1982.
29. Lokota K, Shono F, Yamamoto S, Kominami E, Katunuma N: Transformation of leukotriene D_4 catalyzed by lysosomal cathepsin H of rat liver. J. Biochem. (Tokyo) 94:1173-1178, 1983.

20

EXTRACELLULAR NUCLEOTIDE HYDROLYSIS AND INTEGRATION OF SIGNALLING

L.L. SLAKEY

1. OVERVIEW

The concept to be discussed here is that the reaction sequence:

ATP → ADP → AMP → Adenosine,

occurring extracellularly, is a key regulatory sequence. Adenine nucleotides and adenosine occur extracellularly during many crisis or signalling events, including thrombus formation, neurotransmission, catecholamine release from the adrenal medulla, strenuous exercise, and shock. They are agonists for a wide variety of physiologic responses. The antagonistic and competitive interactions among adenosine and its nucleotides, and the kinetic properties of the ectonucleotidases which carry out the sequential hydrolysis of the phosphate groups, provide for great subtlety in regulating the rate and extent of tissue response to the appearance of nucleotides in the pericellular space.

In my laboratory this concept grew out of an attempt to answer the question: what are the kinetic mechanisms by which normal arterial wall mounts an appropriately timed and appropriately limited response to a breach of its structural integrity? The case of initial response to vascular injury is useful for pointing out briefly the kinds of phenomena that occur during adenine nucleotide signalling.

Two of the specific processes triggered by vascular injury are platelet attachment, aggregation, and release, and vasoconstriction, which is triggered by several of the agonists that promote platelet aggregation and/or are added to the site by the activated platelet. It is well known that the prostaglandins thromboxane and prostacyclin have opposing effects on platelet aggregation and vascular smooth muscle tension (1). However, it is clear that they are not the only mediators in the forming thrombus that provide balance between promotion and inhibition of platelet aggregation.

ATP is released into the extracellular space upon thrombin binding to the vascular cells (2,3). ADP and ATP are released from activated platelets. ADP promotes thromboxane synthesis, platelet aggregation, and release of more ADP. Thus it triggers a positive feedback loop that leads to autoacceleration of platelet activation. ATP and AMP inhibit platelet aggregation, but only by acting as competitive inhibitors of ADP binding. Adenosine, however, can block platelet aggregation induced by ADP and all other stimuli, by causing an elevation of platelet cAMP. The effect of adenosine is modified in mixtures of ADP and adenosine, because ADP can inhibit platelet adenylate cyclase. Similarly, ADP can block the contribution made by prostacyclin toward inhibition of aggregation, since the action of prostacyclin is also mediated by platelet adenylate cyclase. (The effects of ADP and adenosine on platelets have been elegantly reviewed by Haslam and Cusack, 4).

ATP is transiently a vasoconstrictor in some vascular beds (5). ADP, insofar as it causes thromboxane synthesis, promotes vasoconstriction in the immediate area of the thrombus. On the other hand, both will promote vasodilation when acting on the endothelium, since they will trigger endothelial prostacyclin synthesis (6,7). Adenosine is always a vasodilator. The point is, we cannot speak of "the effect" of any component of the adenine nucleotide hydrolysis pathway. Rather we must ask, what are the relative amounts of all the components at a particular moment and microlocation? Further, what secondary mediators is the tissue releasing in response to the adenine nucleotides?

Finally, it is quite striking that the last enzyme in the pathway leading to adenosine, 5'-nucleotidase, is profoundly inhibited by ATP and ADP (8-12). Thus, we can predict that, paradoxically, above a threshold concentration the more ATP or ADP released initially into a pericellular space, the slower will be the appearance of adenosine. This consideration, together with the effects summarized above, suggests that extracellular adenine nucleotides and the ectonucleotidases form a regulatory system that provides multiple switches and quite subtle means of setting the magnitude and net direction of response to the initial stimulus.

In the following sections, this theme will be developed and related to the theme of this volume by: 1) presenting evidence for widespread release of adenine nucleotides into the extracellular space, and estimating extracellular concentrations; 2) describing briefly some ways in which adenosine and adenine nucleotides affect smooth muscle and other tissues, to show how varied the responses can be; 3) summarizing the evidence that some of the effects of adenine nucleotides are mediated by prostaglandin synthesis; 4) describing the kinetics of the ectonucleotidases, especially 5'- nucleotidase, and 5) summarizing the variations in ectonucleotidase activity that have been found in leukemic lymphocytes and various tumor cells.

The concept that the kinetics of extracellular adenine nucleotide hydrolysis play a central role in the integration of signalling developed as I considered diverse findings and points of view. The work of Lowenstein's group on the kinetics of 5'-nucleotidase, Ryan and Ryan's insight that extracellular nucleotide hydrolysis in lung vasculature should be viewed as conversion rather than degradation, Burnstock's description of the properties of the classes of purinergic receptors, and many conversations with my collaborators John Gordon and Jeremy Pearson have contributed especially to its present form.

2. SOURCES OF EXTRACELLULAR ADENINE NUCLEOTIDES

The average intracellular concentration of ATP is in the range of 10 mM. Ordinarily, we think of the extracellular levels as very low in comparison. For example, Forrester and Lind (13) reported the concentration of ATP in venous effluent from resting muscle as 0.5 μg/ml (about 1 μM). They found that this concentration rose 3-fold during and immediately after sustained contraction. In 1965, Douglas, et al. (14) showed that ATP is released together with catecholamine from the chromaffin granule of the adrenal medulla. During the 1960's and early 70's, Burnstock and his collaborators established that a series of non-cholinergic, non-adrenergic autonomic nerve responses were using ATP as a transmitter, and thus must be releasing ATP into the neuromuscular junction (reviewed in reference 15). Stimulated platelets release ATP and ADP. The relative amounts of the two nucleotides are different in different species; in humans, ADP predominates (reviewed in 16 and 17). In 1979, Pearson and Gordon showed that cultured vascular cells from swine release up to 50% of their cellular content of ATP when stimulated by thrombin binding, mechanical stress (2), or neutrophil elastase (18). The cells remained viable, and there was no leakage of cytoplasmic enzymes in these experiments. The observation that thrombin binding causes ATP release has been confirmed and extended with human umbilical vein endothelial cells (3).

For the purposes of investigating the kinetics of extracellular nucleotide hydrolysis, and evaluating the physiologic relevance of cell or tissue responses to exogenously applied nucleotides, it is important to have some estimate of the concentrations of nucleotides that could occur in the immediate pericellular environment of release. I am not aware of direct measurements of such concentrations, but order of magnitude estimates can be made. In human platelets, the ATP and ADP that are released are stored in the dense granule; their combined concentration in the granule is greater than 1 M (17). In chromaffin granules (19) and cholinergic vesicles the ATP content is approximately 0.2 M (20). In the case of an incipient thrombus, if we assume that each platelet releases its granule ATP and ADP into a volume comparable to the platelet's initial volume, which is poorly exchanged with the flowing plasma because of the thrombus forming, the initial concentration in the immediate pericellular space would be millimolar. If thrombin- or elastase-induced release of ATP from vascular cells occurs uniformly around the cells, the interstitial space below a cell to which the protease bound could contain as much as 5 mM initially. (This assumes cellular ATP = 10 mM, and release is into a volume roughly equal to the cell volume.) Similar considerations apply to cells in the immediate neighborhood of a neural junction. The concentrations will change according to the volume into which release occurs, and according to the rate of fluid replacement by flowing plasma or by perfusion of interstitial fluid. These estimations are very rough, but the underlying assumptions are reasonable, indeed conservative. Therefore, observations made using extracellular nucleotide concentrations in the micromolar to millimolar range should be considered potentially relevant to physiologic control situations.

3. PHYSIOLOGIC EFFECTS OF ADENINE NUCLEOTIDES AND ADENOSINE

The physiologic responses to adenosine and its phosphorylated derivatives are extremely diverse. There are several excellent books that provide reviews of the subject, for example, 21, 22, and 23. In writing this section it was necessary to omit many interesting and relevant studies for the sake of brevity. A selection of recent papers was made to illustrate especially the kinds of relationships that can occur among the responses to the components of the sequence ATP → ADP → AMP → adenosine. A difficulty encountered in sorting out these relationships is that only a subset of the published studies have been done in such a way as to determine which compounds act independently as agonists and which act by contributing a compound farther down the sequence. In spite of this difficulty some clear patterns can be discerned.

1) Adenosine may directly oppose the effect(s) of one or more of the phosphorylated species. This is the case with platelet aggregation, as summarized previously. It is also the case with bladder smooth muscle. ATP (and, in some preparations, ADP)

cause contraction of bladder smooth muscle strips (24-28, and reviewed in 29). Adenosine relaxes bladder smooth muscle; both tone and spontaneous contractile activity are reduced (25,29).

2) Adenosine may oppose the effect upon a target cell of catecholamine or acetylcholine co-released with ATP. This pattern appeared in the study of development of neurotransmitter repertoire by Furshpan, et al. (30). Neurons were cocultured with cardiac myocytes, planted on islands of collagen so that each colony of myocytes had only a single neuron. Some of the neurons developed a response of inhibition of myocyte beating. This turned out to be adenosine mediated; the response was inhibited by theophylline, and abolished by adenosine deaminase. All the neurons that had this property also released another transmitter; most were adrenergic, but at least one was cholinergic.

3) The released nucleotide, or a less phosphorylated form, may feed back upon the release process. This also occurs with platelets, in which the binding of ADP to the surface stimulates among other things the release of more ADP, creating a positive feedback loop that makes the process autoaccelerating.

A negative feedback loop of adenosine upon transmitter and nucleotide release occurs in both adrenergic and cholinergic nerves. Burnstock (15), McIlwaine (31) and Paton (32) have recently reviewed the evidence for this.

4) The response of many visceral smooth muscle preparations to ATP is multiphasic. In rabbit bladder, ATP causes a sharp phasic contraction, followed by a prolonged tonic contraction (e.g., 24). ADP gives the tonic response but not the phasic (25). The non-hydrolyzable isostere of ATP, β,γ-methylene ATP (APPCP*) caused the phasic response and not the tonic. In this case it seems reasonable to say that ATP is the agonist for the first response, and ADP for the second.

---

*The following convention will be used to represent analogs of the adenine nucleotides in which one of the bridge oxygens is replaced by carbon or nitrogen: A represents adenosine, P represents phosphate whether or not an oxygen is substituted, and N or C is inserted to represent substitution of the bridge oxygen. Thus unsubstituted ATP would be represented APPP. APNP would be α,β-imido ADP, and so forth. ATPγS represents ATP in which a non-bridge oxygen in the γ-position has been replaced by sulfur.

The methylene and imido isosteres of ATP, ADP, and AMP have been used extensively to try to dissect tissue responses to the adenine nucleotides and adenosine. A difficulty that arises in interpreting this work is that the replacement of oxygen with carbon or nitrogen not only makes the compound resistant to hydrolysis, it also changes the geometry of the polyphosphate chain, and thus may affect binding to a receptor either positively or negatively. This issue has been explored at length by Satchell and MacGuire (29,33). Their observations of the relaxation of guinea pig taenia coli induced by adenine nucleotides and their analogs are consistent with the presence of at least three distinct responses, one to ATP, one to ADP, and one to adenosine. The ultimate response to each of these is relaxation of the tissue, but they occur with quite different time courses and dose response relationships.

The response of rabbit small intestine to adenine nucleotides and adenosine also suggests the presence of distinct receptors for ATP, ADP, and adenosine. ATP and adenosine caused similar responses; a rapid initial relaxation and decrease in spontaneous activity, followed by return to sustained activity of lower force than before treatment. However, APPCP and APPNP caused the same response, and the response to adenosine was sensitive to adenosine deaminase, while the response to ATP, APPCP, and APPNP were not. APCP and APCPP, which would hydrolyze to APCP, also caused a rapid initial relaxation, but the time course of return to contraction was gradual, and the tissue eventually reached the pretreatment force level.

Chapal, et al. (34,35) have shown that ATP, ADP, AMP and adenosine all stimulate insulin release from perfused dog pancreas. The effect of 16 μM ATP or ADP is about five times that of AMP or adenosine at the same concentration. APPCP had no effect and both APCPP and APCP were as effective as ATP or ADP. The authors interpret these data as evidence for a requirement for a particular conformation around the polyphosphate chain. However, the data are equally consistent with the interpretation that ADP is the true agonist for the ATP/ADP response, with rapid hydrolysis of ATP at the cell surface.

Finally, Ford-Hutchinson has reported that UTP and ATP enhance the aggregation of neutrophils (36). UTP, ATP, GTP, TTP, CTP, ADP, AMP, adenosine and APPCP and APCPP were tested in this study. UTP was by far the most effective agonist of the compounds tested (EC_{50} = 2 μM) followed by ATP (EC_{50} = 300 μM). The other nucleotide triphosphates were effective at 300 μM, and adenosine and the methylene isosteres of ATP were not effective at this concentration.

5) There are some cases in which cell or tissue response to adenosine and its nucleotides are strikingly similar in dose response. In these cases, either the response must be truly the same to all the compounds or the conversion at the cell surface is very rapid, and the ectonucleotidases closely associated with the adenosine receptor site. These cases will be taken up in more detail in the context of stimulation of prostaglandin synthesis, below.

4. ADENINE NUCLEOTIDES AND ADENOSINE AS AGONISTS FOR PROSTAGLANDIN SYNTHESIS

As is evident from the summary above, more than one class of receptors for extracellular nucleotides must exist. Burnstock has pointed out that prostaglandins are the probable mediator for many of these receptors (15), although it is clear that this pattern does not apply in all cases. Three kinds of evidence for prostaglandin involvement can be found in the recent literature on adenine nucleotide and adenosine effects. First, there are many studies in which aspirin, indomethacin, or some other inhibitor of prostaglandin synthesis is found to block a particular effect. Second, prostaglandins (PGE_2 and $PGF_{2\alpha}$ were most often tested) were found to mimic some effects. Finally, there is a growing number of papers in which the addition of nucleotide has been shown to cause production of specific prostaglandins, as judged by bioassay or radioimmunoassay.

It has been documented by all of the above criteria that thromboxane A_2 is synthesized by platelets in response to the binding of ADP to the platelet surface, and that thromboxane participates in promoting platelet aggregation and release (37).

Among the cases of multiphasic response to ATP discussed above, there are several in which it is clear that one of the responses is prostaglandin mediated. Husted, *et al.* found that the tonic contraction, but not the phasic, of rabbit detrusor in response to ATP and ADP is blocked by indomethacin. PGE_2 and $PGF_{2\alpha}$ give the tonic response (26). Downie and Larson (27) and Anderson (28) similarly showed that the tonic response to ATP is blocked by indomethacin. Anderson showed in addition that ATP causes this preparation to release PGE into the incubation bath, as shown by radioimmunoassay. Brown and Burnstock (38) observed that ATP causes relaxation, followed by rebound contraction in guinea pig taenia coli, and indomethacin blocked the rebound. They also observed a biphasic response to ATP in guinea pig trachea, a modest contraction followed by profound relaxation. Again indomethacin blocked the second response but not the first.

Advenier, *et al.* (39), working with guinea pig trachea which had been maximally contracted with acetylcholine, found that both ATP and adenosine caused relaxation, with very similar dose response curves. The effect of both ATP and adenosine was enhanced dramatically by dipyridamole, which may suggest that adenosine is the mediator in both cases, since dipyridamole inhibits the reuptake of adenosine. The relaxation caused by ATP and adenosine was not affected by either aspirin or indomethacin. However, if the tissue is at resting tone initially, both ATP and adenosine, again at similar doses, caused a modest contraction. This effect was blocked by aspirin and by nictindole, a TXA_2 synthesis inhibitor.

Jones, *et al.* (40) also studied the responses of guinea pig trachea. They found that when the tissue was initially at resting tone, ATP, ADP, AMP, adenosine, coenzyme A, acetyl-coenzyme A, and 3',5'-ADP all gave a small contraction, followed by relaxation. The relaxing effect of all these compounds was dramatically enhanced by dipyridamole. ATP and ADP were more effective than the other compounds in causing contraction. Indomethacin decreased resting tone, and abolished the contractile response to ATP, ADP, and adenosine. Thus, we see again a dual response to ATP, with one of the actions probably prostaglandin mediated, and again find evidence for prostaglandin mediation of a response to adenosine itself, not a phosphorylated derivative.

Guinea pig uterine smooth muscle strips show a contractile response to ATP, ADP, AMP, and adenosine, again with very similar dose response curves, which is blocked by inhibitors of both the synthesis and the effects of prostaglandins (41). Interestingly, PGE_2 and ATP (or adenosine) have a synergistic effect on this tissue, causing contraction when given together, each at a subthreshold dose. This response is blocked by indomethacin.

In 1974, Needleman, et al.(42), showed that perfusing ATP or ADP through several tissues caused release of material that gave positive response in two bioassays for PGE_2. In 1977, Isakson, et al. (43) showed that perfusion of rat heart with ATP gave rise to release of pre-incorporated ^{14}C-arachidonic acid as PGE_2 and additional radioactive products that were not completely identified at that time, but which probably included PGI_2. Morrison, et al. (44) showed that ATP induced synthesis of both PGE_2 and thromboxane A_2 in perfused kidney that had been ureter obstructed.

In 1981, Schwartzman, et al. (45) described prostaglandin production by perfused kidney and heart from rabbit. The tissues were prelabelled with ^{14}C-arachidonic acid, and the products identified by thin-layer chromatography. In kidney, the principal product was PGE_2. In heart, 50% of the product was PGI_2, 20% was PGE_2, and 20% $PGF_{2\alpha}$. In each tissue, ATP and ADP gave the same profiles. However, they appear to have completely separate receptors, since the effect of adding 95% saturating doses of each of them was additive. Further, the infusion of ADP produced tachyphylaxis to ADP but not ATP. ATP produced tachyphylaxis to ADP and not ATP, perhaps because of rapid hydrolysis at cell surfaces.

Presumably, the vasculature makes a significant contribution to prostaglandin production observed during tissue perfusion. Recently, Boeynaems and Galand (7) observed that rings of rabbit aorta produce PGI_2 and small amounts of PGE_2 and $PGF_{2\alpha}$ in response to ADP. This response required intact endothelium. Pearson, et al. (6) have shown that cultured endothelial cells produce PGI_2, and some PGE_2, when stimulated with ATP. ADP was less effective than ATP, and AMP produced a slight response at high concentration. Adenosine was not an agonist. In this system APPCP gave no response, but this is a case of failure of the altered phosphate chain to fit the receptor, since APPNP and ATPγS, which are also resistant to hydrolysis, were effective substitutes for ATP.

5. KINETICS OF ADENINE NUCLEOTIDE HYDROLYSIS AT CELL SURFACES

It has been shown that adenine nucleotides are hydrolyzed all the way to adenosine in many experimental systems. Hydrolysis of AMP has been studied in perfused heart, liver, and skeletal muscle (46,47). In 1971, Ryan and Smith (48) showed that ATP and AMP are hydrolyzed during passage through the pulmonary circulation, and demonstrated elegantly that the 5'-nucleotidase activity (AMPase) resided on the luminal surface of the vascular endothelium. In 1974, DePierre and Karnovsky (49,50) described an ecto-ATPase, p-nitrophenylphosphatase, and 5'-nucleotidase on guinea pig polymorphonuclear leukocytes. They provided evidence that these activities are due to three separate proteins, and they suggested criteria for establishing that a phosphatase is a true ectoenzyme. Their cells also hydrolyze ADP, but it was not established whether one of the above three enzymes was responsible for ADP hydrolysis. Many isolated cell preparations and cultured cell lines have now been shown to have nucleotidase activities capable of attacking exogenous substrates (9,10,51-59)

Since the extent of phosphorylation can profoundly modify the physiologic effect of extracellular adenine nucleotides, we might predict that the regulation of the time course of their hydrolysis would be complex and finely tuned. Relatively few studies have followed each component of the pathway during the hydrolysis of ATP or ADP. Cooper, et al. (60) measured conversion of ADP to AMP, adenosine, and inosine using several different vascular segment or cell preparations. Crutchley, et al. (61) studied the degradation of ADP to AMP and adenosine by cultured endothelial cells and smooth muscle cells from bovine pulmonary arteries. In both the Cooper and Crutchley studies, "conditioned medium," i.e., ADP solutions that had been incubated with cells, were tested to see what effect they had on platelet aggregation. The details of the two studies are different, but both concluded that the adenosine formed from ADP by the cells significantly inhibited subsequent induction of platelet aggregation by intact ADP. Pearson *et al.*(8) have shown that the ectonucleotidase activity capable of converting ATP to ADP, AMP, and adenosine in cultured aortic endothelial and smooth muscle cells from swine is due to at least three separate enzymes (an ATPase, an ADPase and 5'-nucleotidase). They have shown that the K_m and total activity of each of these enzymes is different in endothelial and smooth muscle cells. They also observed that smooth muscle cells appear to convert ADP to ATP extracellularly.

The kinetic properties of homogeneous or partially purified preparations of 5'-nucleotidase from heart (62), liver (63), lymphocytes (64), and intestinal smooth muscle (10,65) have been described in some detail. In addition, there are descriptions of its properties in several whole cell preparations (9,12). The findings of these studies vary enormously, but some interesting common observations emerge from a comparison of them. All 5'-monoribonucleotides are substrates for this enzyme, but 5'-AMP is hydrolyzed more efficiently than the nucleotides of other bases. The enzyme binds tightly to concanavalin A. Activity is affected by many compounds likely to be present in plasma and interstitial fluid, for example, bicarbonate and free amino acids. It is always found to be inhibited by ATP or ADP; indeed, inhibitability by the ADP analogue APCP has come to be used as a marker to distinguish this enzyme from other phosphatases. Typically the K_I's for ATP and ADP are substantially lower than the K_m of the enzyme for AMP.

This last property, inhibitability by ATP and ADP, establishes the potential for a particularly interesting feature of the regulation of the time course of overall conversion of ATP or ADP to adenosine. That is, when extracellular ATP or ADP concentrations are low, the nucleotides should be quickly degraded all the way to adenosine. When the extracellular concentration of either rises to the level at which it is an effective inhibitor of 5'-nucleotidase, the rate of appearance of adenosine may actually be diminished. Thus, at high extracellular nucleotide concentrations, the length of the time lag created between the extracellular appearance of ATP or ADP and the appearance of a physiologically effective concentration of adenosine would be inversely related to the size of the initial bolus of ATP and/or ADP released.

The overall conversion of ATP to ADP, AMP, and adenosine by cultured endothelial cells from swine aorta behaves according to this prediction. Pearson, et al. showed that when cells are incubated with 1 μM ATP, very little ADP or AMP accumulate, and there is immediate and linear production of adenosine (8). We have repeated these experiments with initial concentrations of ATP in the range from 100 μM to 1 mM, concentrations that we believe would be transiently achieved in the immediate environment of a forming thrombus. As the initial concentration of ATP was raised above 250 μM, a substantial lag in adenosine appearance occurred. When the starting concentration of ATP was 750 μM, over 80% of the ATP was converted to AMP before rapid conversion to adenosine began. The patterns observed were independent of cell number; increasing the number of cells compressed the time for the total conversion to occur, but did not affect the way the lag in adenosine appearance varied with initial ATP concentration. The pattern of hydrolysis of ATP observed with smooth muscle cells is very different from that observed with endothelial cells. Very little ADP or AMP accumulate. ATP is converted with little lag to adenosine. However, the hydrolysis of ATP is retarded at high initial ATP concentrations, so that there is still a paradoxic dependence of the rate of adenosine appearance upon the initial ATP concentration.

We have written a program that simulates the sequential hydrolysis of ATP, and this model has been used (a) to test whether each step in the hydrolysis obeys Michaelis-Menten kinetics, (b) to determine the kinetic constants for each step, and (c) to determine how sensitive the lag in adenosine appearance is to small changes in the V_{max} of 5'-nucleotidase. Matching simulated time courses to experimental data for endothelial cells shows that the ATPase and ADPase obey the Michaelis-Menten relationship in its simplest form. Matching the data for AMP hydrolysis requires the introduction of inhibition by at least ADP into the model.

The kinetic constants for all three reactions found by simulation are very close to those determined by initial velocity measurements. Finally, the lag time in adenosine appearance is very sensitive to small changes in total amount of 5'-nucleotidase activity.

6. VARIATION OF ECTONUCLEOTIDASE ACTIVITY IN MALIGNANT CELLS

Much of the information on ectonucleotidase activity of malignant cells has been gathered in an effort to assess the usefulness of that enzyme as a marker for the success of various chemotherapeutic agents. In patients expressing the common acute lymphocytic leukemia (ALL) antigen, 5'-nucleotidase activity is significantly increased. Activity is low or absent in T cell-ALL (66-70). 5'-nucleotidase is increased in the peripheral leukocytes of some patients in lymphoid blast crisis and depressed in myeloid leukemia (71). 5'-nucleotidase is increased in malignant compared to benign astrocytomas (72).

Sun, *et al*. (73) have recently reported that cell lines derived from lymphocytes of patients with various leukemias contain a protein that is a specific inhibitor of 5'-nucleotidase. Conklyn and Silber (74) investigated the kinetic properties of ecto-5'-nucleotidase in normal and leukemic lymphocytes. Unexpectedly, they found a correlation between K_m and V_{max}, that is, cells with low total activity also had a low K_m for AMP, and cells with high activity had a K_m as much as 10-fold higher than that found with low activity cells. This kind of kinetic pattern could occur if enzyme activity were regulated by binding of a ligand that happened to affect K_m and V_{max} in the same direction. These observations suggest the intriguing possibility that 5'-nucleotidase is ordinarily regulated not only by small ligands in the extracellular milieu but also by a regulatory protein either inside the cell or on its surface.

Less is known about the ecto-ATPase and ADPase activities of tumor cells. In 1981, Wilson, *et al*. (75) studied the cytochemical localization of all three nucleotidases of granulocytes from normal subjects and patients with chronic granulocytic leukemia. Total activities of ATPase and ADPase were the same in patients and controls; 5'-nucleotidase was lower in leukemic granulocytes. ATPase was uniformly distributed over the surface in normal cells and patchy in leukemic cells. 5'-nucleotidase was distributed normally in some leukemic cells and completely absent from others.

Although much of this information was collected for clinical reasons, it has now become apparent that the study of these enzymes will enlarge our basic understanding of regulatory processes in normal as compared with malignant cells. It is clear that extracellular nucleotides play an important role in regulating cellular responses and therefore it should be fruitful to study the kinetics of ectonucleotidases using tumor cells.

ACKNOWLEDGMENTS

A part of this work was supported by NIH grant HL 31854. The author gratefully acknowledges the skillful and patient assistance of Dr. Michael Gross during the literature search and in the preparation of the manuscript.

REFERENCES

1. Lands WEM, Pitt BW, Culp BR: Recent concepts of platelet function and dietary lipids in coronary thrombus, vasospasm and angina. Herz (5): 34-41, 1980.
2. Pearson JD, Gordon JL: Vascular endothelial and smooth muscle cells in culture selectively release adenine nucleotides. Nature (281): 384-386, 1979.
3. Lollar P, Owen WG: Active-site-dependent, thrombin-induced release of adenine nucleotides from cultured human endothelial cells. Ann NY Acad Sci (370): 51-56, 1981.
4. Haslam RJ, Cusack NJ: Recent concepts of platelet function and dietary lipids in coronary thrombus, vasospasm and angina. In: Burnstock G (ed) Purinergic receptors. Chapman and Hall, New York, 1981, pp 221-285
5. Nakatsu K, Bartlett V: Multiple adenine derivative receptors in rat ileum and electrical degradation of purine drugs. In: Baer HP and Drummond GI (eds) Physiological and regulatory functions of adenosine and adenine nucleotides. Raven Press, New York, 1979, pp 79-84.
6. Pearson JD, Slakey LL, Gordon JL: Stimulation of prostaglandin production through purinoceptors on cultured porcine endothelial cells. Biochem J (214): 273-276, 1983.
7. Boeynaems JM, Galand N: Stimulation of vascular prostacyclin synthesis by extracellular ADP and ATP. Biochem Biophys Res Commun (112): 290-296, 1983.
8. Pearson JD, Carleton JS and Gordon JL: Metabolism of adenine nucleotides by ectoenzymes of vascular endothelial and smooth muscle cells in culture. Biochem J (190): 421-429, 1980.
9. Stefanovic V, Mandel P, Rosenberg A: Ecto-5'-nucleotidase of

intact cultured C6 rat glioma cells. J Biol Chem (251): 3900-3905, 1976.

10. Burger RM, Lowenstein JM: 5'-nucleotidase from smooth muscle of small intestine and from brain: inhibition by nucleotides. Biochemistry (14): 2362-2366, 1975.
11. Sullivan JM, Alpers JB: In vitro regulation of rat heart 5'-nucleotidase by adenine-nucleotides and magnesium. J Biol Chem (246): 3057-3063, 1971.
12. Newby AC, Luzio JP, Hales CN: Properties and extracellular location of 5'-nucleotidase of rat fat cell plasma membrane. Biochem J (146): 625-633, 1975.
13. Forrester T, Lind AR: Adenosine triphosphate in the venous effluent and its relationship to exercise. Fed Proc (28): 1280-1283, 1969.
14. Douglas WW, Poisner AM, Rubin RP: Efflux of adenine nucleotides from perfused adrenal glands exposed to nicotine and other chromaffin cell stimulants. J Physiol (179): 130-137, 1965.
15. Burnstock G: Neurotransmitters and trophic factors in the autonomic nervous system. J Physiol (313): 1-35, 1981.
16. DaPrada M, Richards JG, Kettler R: Amine storage granules in platelets. In: Gordon JG (ed) Platelets in biology and pathology. Elsevier, Amsterdam, 1981, pp. 107-146.
17. Ugurbil K, Holmsen H: Nucleotide compartmentalization: radioisotopic and nuclear magnetic resonance studies. In: Gordon, 146-147.
18. LeRoy EC, Ager A, Gordon JL: Effects of neutrophil elastase and other proteases on porcine aortic endothelial prostacyclin production, adenine nucleotide release and responses to vasoactive agents. J Clin Invest (submitted).
19. Winkler H: The composition of adrenal chromaffin granules: an assessment of controversial results. Neuroscience (1): 65-80, 1976.
20. Wagner JA, Carlson SS, Kelly RB: Chemical and physical characterization of cholinergic synaptic vesicles. Biochemistry (17): 1199-1206, 1978.
21. Baer HP, Drummond GI: Physiological and regulatory functions of adenosine and adenine nucleotides. Raven Press, New York, 1979.
22. Berne RM, Rall TW, Rubio R. Regulatory function of adenosine. Martinus Nijhoff, The Hague, 1983.
23. Burnstock G: Purinergic receptors. Chapman and Hall, New York, 1981.
24. Husted S, Sjögren C, Andersson KE: Role of prostaglandins in the responses of rabbit detrusor to non-cholinergic, non-adrenergic nerve stimulation and to ATP. Arch int Pharmacodyn (246): 84-97, 1980.
25. Husted S, Sjogren C, Andersson KE: Mechanisms of responses to non-cholinergic, non-adrenergic nerve stimulation and to ATP in isolated rabbit urinary bladder: evidence for ADP evoked prostaglandin release. Acta pharmacol et toxicol (47): 84-92, 1980.
26. Andersson KE, Husted S, Sjögren C: Contribution of prostaglandins to the adenosine triphosphate-induced contraction of rabbit urinary bladder. Brit J Pharmacol (70): 443-452, 1980.

27. Downie JW, Larsson C: Prostaglandin involvement in contractions evoked in rabbit detrusor by field stimulation and by adenosine 5'-triphosphate. Can J Physiol Pharmacol (59): 253-260, 1980.
28. Anderson GF: Evidence for a prostaglandin link in the purinergic activation of rabbit bladder smooth muscle. J Pharmacol Exp Ther (220): 347-352, 1982.
29. Maguire MH, Satchell DG: Purinergic receptors in visceral smooth muscle. In: Burnstock, 47-92, 1981.
30. Furshpan EJ, Potter DD, Landis C: On the transmitter repertoire of sympathetic neurons in culture. Harvey Lect (76): 149-191, 1981.
31. McIlwain H: Brain: intracellular and extracellular purinergic receptor-systems. In: Burnstock, 163-198, 1981.
32. Paton DM: Presynaptic neuromodulation mediated by purinergic receptors. In: Burnstock, 199-220, 1981.
33. Maguire MH, Satchell DG: Specificity of adenine nucleotide receptor sites: inhibition of the guinea pig taenia coli by adenine nucleotide analogs. In: Baer and Drummond, 33-44, 1979.
34. Chapal J, Loubatieres-Mariani MM, Roye M: Effect of adenosine and phosphated derivatives on insulin release from the newborn dog pancreas. J Physiol, Paris (77): 873-875, 1981.
35. Chapal J, Loubatieres-Mariani MM: Effects of phosphate-modified adenine nucleotide analogues on insulin secretion from perfused rat pancreas. Br J Pharmac (73): 105-110, 1981.
36. Ford-Hutchinson AW: Aggregation of rat neutrophils by nucleotide triphosphates. Br J Pharmac (76): 367-371, 1982.
37. Marcus AJ: The role of lipids in platelet function: with particular reference to the arachidonic acid pathway. J Lipid Res (19): 793-826,1978.
38. Brown CM, Burnstock G: The structural conformation of the polyphosphate chain of the ATP molecule is critical for its promotion of prostaglandin biosynthesis. Eur J Pharmacol (69): 81-86, 1981.
39. Advenier C, Bidet A, Floch-Saint-Aubin A, Renier A: Contribution of prostaglandins and thromboxanes to the adenosine and ATP-induced contraction of guinea pig isolated trachea. Br J Pharmac (77): 39-44, 1982.
40. Jones TR, Lefcoe NM, Hamilton JT: Pharmacological study of adenosine and related compounds on isolated guinea pig trachea: evidence for more than one type of purine receptor. Can J Physiol Pharmacol (58): 1356-1365, 1980.
41. Moritoki H, Takei M, Kasai T, Matsumura Y, Ishida Y: Possible involvement of prostaglandins in the action of ATP on guinea-pig uterus. J Pharmacol Exp Ther (211): 104-111, 1979.
42. Needleman P, Minkes MS, Douglas JR: Stimulation of prostaglandin biosynthesis by adenine nucleotides. Circ Res (34): 455-460, 1974.
43. Isakson PC, Raz A, Denny SE, Pure E, Needleman P: A novel prostaglandin is the major product of arachidonic acid in rabbit heart. Proc Natl Acad Sci USA (74): 101-105, 1977.
44. Morrison AR, Nishikawa K, Needleman P: Thromboxane A_2 biosynthesis in the ureter obstructed isolated perfused kidney of the rabbit. J Pharmacol Exp Ther (205): 1-8, 1978.
45. Schwartzman M, Pinkas R, Raz A: Evidence for different

purinergic receptors for ATP and ADP in rabbit kidney and heart. Eur J Pharm (74): 167-173, 1981.
46. Frick GP, Lowenstein JM: Studies of 5'-nucleotidase in the perfused rat heart. J Biol Chem (251): 6372-6378, 1976.
47. Baer HP, Drummond GI: Catabolism of adenine nucleotides by the isolated perfused rat heart. Proc Soc Exp Med (127): 33-38, 1968.
48. Ryan JW, Smith U: Metabolism of adenosine 5'-monophosphate during circulation through the lungs. Trans Assn Amer Phys (84): 297-306, 1971.
49. DePierre JW, Karnovsky ML: Ecotenzymes of the guinea pig polymorphonuclear leukocyte, I. Evidence for an ectoadenosine monophosphatase, -adenosine triphosphatase, and -p-nitrophenyl phosphatase. J Biol Chem (249): 7111-7120, 1974.
50. DePierre JW, Karnovsky ML: Ectoenzymes of the guinea pig polymorphonuclear leukocyte, II. Properties and suitability as markers for the plasma membrane. J Biol Chem (249): 7121-7129, 1974.
51. Fleit H, Conklyn M, Stebbins RD, Silber R: Function of 5'-nucleotidase in uptake of adenosine from AMP by human lymphocytes. J Biol Chem (250): 8889-8892, 1975.
52. Bruns RR: Adenosine like activity of adenine nucleotides: relation to 5'-nucleotidase activity. Fed Proc (36): 689, 1977.
53. Bingham RW, Burke DC: Isolation of plasma-membrane and endoplasmic-reticulum fragments from chick embryo fibroblasts. Biochim Biophys Acta (274): 348-352.
54. Gurd JW, Evans WH: Distribution of liver plasma-membrane 5'-nucleotidase as indicated by its reaction with anti-plasma membrane serum. Arch Biochem Biophys (164): 305-311, 1974.
55. Trams EG, Lauter CJ: Sidedness of plasma-membrane enzymes. Biochim Biophys Acta (345): 180-197, 1974
56. Lelievre L: Plasma membranes from fibroblastic cells in culture: isolation, morphological and enzymatic identification. Biochim Biophys Acta (291): 662-670, 1973.
57. Dieterle Y, Ody C, Ehrensberger A, Stadler H, Junod AF: Metabolism and uptake of adenosine triphosphate and adenosine by porcine aortic and pulmonary endothelial cells and fibroblasts in culture. Circ Res (42): 869-876, 1978.
58. Dosne AM, Legrand C, Bauvois B, Bodevin E, Caen JP: Comparative degradation of adenyl nucleotides by cultured endothelial cells and fibroblasts. Biochem Biophys Res Commun (85): 183-189, 1978.
59. Glasgow JG, Schade R, Pitlick FA: Evidence that ADP hydrolysis by human cells is related to thrombogenic potential. Thromb Res (13): 255-266, 1978.
60. Cooper DR, Lewis GP, Lieberman GE, Webb H, Westwick J: ADP metabolism in vascular tissue, a possible thrombo-regulatory mechanism. Thromb Res (14): 901-914, 1979.
61. Crutchley DJ, Ryan US, Ryan JW: Effects of aspirin and dipyramidole on the degradation of adenosine diphosphate by cultured cells derived from bovine pulmonary artery. J Clin Invest (66): 29-35, 1979.
62. Naito Y, Lowenstein JM: 5'-nucleotidase from rat heart. Biochemistry (20): 5188-5194, 1981.
63. Widnell CC: Purification of rat liver 5'-nucleotidase as a

complex with sphignomyelin. Meth enzymol (32): part B, 368-375, 1974.

64. Donnand J, Bonnafous JC, Mani JC: Purification and properties of 5'-nucleotidase from lymphocyte plasma membranes. Eur J Biochem (87): 459-465, 1978.

65. Burger RM, Lowenstein JM: Preparation and properties of 5'-nucleotidase from smooth-muscle of small-intestine. J Biol Chem (245): 6274-6280, 1970.

66. Gutensohn W, Thiel E [letter] Blood (60): 536, 1982.

67. Gutensohn W, Wünsch B, Rodt H: Independent expression of the surface markers 5'-nucleotidase and cALLA on leukemic cells. Blut (46): 199-207, 1983.

68. Dornand J, Bonnafous JC, Favero J, Mani JC: Ecto-5'-nucleotidase and adeonsine deaminase activities of lymphoid cells. Biochem Med (28): 144-156, 1982.

69. van Laarhoven JPRM, Spierenburg GT, Bakkeren JAJM, Schouten TJ, De Bruyn CHMM, Geerts SJ, Schretlen EDAM: Purine metabolism in childhood acute lymphoblastic leukemia: biochemical markers for diagnosis and chemotherapy. Leukemia Res (7): 407-420, 1983.

70. Gutensohn W, Thiel E: High levels of 5'-nucleotidase activity in blastic chronic myelogenous leukemia with common ALL-antigen. Leukemia Res (5): 505-510, 1981.

71. Koya M, Kanoh T, Sawada H, Uchino H, Ueda K: Adenosine deaminase and ecto-5'-nucleotidase activities in various leukemias with special reference to blast crisis: significance of ecto-5'-nucleotidase in lympoid blast crisis of chronic myeloid leukemia. Blood (58): 1107-1111, 1981.

72. Amano S, Kreutzberg GW, Reddington M: 5'-nucleotidase activity in human astrocytomas. Acta Neuropathol, Berl (59): 145-149, 1983.

73. Sun AS, Holland JF, Lin K, Ohnuma T: Implications of a 5'-nucleotidase inhibitor in human leukemic cells for cellular aging and cancer. Biochim Biophys Acta (762): 577-584, 1983.

74. Conklyn M, Silber R: Properties of lymphocyte 5'-nucleotidase in normal subjects and patients with chronic lymphocytic leukemia. Leukemia Res (6): 203-210, 1982.

75. Wilson PD, Rustin GJS, Smith GP, Peters TJ: Electronmicroscopic cytochemical localization of nucleoside phosphatases in normal and chronic granulocytic leukaemic human neutrophils. Histochem J (13): 73-84, 1981.

21

EICOSANOIDS AND TUMOR PROMOTION

DONALD A. CHAMBERS AND RHONNA L. COHEN

1. INTRODUCTION

The concept of multistage carcinogenesis dates to observations by Peyton Rous and his colleagues in 1941, who noted a relationship between irritation and tumorigenesis in tar-induced tumors of rabbits (1). Since that time, a large volume of literature has emerged describing multi-stage carcinogenesis in skin and other organs. Some more recent reviews are listed in references 2-5.

As a scientific construct, most investigations have divided multi-stage carcinogenesis into a minimum of two stages, an induction step and a promotion step. At the outset, it should be emphasized that the pathophysiological validity of this construct has yet to be verified or confirmed. Additionally, no definitive mechanism describing multi-stage carcinogenesis has emerged. Nevertheless, recent events suggest that studies of the interactions of specific molecules such as the phorbol esters and protein kinases may yield new mechanistic understanding through which a comprehensive picture of the process of multistage carcinogenesis will emerge (6).

This review will focus on the relationship of eicosanoids to tumor promotion and will be restricted to tumor promotion in the skin. The association of eicosanoids with multi-stage carcinogenesis arises from a number of experimental observations centering on 1) associating increased prostaglandin production in tumors and in cells treated with tumor promoters and 2) correlations of the addition of inhibitors of prostaglandin synthesis with

W.E.M. Lands (ed.), BIOCHEMISTRY OF ARACHIDONIC ACID METABOLISM.

numbers of experimentally induced tumors as well as values for epidermal cell mitogenesis.

In this review, we initially describe the construct of tumor initiation and promotion and review the data associating eicosanoids with this process. An examination of some new insights of molecular mechanisms involving tumor promotion follows. We conclude by considering new avenues through which tumor promotion may be further understood.

In our literature review, we have deliberately not attempted to be all-inclusive but rather have tended to cite more comprehensive papers wherever possible so as to tell a story. In this exercise it was impossible to include all relevant papers.

2. INITIATION AND PROMOTION: THE EXPERIMENTAL SYSTEM

Carcinogenesis appears to be a sequential process involving an initial irreversible alteration of DNA followed by clonal expansion of altered cells. During the proliferative stages of tumor development, additional phenotypic changes may occur until the final properties of the tumor are expressed (7). Two-stage carcinogenesis is the experimental model for skin carcinogenesis which has been widely used for the last 20 years. This model is based on observations that irritation enhances tumor formation (1) and that by alternating application of irritant promoting agents with small doses of chemical carcinogens, greater numbers of tumors will develop more rapidly than if a single large dose or repeated small doses of carcinogen had been applied (8.9). The classic protocol for two-stage carcinogenesis in mouse skin calls for a single subthreshold dose of carcinogen (usually a polycyclic aromatic hydrocarbon) to be followed by repeated applications of an irritant-promoting agent, such as croton oil or TPA, possessing weak, if any, carcinogenic properties. The initial skin tumors to develop are benign papillomas. A minority of papillomas may later become malignant, i.e., squamous cell carcinomas. The mechanism of papilloma

Table 1 Promoting Agents*

croton oil
12-Ø-tetradecanoylphorbol-13-acetate (TPA)
mezerein
12-Ø-retinoylphorbol-13-acetate (RPA)
benzoyl peroxide
lauroyl peroxide
chlorbenzoic acid
ionophore A23187
anthralin
dihydroteleocidin B
cantharidin
1-fluro-2,4,dinitrobenzene
benzo(a)pyrene
iodocaetic acid
Tween 6Ø
repeated skin wounding

*see ref. 2-5 for individual references.

conversion from benign to malignant is obscure.

The sequence of initiator application followed within weeks by repeated applications of promoter is critical to the efficacy of the process. No tumors develop if the promoter is applied before the initiator or if only a few applications of promoter are made over a period of months. Initiation is an irreversible act. The mechanism of initiation is not understood but evidence suggesting that some cells in the treated area undergo alteration of DNA by mutation is generally accepted. In fact, Weinstein and colleagues (1Ø), have suggested that initiation may be characterized by alteration at the level of the gene, and promotion as membrane perturbation. Farber (7) and Fürstenberger and Marks (11) have suggested that a round of cell division is a necessary second step in initiation. Farber has also questioned the dogma that initiating agents always damage DNA by citing evidence that some carcinogens have no known mutagenic or DNA-altering properties (7).

Promoting agents (see Table 1) have the common effects of irritation and epidermal hyperproliferation, but their

Table 2 Responses of Murine Skin to Promoting Agents*

induction of:
- epidermal proliferation
- dark basal keratinocytes
- inflammation
- specific proteins
- ornithine decarboxylase

increases in:
- DNA, RNA
- polyamines
- phospholipid synthesis
- prostaglandins
- histone synthesis and phosphorylation
- cAMP independent protein kinase

decreases in:
- epidermal superoxide dismutase and catalase activities
- isoproterenol stimulation of cAMP

*see ref. 2-5 for details and individual references

mechanism of action is unknown. The pleiotropic response of skin treated with TPA, the classic and most studied promoting agent, has been characterized morphologically and biochemically (2-5) (Table 2). The major morphological responses include inflammation, epidermal hyperproliferation, and appearance of dark cells, i.e., epidermal cells containing large numbers of ribosomes thought to represent a stem cell population (12,13). Key biochemical responses of epidermal cells to TPA include increases in total DNA, RNA and protein synthesis, increases in the synthesis of phospholipids and prostaglandins, increases in ornithine decarboxylase (ODC) activity and polyamine synthesis, decreases in protease, histidase, superoxide dismutase and catalase activities. Twenty years ago, Boutwell suggested that promotion in vivo occurs in two stages (14), the first stage comprised of gene activation and new phenotypic expression, the second consisting of tumor cell proliferation. Recent reports from the laboratories of Slaga (15) and Fürstenberger and Marks (16) have elaborated on this concept.

Slaga's group utilizes the carcinogen-sensitive Sencar mouse strain, originally bred by Boutwell. The two-stage promotion scheme they follow calls for limited application of a complete promoter such as TPA, followed by applications of the weak promoter, mezerein (15). The TPA-induced, first stage is characterized by the appearance of dark cells and elevation of prostaglandins and can be inhibited by fluocinolone acetonide. The second, mezerein-induced stage is characterized by polyamine synthesis and cell proliferation and can be inhibited by retinoic acid. Most recently, Slaga has expanded this concept into a three-stage promotion scheme which further divides the second stage into mezerein-induction of ODC and polyamines followed by application of a non-promoting, hyperplastic agent, e.g., ethylphenylpropiolate (17).

Fürstenberger and Marks propose use of the non-promoting epidermal mitogen, phorbol 12-retinoate 13-acetate (RPA), as the second stage agent (16). RPA combines the structural features of the promoting phorbol esters with the structure of the inhibitor, retinoic acid, thus attempting to separate the characteristics of promotion and proliferation. Both groups, Slaga and Fürstenberger and Marks, point out that mezerein (16) and RPA (17) have biological effects similar to those of TPA, thus complicating clear delineation of the mechanistic events of promotion.

Regardless of the increasing complexity inherent in defining the number of stages in these experimental carcinogenesis systems, it is still not known which, if any, of the biological effects of tumor promoters (Table 2) are necessary and sufficient for tumorigenesis.

Cell culture systems have been extensively employed in the analysis of the biological effects of either initiators or promoters on specific cell types including skin epidermal cells, kidney epithelial cells, lymphocytes, etc. A wealth of data has been acquired from these in vitro experiments (18-20). However, there is no defined end point for

Table 3 Inhibitors of Murine Skin Tumor Promotion*

anti-inflammatory steroids:
- dexamethasone
- fluocinolone acetonide

protease inhibitors:
- tosyl lysine chloromethyl ketone (TLCK)
- tosyl phenylalanine chloromethyl ketone (TPCK)

prostaglandin inhibitors:
- indomethacin
- 5,8,11,14-eicosatetraynoic acid (ETYA)
- phenidone

thromboxane inhibitors:
- imidazolacetophenone
- imidazolphenol

arachidonic acid
retinoic acid

see ref. 2-5 for details and individual references.

promotion in in vitro systems. Definition of two-stage carcinogenesis is based on the number of tumors developed in vivo. The morphological and biochemical parameters listed in Table 2 do not by themselves define promotion. Although conventional wisdom supports the value of promoter-treated cell cultures which reproduce some findings of in vivo experiments, e.g., mitogenesis and ODC levels, the two systems are not equivalent a priori and the precise mechanisms of TPA action may differ.

3. EICOSANOIDS AND TUMOR PROMOTION

3.1 Prostaglandins

It is logical that in the search for mechanism(s) of tumor promotion, potential regulatory roles for eicosanoids would be intensively investigated because of the following evidence: 1) reports of elevated prostaglandin levels in spontaneous and experimentally induced tumors (21, 22), 2) interaction between TPA and cell membranes results in altered phospholipid metabolism (14) and prostaglandin elevation (23,24) and 3) observations that certain inhibitors of promotion (e.g., indomethacin, fluocinolone acetonide) (Table 3) also inhibit arachidonic acid metabolism (25-27). Three of the biologic responses considered to be hallmarks of promotion, i.e., epidermal hyperplasia, induction of ODC and inflammation, have also been associated with prostaglandins.

Early biochemical evidence of the interaction of TPA with cell membranes, including direct binding studies (28) and choline incorporation into cell membranes (29) prompted Rohrschneider, O'Brien and Boutwell to examine phospholipid metabolism in croton oil or TPA-treated mouse skin (30). Their finding of increased metabolism in whole skin of phosphatidyl ethanolamine and phosphatidylcholine within hours of treatment was later confirmed in pure epidermal preparations (31). In 1977, Verma, Rice and Boutwell provided more direct evidence of a link between prostaglandins and promotion when they showed that the induction of ODC could be inhibited by pretreating mouse skin with indomethacin or other agents which inhibit both inflammation and prostaglandin synthesis prior to TPA application (25). Addition of exogenous PGE_2 but not $PGF_{2\alpha}$, simultaneously with TPA, restored ODC induction. Utilizing the same type of treatment protocol, epidermal cell division (26) and papilloma formation (27) were also linked with PGE_2 increases. Similar studies measuring the in vitro effect of tumor promoters on arachidonic acid metabolism, effects of prostaglandin inhibitors on ODC-induction and DNA synthesis have been reported (32-36).

The effect of prostaglandins on the classic two-stage carcinogenesis protocol was studied by Fischer et al. (37). Sencar mouse skin was treated with a subthreshold dose of the carcinogen dimethylbenzanthracene followed by 15 weeks of TPA treatment with simultaneous application of prostaglandins E_1, E_2, $F_{2\alpha}$, arachidonic or linoleic acid. Papilloma number was increased with $PGF_{2\alpha}$ treatment, decreased by PGE_1 and unaffected by arachidonic or linoleic acids. Experiments varying the time of PGE_2 treatment revealed that PGE_2 inhibited papilloma formation when applied 15 minutes before to 45 minutes after TPA, slightly enhanced papilloma formation when applied 90 or 45 minutes before TPA and had no effect 90 minutes after TPA treatment. Although these experiments suggest a role for prostaglandins in cell proliferation, there is no direct evidence that elevated prostaglandin levels are sufficient to cause either ODC induction (38), epidermal proliferation (39) or tumor promotion (40,41).

Recent work by Fürstenberger, Gross and Marks (40) indicates that both PGE_2 and $PGF_{2\alpha}$ have comitogenic effects when concomitantly administered with promoting agents. They propose that promotion occurs in two stages. In the first, irreversible stage, a single treatment of TPA to NMRI mouse skin results in a rapid PGE_2 elevation followed by a transient wave of epidermal hyperplasia. A wave of $PGF_{2\alpha}$ increase, which occurs hours later, appears to be the critical step in the development of promotion in this system. The second stage of promotion, which follows after repeated applications of RPA, a non-promoting agent (16), is reversible and consists of sustained epidermal hyperplasia. Fürstenberger and colleagues conclude from these studies that different populations of epidermal cells are triggered by promoting agents and that prostaglandins may serve as regulatory molecules either directly or in concert with endogenous growth factors.

During the course of in vivo experiments in which the multistage TPA-mezerein promotion protocol was used with the selective addition of appropriate inhibitors, Fischer and coworkers found that decreased numbers of papillomas could be related to inhibition of phospholipase 2, cyclooxygenase, lipoxygenase and thromboxane synthetase (42). Thus, for the first time, implicating metabolites of arachidonic acid other than prostaglandins in tumor promotion.

The effects of different classes of prostaglandins on skin tumor promotion are difficult to dissect, being dependent on variables such as the specific type and dose of promoting agent, the protocol followed, strain of mouse

Table 4 Free Radical Generating Promoting Agents*

benzoyl peroxide
m-chlorobenzoic acid
p-nitroperbenzoic acid
lauroyl peroxide
decanoyl peroxide
cumene peroxide
2,2'-azobis-2-methylpropionitrile
tert-butyl hydroperoxide (weak promoter)
methyl ethyl ketone hydroperoxide (weak promoter)

*see ref. 44-52 for additional agents and details.

treated and method of assay. However confusing the reports in the literature may seem to be, two things are clear. First, topical application of prostaglandins does not by itself lead to tumor promotion (24,37) and second, prostaglandin elevation is not by itself sufficient for promotion (27,33,41).

3.2 Free Radicals

The possibility that free radicals may act as mediators of promotion is currently a hot area for speculation but at present there is little hard data to support this concept. Suggestive evidence for a role for free radicals comes from four sources: 1) free radical generating compounds can serve as promoters, 2) inhibitors of promotion also inhibit free radical generation, 3) arachidonic acid and hydroperoxides can act as clastogenic agents, 4) endoperoxides of the lipoxygenase pathway have been implicated in epidermal proliferation.

Certain free radical generating compounds (Table 4) can serve as incomplete promoters, capable of eliciting some, but not all, of the biological effects characteristic of tumor promotion (Table 2). A correlation between degree of promoter efficacy and free radical generating capacity has been reported (44).

The cellular enzymes, superoxide dismutase (SOD) and

catalase, act in concert to reduce free radicals and H_2O_2 produced by actively metabolizing cells. High levels of free radicals and H_2O_2 can lead to DNA damage by direct action or in association with hydroperoxide formation. Inhibition of promoter-induced superanion production by inhibitors of tumor promotion, such as retinoids, protease inhibitors and anti-inflammatory steroids has been demonstrated. Solanki et al. (45) reported that TPA or mezerein treatment of Sencar mouse skin appeared to significantly reduce intracellular levels of SOD and catalase. However, inhibitors of tumor promotion, e.g., retinoic acid and fluocinolone acetonide, did not restore SOD and catalase levels in these experiments. Kensler et al. (46) replaced the anti-promotion activity of SOD by pretreating CD-1 mouse skin with an SOD-memetic agent before TPA application, resulting in effective inhibition of ODC induction and papilloma formation. Taken together, these observations suggest that increased free radical concentration may play a role in promotion, but free radical-mediated effects on promoter treated skin are not solely sufficient as a mechanism for promotion.

Inflammation is a hallmark of promotion. A number of investigators (44,47-49) have reasoned that free radicals generated by inflammatory cells may play a key role in altering epidermal DNA by causing chromosome breaks or affecting specific transcription by direct or indirect means. In a recent workshop on the role of free radicals in promotion (50) Birnboim suggested that promoters can be classified into those that cause direct target cell DNA (50) strand breaks, e.g., benzoyl peroxide (48) and those that act on effector (inflammatory) cells. Inflammatory cells such as polymorphonuclear leucocytes (44,47), macrophages (51) and lymphocytes (49,52) treated with promoters have been demonstrated to produce membrane-associated clastogenic factors. This group of low molecular weight factors including oxygen radicals, peroxides, aldehyde breakdown products and free arachidonic acid, has been linked to chromosome damage seen after exposure to ionizing radiation, as well as in chronic inflammatory diseases and genetic diseases associated with high cancer rates. Emerit and Cerutti suggest that clastogenic factors may be transferred from inflammatory cells into adjacent target cells (49). This indirect mode of chromosome damage may provide a mechanistic role in tumor promotion for the inflammatory cells elicited by promoting agents. DNA strand breaks, sister chromatid exchanges and structural alterations have been reported in promoter treated epidermal cells in vitro (53), but no direct correlation to mechanism has as yet been suggested.

3.3 Leukotrienes

The lipoxygenase pathway of arachidonic acid metabolism is believed by some investigators to be of critical importance in tumor promotion (43). Supportive data comes from the work of Fischer et al. (42) and Kato et al. (54). The Fischer group compared the number of papillomas resulting from concomitant treatment of Sencar mouse skin with TPA and inhibitors of phospholipase A_2 (dibromoacetophenone) or cyclooxygenase (indomethacin, flurbiprofen) (42). The greatest inhibition of papilloma formation occurred with phospholipase A_2 and cyclooxygenase/lipoxygenase inhibitors (5,8,11,14-eicosatetrayonic acid, phenyl-3-pyrazolidinone). Parallel experiments in epidermal cell culture assaying DNA synthesis as the end point, showed inhibition in cultures treated with antiphospholipase A_2 or anti-cyclooxygenase/lipoxygenase agents. Kato and his colleagues reported similar findings of reduced papilloma induction in vivo after combined treatment with TPA and inhibitors of phospholipase A_2 (BPB, mepacrine) or lipoxygenase (NDGA, quercetin, morin) (54).

Additional suggestive evidence for a role for leukotrienes in tumor promotion and epidermal hyperplasia comes from recent investigations of eicosanoids in the etiology of the benign hyperproliferative skin disease, psoriasis. At least two research groups have noted a marked increase of the leukotriene LTB4 in active psoriatic lesions which is not found in uninvolved or normal skin (55,56).

4. MOLECULAR MECHANISMS OF TUMOR PROMOTION

In the past, hypotheses of the molecular mechanism of TPA action centered around induction of the enzyme ornithine decarboxylase, leading to increased cellular levels of polyamines (38). However, that polyamine increases could not directly account for TPA effects was demonstrated when Yuspa and his colleagues showed that skin subjected to TPA in the presence fo the anesthetic, lidocaine, failed to induce ODC but did respond mitogenically to the promoter (57). Similar results have been found by others (38, 58-60). Thus, although it is probable that polyamines and ODC induction play a role in TPA action, that role does not appear to be obligatory for the mitogenic properties of the compound.

Recent excitement in the problems of mechanisms of tumor promotion has centered on the characterization of the TPA receptor and its association with a Ca++, phospholipid-dependent protein kinase, protein kinase C (6). Thus, three separate research groups: Blumberg and his collaborators (61,62), Slaga and his colleagues (63-65), and Sando and her associates (66,67) have been able to identify the TPA receptor in cytosol, particulate fractions and whole cells by using the phorbol derivative (20-^{3}H) phorbol 12, 13 dibutyrate to specifically label the receptor in several tissues including brain, skin and thymus cells (68).

Some investigators have suggested that TPA binding sites can be located in the nucleus. In 1974, data reported by Slaga and colleagues (69) provided evidence of TPA binding to chromatin fractions derived from CD-1 mouse skin treated with (^{3}H)TPA. The level of binding was equivalent to 16-24% of the total subcellular distribution and could be suppressed in both chromatin and cytosol fractions by dexamethasone, a promotion inhibitor. More recently, Perella and coworkers (7Ø) demonstrated two (^{3}H) TPA binding sites in epidermal nuclei isolated from CD-1 mouse skin, one of high affinity and a second low affinity component which may represent non-specific binding to the nuclear membrane or nuclear protein matrix. This data suggests that TPA could act directly on the cellular genome.

Most receptor-ligand binding studies have revealed Scatchard plot analysis of at least two classes of binding sites; high affinity and low affinity (68). Such heterogenous binding sites could be present on the same cells or could reflect cellular heterogeneity. Even apparently homogeneous cell lines show curvilinear binding of TPA, suggesting multiple binding sites (67). It will be important to resolve this problem because it is possible that different classes of binding sites may be functionally related to different effector molecules. Competitive binding studies with derivatives of TPA showed a close parallelism between the affinity of binding of the specific derivative and its effect as a tumor promoter in tissues previously shown to be responsive. Binding required the presence of Ca++, and diacylglycerol competed effectively for TPA (68). This observation suggests that the endogenous co-factor is diacyl glyceride.

Recent studies on the turnover of the TPA receptor, reviewed by Solanki and Slaga (65), reveal that this receptor is subject to down modulation. In epidermal cells, recovery of the receptor to normal levels occurs rapidly within one hour and appears to be independent of a requirement for concomitant protein synthesis because the reappearance of the receptor is not affected by the protein synthesis inhibitor, cycloheximide (64). Some evidence exists that in cell culture the TPA receptor is stabilized by the presence of serum (64,65).

TPA binding is not obligatory to activation of the phenotypic TPA-mediated response. Thus, although Estensen and coworkers have been able to correlate TPA binding with mitogenic activity in human lymphocytes (72), Shoyab et al. observed that TPA can bind to two cell types which lack promotability (73) and Chambers et al. have shown that TPA can bind to epidermal cells derived from TPA-resistant nude mice (74-76).

Evidence is accumulating that as a function of membrane perturbation the enzyme, phospholipase C,found within the cell membrane, specifically releases diacyl glycerides from phosphatidyl inositides (77). The generation of the diacyl glycerol is coupled with the release of bound arachidonic acid, which can then be transformed into a variety of eicosanoids (78), both of the prostaglandin and leukotriene series.

This complex series of reactions involving the turnover of phospholipids also involves, in addition to phospholipase C, a diacyl glycerol kinase (79). An alternate means of releasing membrane arachidonic acid is through phospholipase A_2 (8Ø). The relative contributions of each pathway are unknown. However, it is assumed by many that the phospholipase C-mediated pathway is the main one involved in regulatory events. It should be noted that the activation of the arachidonic acid cascade may well be a by-product of the reactions accompanying the binding of the tumor promoter to the cell, rather than a causal event in the primary mechanisms of tumor promotion.

A further mechanistic breakthrough occurred when it was shown that the TPA receptor couples with protein kinase C, suggesting that at least some portion of the pleiotropic effects of TPA, could arise from the post-translational modification of proteins by the TPA-mediated activation of protein kinase C (63,67,68,81-83). Protein kinase C (PKC) was first discovered by Nishizuka and his colleagues in 1977 (84). It is a ubiquitous enzyme, being widely found in many different tissues of the body (85). Often its activity far exceeds the cyclic AMP or cyclic GMP-dependent protein kinases. Protein kinase C has been purified from brain (86) and platelets (87) and where examined, appears to have similar characteristics independent of tissue source. The enzyme consists of a single polypeptide of 77,000 daltons containing a hydrophilic domain and a hydrophobic domain (86,88). Presumbly the role of the hydrophobic domain is to promote membrane interactions whenever the hydrophilic portion of the molecule contains the active site. Protein kinase C is maximally active in the presence of diacyl glycerol (83,89) containing an unsaturated fatty acid, i.e., arachidonic acid, phosphatidyl serine (90) and Ca++.

The enzyme has been found both in cytosol and in association with cell membranes (6,81,86,87,91). Apparently, an equilibrium between soluble and particulate species of PKC exists. In the presence of increased concentrations of calcium, the particulate form of the enzyme appears to be favored. TPA appears also to stabilize the particulate form of the enzyme (91).

A class of Ca++-dependent thiol proteases have been identified which are capable of cleaving PKC to a fragment active in the absence of co-factors (89). The physiological significance of these proteases has yet to be established.

Studies of substrate specificity of PKC show that it phosphorylates proteins at serine and threonine (92,94). Unlike the situation found with the EGF kinase, the platelet derived growth factor kinase or kinases associated with oncogenes (93,94), no tyrosine phosphorylation has been associated with PKC. The recent observation that the TPA receptor copurifies with PKC has excited many investigators and led to the intriguing postulate that TPA may exert its pleiotropic effects through the action of PKC, somewhat analogous to the mechanisms of action of protein hormones mediated through the cAMP-dependent protein kinases. Thus, TPA appears to replace the role of diacyl glycerol as a PKC cofactor as well as to diminish the need of the enzyme for calcium by increasing the enzyme affinity for Ca++ to 10^{-7}M. TPA effectively competes with diacyl glycerol for binding to the receptor (95) and stoichiometric studies suggest that one molecule of TPA binds per molecule of enzyme, probably through the phospholipid cofactor (96,97). In fact, analysis of the chemical structure of the two moieties suggests structural similarities (6). Some evidence suggests that TPA can directly activate PKC. Thus, when platelets are incubated in the presence of TPA, a specific 40,000 dalton protein is phosphorylated to the same extent as it is when platelets are activated by the natural activator, thrombin. Although analysis of the labeled ^{32}P protein suggested that the phosphorylation was mediated by PKC, no concomitant breakdown of phospholipid or synthesis of diacyl glycerol could be observed suggesting that TPA could substitute for the enzyme's requirement of diacyl glycerol (98).

Other tumor promoters which have chemical structures differing from TPA or diacyl glycerol, e.g., mezerein and teleocidin, can also activate PKC (6,99). The mechanism of their activation is obscure but suggestive of a membrane perturbation, resulting in an active enzyme species with a conformation similar to that induced by the diacyl glycerol-like factors.

Although the specific substrates for PKC have yet to be fully identified, a 40,000 dalton protein has been associated with PKC activity in platelets under in vitro conditions. Protein kinase C has mediated the phosphorylation of a number of proteins including vinculin, microtubule-associated protein and myosin-light chains (6). Recent studies by Gilmore and Martin utilizing cultured fibroblasts have suggested an association of TPA with tyrosine phosphorylation (100). Other studies in A431 cells, by King and her collaborators, have revealed that when membranes are incubated with TPA in the absence of PKC, protein phosphorylation occurs at tyrosine moieties (101). However, when these membranes were specifically reconstituted with PKC, protein phosphorylation occurred at threonine and serine (102). These results suggest the possibility of feedback loops involving the interaction of

PKC with tyrosine kinase and may begin to illuminate an even more complex mechanism.

The importance of protein kinases is their role in post-translational modification of their specific substrate proteins. Thus, to gain greater understanding of the ultimate effects of protein kinase, we need to identify specific protein phosphorylation which occurs as a function of activation of the protein kinase in question, e.g., protein kinase C. To date, few studies have been carried out in skin.

Recently, three different research groups have noted that addition of TPA to A431 cells leads to specific phosphorylation of the EGF receptor to produce phosphoserine residues (103-105). In addition, Belman and Garte (106) have shown that TPA uncouples β-adrenergic receptors from adenylate cyclase in mouse epidermis and Sibley et al (107) have shown that the β-adrenergic receptor-associated activation of adenylate cyclase can be desensitized by a TPA-mediated kinase event. The latter group have suggested that this data supplies important evidence for the postulate that protein kinase C regulates β-adrenergic function.

In parallel studies, with the aim of associating the appearance of proteins in the skin with TPA treatment, a number of different investigators have reported such changes (76,108,109). Slaga and his collaborators (108) have examined a spectrum of proteins which change as a function of TPA addition as have Yuspa et al. (109) and Chambers et al. (76). Taken in total, all these experiments can be summarized by observing that TPA mediates changes in cellular proteins and protein phosphorylation, but for the present, the meaning of these changes eludes understanding.

This section has centered on the association of TPA with protein kinase C. However, it is important to point out that cells contain other protein kinases as well, including calmodulin-dependent protein kinase, cyclic AMP-dependent protein kinase, cyclic GMP-dependent protein kinase, growth factor-protein kinase (93,110). Among these protein kinases, some have been shown to phosphorylate protein at serine and threonine amino acids while others are specific for tyrosine. The mechanisms of how protein kinases work in concert to regulate cell physiology has barely been touched upon. Further understanding of these events is likely to be necessary for the solution of the tumor promotion problem. However, the association of a specific kinase, protein kinase C, with the TPA receptor may be a milestone in the path towards solution.

5. CONCLUSIONS

In this chapter, we have reviewed the data suggesting a relationship between eicosanoids and tumor promotion. That eicosanoids are involved in the processes of cell proliferation, differentiation, hyperplasia and neoplasia there can be no doubt. However, as of the date of this review, no causal evidence exists to suggest that eicosanoids play a major role in the mechanisms of tumor promotion.

In the construct of tumor promotion, a major problem exists relating to definitions. Thus, since tumor promotion has as its real in vivo endpoint, the producton of tumors, it is difficult, if not impossible, to define in vitro equivalents for the in vivo situation. The functional in vitro equivalent of promotion has become mitogenesis but studying mitogenesis and the factors leading to it as a function of promotion can be fraught with peril. One example will serve to illustrate this point. Prostaglandins of the E series have been associated with increases of intra-cellular cyclic AMP (78). If one then tries to hypothesize a general promoter effect resulting in mitogenesis mediated through a PGE-cAMP process, the following dilemma occurs. Cyclic AMP is a mitogen for epidermal cells (111,112), but an inhibitor of lymphocyte proliferation (113,114). Thus, it may be likely that specific mechanisms for promotion may require definition for each cell type. A correlate of this argument is that careful attention has to be paid to the appropriateness of the model system.

A great mechanistic advance has come from the studies cited herein of the characterization of the TPA receptor and its association with protein kinase C. This research heralds a new examination of the interrelationship of all protein kinases, their substrate proteins and the means through which phosphoprotein modifications govern bio-regulatory events. We believe that additional advances will come from detailed in vivo and in vitro studies of the interaction of promoters with cells and tissues derived from normal and promoter-hypersensitive, i.e., Sencar mice and promoter-resistant animals, i.e., nude mice (74,75,115-118). Such studies may bring to the field of tumor promotion, the analytic systems capable of resolving the mechanistic mysteries that now prevail. It is our hope that future reviewers will, as a result of continued innovative research, be able to give clearer expositions of the promoter problem than we have.

ACKNOWLEDGEMENTS

The authors would like to thank B. Coyne, M. Ostrega, J. Ziomek and B. Whittington for their invaluable assistance in assembling this manuscript. In addition, we express our appreciation to inenumerable colleagues who have discussed the "tumor promotion problem" with us.

REFERENCES

1. Rous P, Kidd JG: Conditional neoplasms and subthreshold neoplastic states: A study of the tar tumors of rabbits. J Exp Med (73): 365-389, 1941.

2. Boutwell RK: The function and mechanism of promoters of carcinogenesis. CRC Crit Rev Toxicol (2): 419-443, 1974.

3. Blumberg PM: In vitro studies on the mode of action of the phorbol esters, potent tumor promoters: Part I. CRC Crit Rev Toxicol (8):

153-196, 1980.

4. Diamond L, O'Brien TG, Baird WM: Tumor promoters and the mechanism of tumor promotion. Adv Cancer Res (32): 1-74, 1980.

5. Slaga TJ, Fischer SM, Weeks CE, Klein-Szanto AJP, Reiners J: Studies on the mechanisms involved in multistage carcinogenesis in mouse skin. J Cell Biochem (18): 99-119, 1982.

6. Nishizuka Y: The role of protein kinase C in cell surface signal transduction and tumour promotion. Nature (308): 693-698, 1984.

7. Farber E: Chemical carcinogenesis: A biologic perspective. Amer J Pathol (106): 271-296, 1982.

8. Berenblum I: The cocarcinogenic action of croton resin. Cancer Res (1): 44-48, 1941.

9. Mottram JC: A developing factor in experimental blastogenesis. J Pathol Bacteriol (56): 181-187, 1944.

10. Weinstein IB, Gattoni-Celli S, Kirschmeier P, Lambert M, Hsiao W, Backer J, Jeffrey A: Multistage carcinogenesis involves multiple genes and multiple mechanisms. In: Levine AJ, Vande Woude GF, Topp

11. Fürstenberger G, Gross M, Marks F: Involvement of prostaglandins in the process of skin tumor promotion. In: Thaler-Dao H, Craster de Paulet A, Paoletti R, (eds) Icosanoids and cancer. Raven Press, New York, 1984, pp 91-100.

12. Raick AN: Ultrastructural, histological and biochemical alterations produced by 12-0-tetradecanoyl-phorbol-13-acetate on mouse epidermis and their relevance to skin tumor promotion. Cancer Res (33): 269-286, 1973.

13. Klein-Szanto AJP: Morphological evaluation of tumor promoter effects on mammalian skin. In: Slaga TJ (ed) Mechanisms of tumor promotion. v.II Tumor promotion and skin carcinogenesis, CRC Press, Inc., Boca Raton, Florida, 1984, pp 41-72.

14. Boutwell RK: Some biological aspects of skin carcinogenesis. Prog Exp Tumor Res (4): 207-250, 1964.

15. Slaga TJ, Fischer SM, Nelson K, Gleason GL: Studies on the mechanism of skin tumor promotion: Evidence for several in stages promotion. Proc Natl Acad Sci

USA (77): 3659-3633, 1980.

16. Fürstenberger G, Berry DL, Sorg B, Marks F: Skin tumor promotion by phorbol esters is a two-stage process. Proc Natl Acad Sci USA (78): 7722-7726, 1981.

17. Slaga TJ: Multistage skin tumor promotion and specificity of inhibition. In: Mechanisms of tumor promotion. v.II. Tumor promotion and skin carcinogenesis. CRC Press, Inc., Boca Raton, Florida, 1984, pp 189-196.

18. Diamond L, O'Brien TG, Rovera G: Tumor promoters: Effects on proliferation and differentiation of cells in culture. Life Sciences (23): 1979-1988, 1978.

19. Fusenig NE, Dzarlieva RT: Phenotypic and chromosomal alterations in cell cultures as indicators of tumor-promoting activity. In: Hecker E, Kunz W, Theilman HW, Fusenig NE, Marks F (eds) Carcinogenesis v.7 Cocarcinogenesis and biological effects of tumor promoters. Raven Press, New York, 1982, pp 201-216.

20. Yuspa SH, Hennings H, Kulesz-Martin M, Lichti U: The study of tumor promotion in a cell culture model for mouse skin-a tissue that exhibits multistage carcinogenesis in vivo. In: Hecker E, Kunz W, Theilman HW, Fusenig NE, Marks F (eds) Carcinogenesis v.7 Cocarcinogenesis and biological effects of tumor promoters. Raven Press, New York, 1982, pp 217-230.

21. Levine L: Arachidonic acid transformation and tumor production. Adv Cancer Res (35): 49-79, 1981.

22. Honn KV, Bockman RS, Marnett LJ: Prostaglandins and cancer: A review of tumor initiation through tumor metastasis. Prostaglandins (21): 833-864, 1981.

23. Bresnick E, Meunier P, Lamden M: Epidermal prostaglandins after topical application of a tumor promotor. Cancer Letters (7): 121-125, 1979.

24. Ashendel CL, Boutwell RK: Prostaglandin E and F levels in mouse epidermis are increased by tumor-promoting phorbol esters. Biochem Biophys Res Commun (90): 623-627, 1979.

25. Verma AK, Rice HM, Boutwell RK: Prostaglandins and skin tumor promotion: Inhibition of tumor promoter-induced ornithine decarboxylase activity in epidermis by inhibitors of prostaglandin synthesis. Biochem Biophys Res Commun (79): 1160-1166, 1977.

26. Fürstenberger G, Marks F: Indomethacin inhibition

of cell proliferation induced by the phorbolester TPA is reversed by prostaglandin E_2 in mouse epidermis in vivo. Biochem Biophys Res Commun (84): 1103-1111, 1978.

27. Verma AJ, Ashendel CL, Boutwell RK: Inhibition by prostaglandin synthesis inhibitors of the induction of epidermal ornithine decarboxylase activity, the accumulation of prostaglandins, and tumor promotion caused by 12-0-tetradecanoylphorbol-13-acetate. Cancer Research (40): 308-315, 1980.

28. Sivak A, van Duuren BL: Cellular interactions of phorbol myristate acetate in tumor promotion. Chem Biol Interactions (3): 401-411, 1971.

29. Süss R, Kreibich G, Kinzel V: Phorbol esters as a tool in cell research? Europ J Cancer (8): 299-304, 1972.

30. Rohrschneider LR, O'Brien DH, Boutwell RK: The stimulation of phospholipid metabolism in mouse skin following phorbol ester treatment. Biochem Biophys Acta (280): 57-70, 1972.

31. Balmain A, Hecker E: On the biochemical mechanism of tumorigenesis in mouse skin. VI. Early effects of growth-stimulating phorbol esters on phosphate transport and phospholipid synthesis in mouse epidermis. Biochem Biophys Acta (362): 457-468, 1974.

32. Ganss M, Seeman D, Fürstenberger G, Marks F: Calcium-dependent release of arachidonic acid from a murine epidermal cell line induced by the tumor promoter TPA or ionophore A 23187. FEBS Letters (142): 54-58, 1982.

33. Levine L: Effects of tumor promoters on arachidonic acid metabolism by cells in culture. In: Hecker E, Kunz W, Theilman HW, Fusenig NE, Marks F (eds) Carcinogenesis v.7 Cocarcinogenesis and biological effects of tumor promoters. Raven Press, New York, 1982, pp 477-494.

34. Fürstenberger G, Richter H, Fusenig NE, Marks F: Arachidonic acid and prostaglandin E_2 release and enhanced cell proliferation induced by the phorbol ester TPA in a murine epidermal cell line. Cancer Letters (11): 191-198, 1981.

35. Fürstenberger G, Delescluse C, Fischer SM, Richter H, Marks F: Early induction of the arachidonic acid cascade and stimulation of DNA synthesis by TPA in murine and guinea pig epidermal cells in culture.

In: Hecker E, Kunz W, Thielman HW, Fusenig NE, Marks F (eds) Carcinogenesis v.7 Cocarcinogenesis and biological effects of tumor promoters. Raven Press, New York, 1982, pp 325-329.

36. Levine L, Goldstein SM, Snoek GT, Rigas A: Arachidonic acid metabolism by cells in culture: Effects of tumor promoters. In: Thaler-Dao H, Craster de Paulet A, Paoletti R (eds) Icosanoids and cancer. Raven Press, New York, 1984, pp 115-125.

37. Fischer SM, Gleason GL, Hardin LG, Bohrman JS, Slaga J: Prostaglandin modulation of phorbol ester skin tumor promotion. Carcinogenesis (1): 245-248, 1980.

38. Weeks CE, Slaga TJ, Boutwell RK: The role of polyamines in tumor promotion. In: Slaga TJ (ed) Mechanisms of tumor promotion v.2 Tumor promotion and skin carcinogenesis. CRC Press, Inc., Boca Raton, Florida, 1984, pp 127-142.

39. Fürstenberger G, Marks, F: Early prostaglandin E synthesis is an obligatory event in the induction of cell proliferation in mouse epidermis in vivo by the phorbol ester TPA. Biochem Biophy Res Commun (92): 749-756.

40. Fürstenberger G, Gross M, Marks F: On the role of prostaglandins in the induction of epidermal proliferation, hyperplasia and tumor promotion in mouse skin. In: Powles TJ, Bockman RS, Honn KV, Ramwell P (eds) Prostaglandins and cancer: First international conference, Alan R Liss, Inc., New York, 1982, pp 239-254.

41. Fischer SM: The role of prostaglandins in tumor production. In: Slaga TJ, (ed) Mechanisms of tumor promotion v.2 Tumor promotion and skin carcinogenesis. CRC Press, Inc, Boca Raton, Florida, 1984, pp 113-126.

42. Fischer, SM, Mills GD, Slaga TJ: Inhibition of mouse skin tumor promotion by several inhibitors of arachidonic acid metabolism. Carcinogenesis (3): 1243-1245, 1982.

43. Fischer SM: Arachidonate cascade and skin tumor promotion. In: Thaler-Dao H, Craster de Paulet A, Paoletti R (eds) Icosanoids and cancer. Raven Press, New York, 1984, pp 79-90.

44. Troll W, Witz G, Goldstein B, Stone D, Sugimura T: The role of free oxygen radicals in tumor promotion and carcinogenesis. In: Hecker E, Kunz W, Theilman HW, Fusenig NE, Marks F (eds)

Carcinogenesis v.7 Cocarcinogenesis and biological effects of tumor promoters. Raven Press, New York, 1982, pp 593-597.

45. Solanki V, Rana RS, Slaga TJ: Diminution of mouse epidermal superoxide dismutase and catalase activities by tumor promoters. Carcinogenesis (2): 1141-1146, 1981.

46. Kensler TW, Bush DM, Kozumbo WJ: Inhibition of tumor promotion by a biomimetic superoxide dismutase. Science (221): 75-77, 1983.

47. Goldstein BD, Witz G, Amoruso M, Stone DS, Troll W: Stimulation of human polymorphonuclear leukocyte superoxide anion radical production by tumor promoters. Cancer letters (11): 257-262, 1981.

48. Slaga TJ, Klein-Szanto AJP, Triplett LL, Yotti LP: Skin tumor-promoting activity of benzoyl peroxide, a widely used free radical-generating compound. Science (213): 1023-1025, 1981.

49. Emerit I, Cerutti PA: Tumor promoter phorbol 12-myristate 13-acetate induces a clastogenic factor in human lymphocytes. Proc Natl Acad Sci USA (79): 7509-7513, 1982.

50. Copeland ES: Free radicals in promotion - a Chemical Pathology Study Section workshop. Cancer Res (43): 5631-5637, 1983.

51. Emerit I, Cerutti P: Icosanoids and chromosome damage. In: Thaler-Dao H, Craster de Paulet A, Paoletti R (eds) Icosanoids and cancer. Raven Press, New York, 1984, pp 127-138.

52. Emerit I, Levy A, Cerutti P: Suppression of tumor promoter phorbolmyristate acetate-induced chromosome breakage by antioxidants and inhibitors of arachidonic acid metabolism. Mut Res (110): 327-335, 1983.

53. Dzarlieva RT, Fusenig NE: Tumor promoter 12-0-tetradecanoyl-phorbol-13-acetate enhances sister chromatid exchanges and numerical and structural chromosome aberrations in primary mouse epidermal cell cultures. Cancer Letters (16): 7-17, 1982.

54. Kato R, Nakadate T, Yamamoto S: Involvement of lipoxygenase products of arachidonic acid in tumor-promoting activity of TPA. In: Thaler-Dao H, Craster de Paulet A, Paoletti R (eds), Icosanoids and cancer. Raven Press, New York, 1984, pp 101-

113.

55. Brain SD, Camp RDR, Dowd PM, Black AK, Woolard PM, Mallet AI, Greaves MW: Psoriasis and leukotriene B_4. Lancet (2): 762-763, 1982.

56. Marcelo CL, Voorhees JJ: Cyclic nucleotides, prostaglandins and polyamines in psoriasis. In: Baden HP (ed) The chemotherapy of psoriasis. Pergamon Press, Oxford, 1984, pp 111-124.

57. Yuspa SH, Lichti U, Ben T: Local anesthetics inhibit induction of ornithine decarboxylase by the tumor promoter 12-0-tetradecanoylphorbol 13-acetate. Proc Natl Acad Sci (USA) (77): 5312-5316, 1980.

58. Lanz R, Brune K: Tumour-promoter-induced proliferation of 3T3 cells is independent of prostaglandin release and ornithine decarboxylase induction. In: Hecker E, Kunz W, Theilman HW, Fusenig NE, Marks F (eds) Carcinogenesis v.7 Cocarcinogenesis and biological effects of tumor promoters. Raven Press, New York, 1982, pp 495-498.

59. Marks F, Fürstenberger G, Kownatzki E: Prostaglandin E-mediated mitogenic stimulation of mouse epidermis in vivo by divalent cation ionophore A 23187 and by tumor promoter 12-0-tetradecanoylphorbol-13-acetate. Cancer Res (41): 696-702, 1981.

60. O'Brien TG, Lewis MA, Diamond L: Ornithine decarboxylase activity and DNA synthesis after treatment of cells in culture with 12-0-tetradecanoyl-phorbol-13-acetate. Cancer Res (39): 4477-4480, 1979.

61. Leach KL, James ML, Blumberg PM: Characterization of a specific phorbol ester aporeceptor in mouse brain cytosol. Proc Natl Acad Sci USA (80): 4208-4212, 1983.

62. Dunn JA, Blumberg PM: Specific binding of (20-^{3}H) 12-deoxyphorbol 13-isobutyrate to phorbol ester receptor subclasses in mouse skin particulate preparations. Cancer Res (43): 4632-4637, 1983.

63. Solanki V, Slaga TJ: Specific binding of phorbol ester tumor promoters to intact primary epidermal cells from Sencar mice. Proc Natl Acad Sci USA (78): 2549-2553, 1981.

64. Solanki V, Slaga TJ: The down modulation of receptors for phorbol ester tumor promoter in

primary epidermal cells. Carcinogenesis 3: 993-998, 1982.

65. Solanki V, Slaga TJ: Phorbol ester tumor promoter receptors and their down modulation. In: Slaga TJ (ed) Mechanisms of tumor promotion v.II Tumor promotion and skin carcinogenesis. CRC Press, Inc., Boca Raton, Florida, 1984, pp 97-111.

66. Sando JJ, Hilfiker ML, Salomon DS, Farrar JJ: Specific receptors for phorbol esters in lymphoid cell populations: Role in enhanced production of T-cell growth factor. Proc Natl Acad Sci USA (78): 1189-1193, 1981.

67. Sando JJ, Young MC: Identification of high affinity phorbol ester receptor in cytosol of EL4 thymoma cells: Requirement for calcium, magnesium, and phospholipids. Proc Natl Acad Sci USA (80): 2642-2646, 1983.

68. Blumberg PM, Sharkey NA, Konig B, Jaken S, Leach KL, Jeng AY: Membrane and cytosolic receptors for the phorbol ester tumor promoters. In: Levine AJ, Vande Woude GF, Topp WC, Watson JD (eds) Cancer cells. The transformed phenotype. Cold Spring Harbor Laboratory, Cold Spring Harbor New York, 1984, pp 245-251.

69. Slaga TJ, Scribner JD, Rice JM, Das SB, Thompson S: Inhibition by dexamethasone of intracellular binding of phorbol esters to mouse skin. J Natl Cancer Inst (52): 1611-1618, 1974.

70. Perrella FW, Ashendel CL, Boutwell RK: Specific high-affinity binding of the phorbol ester tumor promoter 12-0-tetradecanoylphorbol-13-acetate to isolated nuclei and nuclear macromolecules in mouse epidermis. Cancer Res (42): 3496-3501, 1982.

71. Hecker E, Kunz W, Theilman W, Fusenig NE, Marks F (eds) Carcinogenesis v.7 Cocarcinogenesis and biological effects of tumor promoters. Raven Press, New York, 1982.

72. Estensen RD, DeHoogh DK, Cole CF: Binding of (^{3}H) 12-0-tetradecanoylphorbol-13-acetate to intact human peripheral blood lymphocytes. Cancer Res (40): 1119-1124, 1980.

73. Shoyab M, Warren TC, Todaro GJ: Phorbol-12 13-diester 12-ester hydrolase may prevent tumour promotion by phorbol diesters in skin. Nature (295): 152-154.

74. Chambers DA, Sando JJ, Cohen RL, Krueger GG: Innate resistance to tumour promotion in nude mice. In: Wright NA (Camplejohn RS (eds) Psoriasis: Cell proliferation. Churchill Livingstone, London, 1983, pp 306-313.

75. Chambers, DA, Cohen RL, Sando JJ, Krueger GG: Resistance to 12-0-tetradecanoyl phorbol 13-acetate-induced mitogenesis in the nude mouse. In: Sordat B (ed) Immune-deficient animals. S. Karger, Basel, 1984, pp 125-128.

76. Chambers DA, Cohen RL: The direct effects of biological response modifiers on epidermal cells. Brit J Derm (Supple 27): 114-122, 1984.

77. Farese RV: Phospholipids as intermediates in hormone action. Molec Cell Endocrin (35): 1-14, 1984.

78. Samuelsson B: Leukotrienes: Mediators of immediate hypersensitivity reactions and inflammation. Science (220): 568-575, 1983.

79. Kaibuchi K, Takai Y, Sawamura M, Hoshijima M, Fujikura T, Nishizuka Y: Synergistic functions of protein phosphorylation and calcium mobilization in platelet activation. J Biol Chem (258): 6701-6704, 1983.

80. Lands WEM, Samuelsson B: Phospholipid precursors of prostaglandins. Biochim Biophys Acta (164): 426-428, 1968.

81. Niedel JE, Kuhn LJ, Vandenbark GR: Phorbol diester receptor copurifies with protein kinase C. Proc Natl Acad Sci USA (80): 36-40, 1983.

82. Ashendel CL, Staller JM, Boutwell RK: Protein kinase activity associated with a phorbol ester receptor purified from mouse brain. Cancer Res (43): 4333-4337, 1983.

83. Takai Y, Kishimoto A, Kikkawa U, Mori T, Nishizuka Y: Unsaturated diacylglycerol as a possible messenger for the activation of calcium-activated, phospholipid-dependent protein kinase system. Biochem Biophys Res Commun (91): 1218-1224, 1979.

84. Inoue M, Kishimoto A, Takai Y, Nishizuka Y: Studies on a cyclic nucleotide-independent protein kinase and its proenzyme in mammalian tissues. II. Proenzyme and its

activation by calcium-dependent protease from rat brain. J Biol Chem (252): 7610-7616, 1977.

85. Kuo JF, Andersson RGG, Wise BC, Mackerlova L, Salomonsson, I, Brackett NL, Katoh N, Shoji M, Wrenn RW: Calcium-dependent protein kinase: Widespread occurrence in various tissues and phyla of the animal kingdom and comparison of effects of phospholipid, calmodulin and trifluoperazine. Proc Natl Acad Sci USA (77): 7039-7043, 1980.

86. Kikkawa U, Takai Y, Minakuchi R, Inohara S, Nishizuka Y: Calcium-activated, phospholipid-dependent protein kinase from rat brain. Subcellular distribution, purification and properties. J Biol Chem (257): 13341-13348, 1982.

87. Nishizuka Y, Takai Y: Calcium and phospholipid turnover in a new receptor function for protein phosphorylation. Cold Spring Harbor. Conf Cell Prolif (8): 237-249, 1981.

88. Kishimoto A, Kajikawa N, Shiota M, Nishizuka Y: Proteolytic activation of calcium-activated, phospholipid-dependent protein kinase by calcium-dependent neutral protease. J Biol Chem (258): 1156-1164, 1983.

89. Kishimoto A, Takai Y, Mori T, Kikkawa U, Nishizuka Y: Activation of calcium and phospholipid-dependent protein kinase by diacylglycerol, its possible relation to phosphatidylinositol turnover. J Biol Chem (255): 2273-2276, 1980.

90. Kaibuchi K, Takai Y, Nishizuka Y: Cooperative roles of various membrane phospholipids in the activation of calcium-activated, phospholipid-dependent protein kinase. J Biol Chem (256): 7146-7149, 1981.

91. Kraft AS, Anderson WB: Phorbol esters increase the amount of Ca^{2+}, phospholipid-dependent protein kinase associated with plasma membrane. Nature (301): 621-623, 1983.

92. Nishizuka Y: Three multifunctional protein kinase systems in transmembrane control. In: Chapeville F, Haenni A-L (eds) Molec Biol Biochem Biophys v.32 Chemical recognition in biology. Springer-Verlag, Berlin, 1980, pp 113-135.

93. Hunter T, Cooper JA: Tyrosine protein kinases and their substrates: An overview. Adv Cyclic Nucleotide and Protein Phosphorylation Res (17): 443-455, 1984.

94. Nishizuka Y: Phospholipid turnover in hormone action. Rec Prog Horm Res (40): 301-345, 1984.

95. Castagna M, Takai Y, Kaibuchi K, Sano K, Kikkawa U, Nishizuka Y: Direct activation of calcium-activated, phospholipid-dependent protein kinase by tumor-promoting phorbol esters. J Biol Chem (257): 7847-7851, 1982.

96. Kikkawa U, Takai Y, Tanaka Y, Miyake R, Nishizuka Y: Protein kinase C as a possible receptor protein of tumor-promoting phorbol esters. J Biol Chem (258): 11442-11445, 1983.

97. Delclos KB, Yeh E, Blumberg PM: Specific labeling of mouse brain membrane phospholipids with (20-^{3}H) phorbol 12-p-azidobenzoate 13-benzoate, a photolabile phorbol ester. Proc Natl Acad Sci USA (80): 3054-3058, 1983.

98. Kikkawa U, Kaibuchi K, Castagna M, Yamanishi J, Sano K, Tanaka Y, Miyake R, Takai Y, Nishizuka Y: Protein phosphorylation and mechanism of action of tumor-promoting phorbol esters. Adv in Cyclic Nucleotide and Protein Phosphorylation Res (17): 437-441, 1984.

99. Kikkawa U, Miyake R, Tanaka Y, Takai Y, Nishizuka Y: Protein kinase C and the mechanism of action of tumor promoters. In: Levine AJ, Vande Woude GF, Topp WC, Watson JD (eds) Cancer cells v.1 The transformed phenotype. Cold Spring Harbor Laboratory, Cold Spring Harbor, New York, 1984, pp 239-244.

100. Gilmore T, Martin GS: Phorbol ester and diacyl-glycerol induce protein phosphorylation at tyrosine. Nature (306): 487-490, 1983.

101. Moon SO, Palfrey HC, King AC: Phorbol esters potentiate tyrosine phosphorylation of epidermal growth factor receptors in A431 membranes by a calcium-independent mechaism. Proc Natl Acad Sci USA (81): 2298-2302, 1984.

102. Fearn JC, King AC: EGF receptor affinity is regulated by intracellular calcium and protein kinase C. (submitted for publication).

103. Davis RJ, Czech MP: Tumor-promoting phorbol diesters mediate phosphorylation of the epidermal growth factor receptor. J Biol Chem (259): 8545-8549, 1984.

104. Cochet C, Gill GN, Meisenhelder J, Cooper JA,

Hunter T: C-kinase phosphorylates the epidermal growth factor receptor and reduces its epidermal growth factor-stimulated tyrosine protein kinase activity. J Biol Chem (259): 2553-2558, 1984.

105. Iwashita S, Fox CF: Epidermal growth factor and potent phorbol tumor promoters induce epidermal growth factor receptor phosphorylation in a similar but distinctively different manner in human epidermoid carcinoma A431 cells. J Biol Chem (259): 2559-2567, 1984.

106. Belman S, Garte SJ: Phorbol myristate acetate uncouples the relationship between B-adrenergic receptors and adenyl cyclase in mouse epidermis: A phenotypic trait in papillomas. In: Hecker E, Kunz W, Theilman HW, Fusenig NE, Marks F (eds) Carcinogenesis v.7 Cocarcinogenesis and biological effects of tumor promoters. Raven Press, New York, 1982, pp 561-563.

107. Sibley DR, Nambi P, Peters JR, Lefkowitz RJ: Phorbol diesters promote B-adrenergic receptor phosphorylation and adenylate cyclase desensitization in duck erythrocytes. Biochem Biophys Res Commun (121): 973-979, 1984.

108. Slaga TJ, Nelson KG: Critical protein modifications during skin tumor promotion in mice. In: Slaga TJ (ed) Mechanisms of tumor promotion v.II Tumor promotion and skin carcinogenesis. CRC Press, Inc., Boca Raton, Florida, 1984, pp 159-188.

109. Cabral F, Gottesman MM, Yuspa SH: Induction of specific protein synthesis by phorbol esters in mouse epidermal cell culture. Cancer Res (41): 2025-2031, 1981.

110. Mamrack MD: Protein phosphorylation and tumor promotion. In: Slaga TJ (ed) Mechanisms of tumor promotion v.II Tumor promotion and skin carcinogenesis. CRC Press, Inc., Boca Raton, Florida, 1984, pp 143-158.

111. Green H: Cyclic AMP in relation to proliferation of the epidermal cell: a new view. Cell (15): 801-811, 1978.

112. Marcelo CL: Differential effects of cAMP and cGMP on in vitro epidermal cell growth. Exp Cell Res (120): 201-210, 1979.

113. Bourne HR, Lichtenstein LM, Melmon KL, Henney CS, Weinstein Y, Shearer GM: Modulation of inflammation and immunity by cyclic AMP. Science (184): 19-28,

1974.

114. Chambers, DA, Martin, DW, Weinstein Y: The effect of cyclic nucleotides on purine biosynthesis and the induction of PRPP synthetase during lymphocyte activation. Cell (3): 375-380, 1974.

115. Outzen HC: A preliminary analysis of skin tumor induction in immunomodulated mice. In: Houchens DP, Ovejera AA (eds) Proceedings of the Symposium on the Use of Athymic (nude) Mice in Cancer Research. Gustav Fischer, New York, 1978, pp 93-99.

116. Yuspa SH: Maintenance of human skin on nude mice for studies of chemical carcinogenesis. Cancer Lett (6): 301-310, 1979.

117. Holland JM, Perkins EH: Resistance of germfree athymic nude mice to two-stage carcinogenesis. In: Reed ND (ed) Proceedings of the Third International Workshop on Nude Mice. Gustav Fischer, New York, 1982, pp 423-433.

118. Krueger GG, Chambers, DA, Shelby J: Epidermal proliferation of nude mouse skin, pig skin, and pig skin grafts. Failure of nude mouse skin to respond to the tumor promoter 12-0-tetradecanoyl phorbol 13-acetate. J Exp Med (152): 1329-1339, 1980.

22

HIGH PRESSURE LIQUID CHROMATOGRAPHY OF EICOSANOIDS

WILLIAM S. POWELL

1. INTRODUCTION

Arachidonic acid (20:4) and other polyunsaturated fatty acids can be oxygenated in many tissues by cyclooxygenase or a variety of lipoxygenases. The intermediates formed in these reactions are then converted enzymatically to a variety of other products. In many cases large numbers of closely-related substances are formed in these reactions, necessitating the use of sophisticated methods for their separation and analysis. High pressure liquid chromatography (HPLC) is ideally suited for this purpose. The use of this technique has been very important for the identification and analysis of a large number of recently discovered 20:4 metabolites, including leukotrienes (1,2) and epoxy and dihydroxy products formed by the actions of epoxygenases related to cytochrome P-450 (3,4). The utility of HPLC is due to the improved chromatography achieved by the use of stationary phases composed of very small and uniform particles. This minimizes peak broadening due to differences in the paths followed by solute molecules as they pass through the column (Eddy diffusion). Consequently, the efficiency of a column increases as the particle size decreases. The high back pressures encountered with columns of very small particle sizes often makes it necessary to use shorter columns. In the past, stationary phases with particle sizes of 10 μm were most commonly used, but now particle sizes of 3 and 5 μm are becoming increasingly popular.

One general problem with chromatography of eicosanoids is that some of these products exist in several tautomeric forms which are in equilibrium with one another. This is true of 6-oxoPGF$_{1\alpha}$ and TXB$_2$, which exist both as uncyclized oxo derivatives and cyclic hemiketals or hemiacetals, respectively. This causes very broad peaks for TXB$_2$ with both normal-phase (NP) and reversed-phase (RP) HPLC, and for 6-oxoPGF$_{1\alpha}$ with normal-phase HPLC.

This chapter deals with the separation of both cyclooxygenase and lipoxygenase products by normal-phase, reversed-phase, and argentation HPLC. We have chromatographed mixtures containing both types of products so that their relative chromatographic behaviors in several different systems can be compared. In many cases we have used somewhat complicated combinations of both isocratic and gradient elution, since the samples contained a complex mixture of products. These conditions can often be simplified, depending on the specific products which it is desired to separate.

2. PREPARATION OF STANDARDS

In order to compare a number of different HPLC methods, we have biosynthetically prepared mixtures of cyclooxygenase products [12-hydroxy-5,8,10-heptadecadienoic acid (HHT), PGD$_2$, PGE$_2$, PGF$_{2\alpha}$, 6-oxoPGF$_{1\alpha}$, and TXB$_2$] from bovine lung homogenates (5), and 5- and 12-lipoxygenase products [12-hydroxy-5,8,10,14-eicosatetraenoic acid (12-HETE, 12h-20:4), 5-hydroxy-6,8,11,14-eicosatetraenoic acid (5-HETE, 5h-20:4), LTB$_4$ (5S,12R-dihydroxy-6,8,10,14-ZEEZ-eicosatetraenoic acid), 5S,12S-dihydroxy-6,8,10,14-EZEZ-eicosatetraenoic acid (5S,12S-dh-20:4), 5S,12R-dihydroxy-6,8,10,14-EEEZ-eicosatetraenoic acid (isomer-1), 5S,12S-dihydroxy-6,8,10,14-EEEZ-eicosatetraenoic acid (isomer-2), 5,6-dihydroxy-7,9,11,14-eicosatetraenoic acid (5,6-dh-20:4), and 5,15-dihydroxy-6,8,11,13-eicosatetraenoic acid (5,15-dh-20:4)] from human polymorphonuclear leukocytes contaminated with some

platelets (6). Soybean lipoxygenase was used to synthesize 15-hydroxy-5,8,11,13-eicosatetraenoic acid (15-HETE, 15h-20:4) (7), whereas 11-hydroxy-5,8,12,14-eicosatetraenoic acid (11-HETE, 11h-20:4) was prepared by chemical oxidation of 20:4 (8).

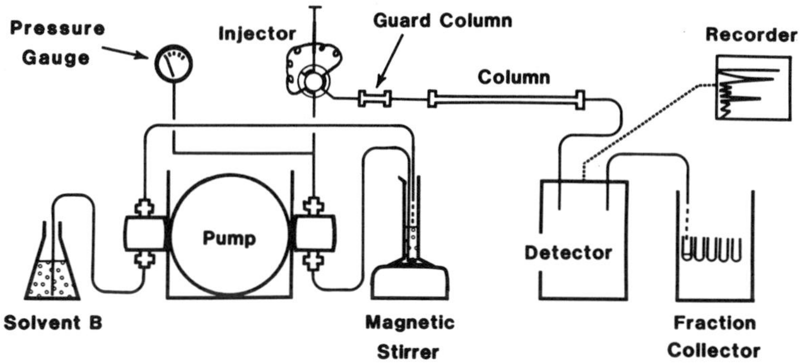

FIGURE 1. Inexpensive HPLC system utilizing a Milton Roy duplex minipump to deliver a linear gradient between the weaker solvent in a graduated cylinder and a stronger solvent (solvent B) in a reservoir.

3. EQUIPMENT

The minimum equipment required for HPLC consists of a pump, a pressure gauge, an injector, and a column. Although it is possible to monitor the column eluate by liquid scintillation counting or radioimmunoassay, it is highly desirable to have an ultraviolet or radioactivity detector.

An inexpensive HPLC system capable of making linear solvent gradients, which we have used in our laboratory (9,10), is illustrated in Figure 1. This consists of a Milton Roy duplex minipump coupled to a pressure gauge and a Valco injector. The column is first equilibrated (e.g. for 10 min at a flow rate of 2 ml/min) with the weaker starting solvent (solvent A) and the pump is then shut off. Linear solvent gradients can be made by placing solvent A (volume of solvent A = flow rate x time/2) into a graduated cylinder. The sample is then injected, and the pump is turned on so that one of the pump heads pumps the contents of the graduated cylinder on to the column, and the second pump head pumps the stronger solvent B into the graduated cylinder at a flow rate equal to half the flow rate through the column. Although there is considerable pressure pulsation with this system, this appears to have a minimal effect on the chromatography, and the lifetime of the column is not appreciably shortened. The disadvantage of this system is that it requires a considerable amount of attention, since it is necessary to shut the pump off precisely at the end of the run. Nevertheless, this may be a good introduction to HPLC for those who do not want to make a large initial investment in more expensive equipment.

A more sophisticated system would consist of two solvent delivery systems, each consisting of two pump heads and a pulse dampener, along with a gradient controller, which will allow the preprogramming of complex solvent gradients. We have used such a system for the chromatograms illustrated in this chapter. Alternatively, a single solvent delivery system connected to a solenoid valve, which switches the solvent intake from one reservoir to another, can be used to deliver gradients made from 2 or 3 different solvent mixtures.

4. DETECTION OF EICOSANOIDS

Both leukotrienes (LTs) and monohydroxyeicosatetraenoic acids (HETEs) can readily be detected using an ultraviolet detector. Leukotrienes and their isomers each have three λmax values between 260 and 290 nm, with extinction coefficients of about 40,000 to 50,000, whereas HETEs have λmax values at about 235 nm with extinction coefficients of about 30,000 (Table 1). Unlike many other other dihydroxyeicosatetraenoic acids, 5,15-dihydroxyeicosatetraenoic acid has two pairs of conjugated double bonds, and has a max at 243 nm.

Underivatized prostaglandins and TXB_2 can only detected at low wavelengths (ca. 200 nm or less) using a UV detector. This can be useful in cases where there are relatively few

Table 1. Ultraviolet spectra of some eicosanoids

Compound	λ max (extinction coefficient) (nm)			Reference
LTB_4	260(38,000)	270(51,000)	281(39,500)	11
Isomer-1	258(42,000)	268(56,000)	280(44,000)	12
Isomer-2	258(42,000)	268(56,000)	280(44,000)	12
5S,12S-dh-20:4	259;	268;	278	13
5,15-dh-20:4	243(30,000)			14
LTC_4	270	280(40,000)	292	2
LTD_4	270	280(40,000)	292	15
LTE_4	270	280(40,000)	290	16
5-HETE	235(30,500)			17
12-HETE	237(30,500)			18
HHT	232(33,400)			18
PGB_1	280(26,800)			19

interfering substances, but unfortunately, many biological samples contain many components which absorb at low wavelengths, making it very difficult to detect low concentrations of prostaglandins. A similar problem would be encountered with a refractive index detector, which would detect all compounds. One way to get around this problem is to esterify prostaglandins with a reagent which absorbs light at a higher wavelength. Phenacyl bromide and its *p*-bromo and *p*-nitro derivatives have been used to convert prostaglandins to phenacyl ester derivatives, which can be detected using a fixed wavelength UV detector at 254 nm (20,21). Eicosanoids can also be converted to fluorescent derivatives by treatment with 4-bromomethyl-7-methoxycoumarin (22) or 9-anthryldiazomethane (23). The latter derivatives can be detected using a fluorescence detector (fluorescence monitored at 418 nm with excitation at 365 nm) with sensitivities below 1 ng (23). These methods also suffer from lack of specificity, however, since all carboxylic acids in the sample will react to form UV-absorbing or fluorescent derivatives.

If it is desired to investigate the biosynthesis of arachidonic acid metabolites by tissues or cells, tritium- or ^{14}C-labeled arachidonic acid can be used, and the radioactive products can be detected using a radioactivity monitor or by liquid scintillation counting. Radioactivity can be detected with a monitor either by passing the column eluate through a cell containing a solid scintillator, such as cerium-coated glass beads (heterogeneous counting), or by mixing the eluate with a liquid scintillator and passing the mixture through the monitor (homogeneous counting). One advantage of cells containing solid scintillators is that all of the eluate can be collected after it has passed through the detector, and the products of interest can be recovered after evaporation of the solvent. This type of detector is fairly sensitive for ^{14}C (ca. 30 to 40% efficiency), but much less sensitive for tritium (often less than 1% efficiency). Recently, more sensitive solid scintillators have become available, which have efficiencies for tritium in excess of 5%, but unfortunately carboxylic acids may bind to these materials, causing increases in background radioactivity. Thus, in order to take advantage of these scintillators, samples would have to be esterified prior to HPLC. The most sensitive means of detecting tritium is by homogeneous counting, which has efficiencies of over 50%. If it is desired to collect fractions, the column eluate can be split prior to the addition of liquid scintillant, so that only a small portion of the total eluate is lost. Homogeneous counting is more costly than heterogeneous counting, however, since relatively large amounts of liquid scintillant are required, and it is necessary to have pumps for the scintillant and for splitting the column eluate.

5. **PREPARATION OF SAMPLES FOR HPLC**

Prior to HPLC, eicosanoids are usually extracted from biological samples in order to remove unrelated contaminating substances. This can be accomplished by extraction of acidified aqueous solutions with diethyl ether or ethyl acetate. More polar substances (e.g. PG metabolites in urine) can be extracted using columns of XAD resin (24).

We have developed a novel method to extract most eicosanoids (except for those containing amino acids) using octadecylsilyl (ODS) silica (Figure 2) (5,25). Either cartridges (Waters C18 Sep-Paks) or syringes (Baker octadecyl disposable extraction columns) containing this substance can be used, or alternatively, columns can be prepared from ODS silica purchased in bulk or prepared as described in the literature (26). The ODS silica is first washed with ethanol (20 ml) and water (20 ml). Biological fluids or tissue extracts in 15% aqueous ethanol are then acidified and passed through the stationary phase. The column is washed with 20 ml each of 15% aqueous ethanol, water (to remove the ethanol), and petroleum ether (to remove neutral lipids and water). Arachidonic acid metabo-

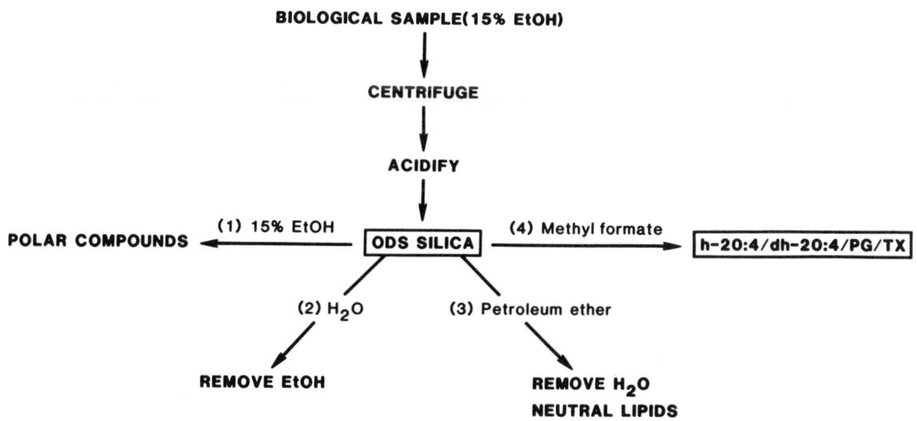

FIGURE 2. Scheme for the extraction of eicosanoids using octadecylsilyl silica. The ODS silica should first be washed with ethanol (20 ml) and water (20 ml) before use.

lites are then eluted (80 to 100% recoveries) with methyl formate (10 ml). The latter fraction contains HETEs; dh-20:4s such as leukotriene B_4, along with its omega-oxidation products; prostaglandins and their metabolites; and thromboxane B_2. Arachidonic acid is also present, but its recovery would not be optimal with these conditions. In order to increase the recovery of 20:4, the sample should be loaded on to the ODS silica in 30% ethanol, but this could reduce the recovery of polar metabolites. For extraction of unstable metabolites, it is not necessary to acidify the sample, but in this case, it might be necessary to reduce the concentration of ethanol to 10% (25). We normally purchase the methyl formate (practical grade, Eastman Kodak) in 5 gal quantities, and redistill it before use, since it contains some UV-absorbing materials.

Extraction of eicosanoids using this method involves adsorption from the initial solution by reversed-phase chromatography using aqueous ethanol, followed by elution by normal-phase chromatography with water-immiscible organic solvents. The main advantages of this procedure are that it is rapid and gives an extract in a highly volatile organic solvent, which can readily be removed under a stream of nitrogen. The extracts contain relatively small amounts of contaminants, and can be analyzed directly by HPLC without any further purification, even when large amounts of tissue have been extracted. The ODS silica can be regenerated by washing with 80% aqueous ethanol and 100% ethanol after use.

This method does not work very well for leukotrienes containing cysteine residues (i.e. leukotrienes C_4, D_4, E_4 and F_4). These compounds can be extracted on columns of XAD-7 resin (27), silica (28), or alternatively, media from cells can be analyzed directly without extraction by RP-HPLC (29,30).

6. NORMAL-PHASE HPLC

Normal-phase (straight-phase) HPLC is usually carried out on columns of silica. We used a column of 300 x 4.6 mm of RoSiL (particle size, 5 μm; Alltech Associates) for the chromatograms shown in this section (Figures 3-5). Some advantages of NP-HPLC are that (i) the capacity is quite high; (ii) the solvents used are volatile, and thus easily removed, and (iii) the solvents generally have relatively low viscosities, resulting in low back pressures. A disadvantage is that many nonpolar organic solvents are not transparent to UV light, so that if it is necessary to monitor UV absorbance at low wavelengths, the mobile phase will probably have to be composed mainly of hexane or some other saturated hydrocarbon. This in turn restricts the polar components of the mobile phase to those which are miscible with hexane, thus limiting the use of methanol, ethanol, and acetonitrile (the addition of 5% propanol to hexane will render the above solvents miscible). Other disadvantages of NP-HPLC are that recoveries are sometimes rather low, and that retention times can be altered markedly by the solvent used to inject the sample as discussed below.

One system which has often been used for the purification of both the free acids and methyl esters of h-20:4s (31) and dh-20:4s (13) is hexane/isopropanol/acetic acid. This system gives excellent separation of the 5, 8, 9, 11, 12, and 15 monohydroxy metabolites of 20:4 (31), as well as of LTB_4 and its 5,12-dh-20:4 isomers (13). Figure 3 shows the separation of a mixture of standards, prepared as described above, using isocratic and gradient elution with hexane/isopropanol/acetic acid. The sample was injected in 4 μl of isopropanol, and under these conditions, the peaks for 12-HETE and 15-HETE were quite close together (see below). LTB_4 was well separated from 5S,12S-dh-20:4 and its 6-trans isomers (isomers 1 and 2), but 6-oxo$PGF_{1\alpha}$ and TXB_2 appeared as very broad peaks and were not resolved from PGE_2. Although the mobile phase used for this chromatogram is transparent to UV light at 235 nm, we got a very poor baseline at high sensitivity, especially at the beginning of the first gradient segment of the run. If it is necessary to monitor absorbance at this wavelength using high sensitivities, it would probably be necessary to redistill all the solvents.

The medium used for the injection of samples in NP-HPLC can have considerable effects on the results obtained,

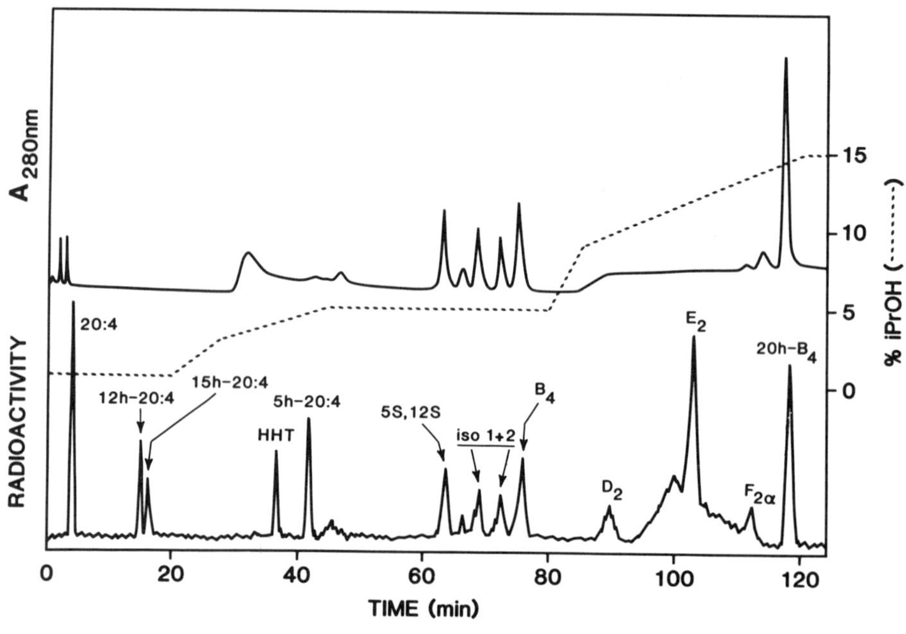

FIGURE 3. Separation by NP-HPLC of a mixture consisting of (i) metabolites of 20:4 formed by human polymorphonuclear leukocytes, (ii) metabolites of 20:4 formed by a bovine lung homogenate, and (iii) 15-HETE (15h-20:4). The mobile phase consisted of a gradient between solvent A (hexane/isopropanol/acetic acid, 99.4:0.6:0.1) and hexane/isopropanol/acetic acid, 85:15:0.1) as follows: 0 min, 0% B; 20 min, 0% B; 28 min, 17% B; 45 min, 31% B; 80 min, 31% B; 85 min, 51% B; 120 min, 100% B. The flow rate was 2 ml/min. 5S,12S, 5S,12S-dh-20:4; iso 1+2, isomers 1 and 2 (i.e. the two trans isomers of LTB_4); B_4, LTB_4; D_2, PGD_2; E_2, PGE_2; $F_{2\alpha}$, $PGF_{2\alpha}$; 20h-B_4, 20-hydroxyLTB_4.

especially with the hexane/isopropanol/acetic acid system. In general, the best results are obtained if the sample is injected in a small volume of mobile phase. This presents a problem if the sample contains both polar and nonpolar materials. Nonpolar eicosanoids such as HETEs are not lost when residues in tubes are dissolved in nonpolar solvents (e.g. 20 µl of hexane/isopropanol/acetic acid, 99.4/0.6/0.1, cf. Fig. 3), but there are considerable losses for more polar products such as prostaglandins. Thus, a more polar solvent such as isopropanol is often required to dissolve samples containing mixtures of 20:4 metabolites. We normally dissolve the sample in 4 or 5 µl of isopropanol, and then add a few µl of mobile phase. The problem with this procedure is that the isopropanol (or other polar organic solvent) is retained by the column and passes through as a band (which was detected as a negative peak in absorbance at 235 nm). The composition of the mobile phase is altered within this band, and this can cause marked alterations in the chromatographic behavior of nonpolar solutes. This is illustrated in Figure 4, which shows the separation of 12-HETE and 15-HETE with hexane/isopropanol/acetic acid (99.4/0.6/0.1) as the mobile phase. The sample was dissolved in 20 µl of mobile phase, to which were added between 0 and 6 µl of isopropanol. When the sample was dissolved in mobile phase alone, the retention times of 12-HETE and 15-HETE were 19.5 min and 24.2 min, respectively. When only 1 µl of isopropanol was added, there was a small negative peak at 235 nm with a retention time of 17.4 min, followed by a very sharp peak at 17.8 min for 12-HETE. In addition, the peak for 15-HETE was split into several smaller peaks. As increasing amounts of isopropanol were added, the peak for 12-HETE appeared just before the negative peak, whereas that for 15-HETE gradually moved closer, and, with the addition of 6 µl of isopropanol, became very sharp. The retention times of more polar products such as LTB_4 and prostaglandins were not significantly affected by the addition of isopropanol.

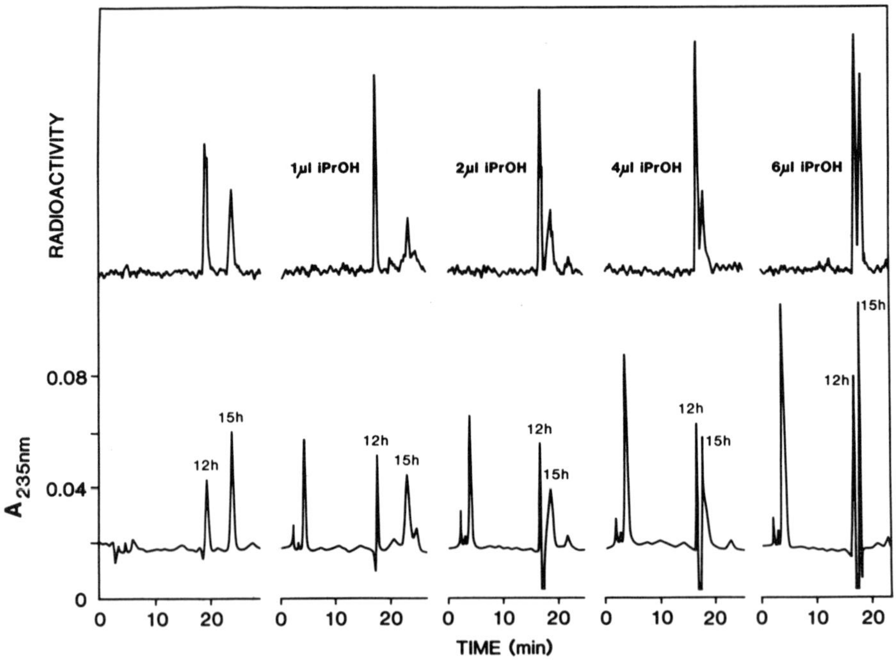

FIGURE 4. Separation by NP-HPLC of 12-HETE (12h) and 15-HETE (15h) using hexane/isopropanol/acetic acid (99.4:0.6:0.1) at a flow rate of 2 ml/min as the mobile phase. The samples were injected in 20 µl of mobile phase (extreme left) or 20 µl of mobile phase containing different amounts of isopropanol.

Another type of mobile phase (5) which can be used for separating 20:4 metabolites by NP-HPLC is a gradient between hexane/toluene/acetic acid (50/50/0.5) and toluene/ethyl acetate/acetonitrile/methanol/acetic acid (30/40/30/2/0.5) as shown in Figure 5. This system is not as good as hexane/ isopropanol/acetic acid for separating monohydroxy 20:4 metabolites, but, unlike the latter system, it completely separates PGD_2, TXB_2, PGE_2, 6-oxo$PGF_{1\alpha}$, and $PGF_{2\alpha}$. This mobile phase is therefore very useful for the separation of cyclooxygenase products.

Methyl esters of 20:4 metabolites often give better results than free acids with NP-HPLC, especially mono- and di- hydroxy metabolites. The chromatographic profile of prostaglandins can be rather complex, however, due to the presence of multiple sharp peaks for 6-oxoPGF$_{1\alpha}$, which exists in several tautomeric forms. The mobile phases used for the free acids can usually also be used for methyl esters, although the acetic acid can often be omitted (cf. 10, 32).

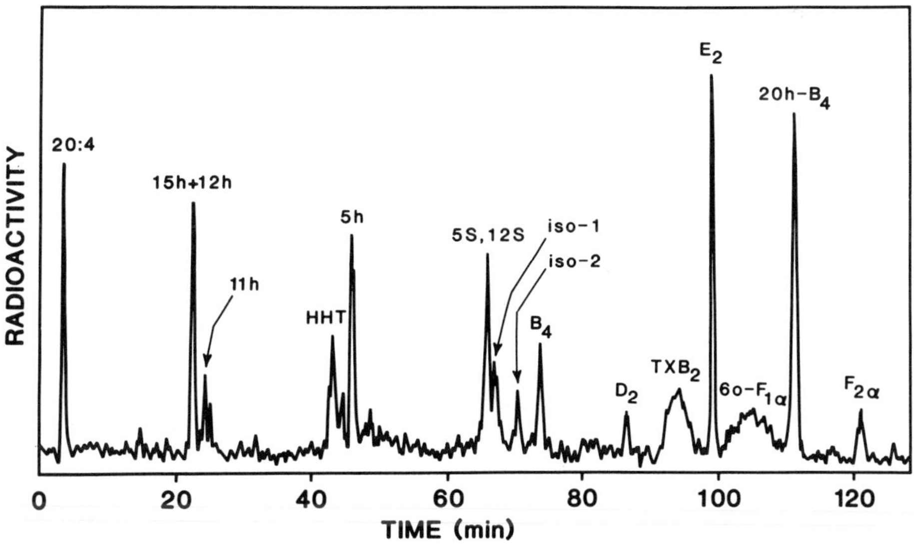

FIGURE 5. Separation by NP-HPLC of a mixture of eicosanoids similar to that described in the legend to Figure 3, but also containing 11-HETE (11h). The mobile phase consisted of a gradient formed from solvent C (hexane/toluene/acetic acid, 50:50:0.5) and solvent D (toluene/ethyl acetate/acetonitrile/methanol/acetic acid, 30:40:30:2:0.5) as follows: 0 min, 4%D; 25 min, 4% D; 40 min, 25% D; 76 min, 25% B; 81 min, 47% B; 131 min, 80% B. The flow rate was 2 ml/min. The abbreviations are defined in Figures 3 and 4. 6o-F$_{1\alpha}$, 6-oxoPGF$_{1\alpha}$.

7. REVERSED-PHASE HPLC

Reversed-phase HPLC can be carried out on a variety of stationary phases composed of a hydrophobic group such as octylsilyl, octadecylsilyl, or phenylsilyl bound chemically to silica. Recoveries of solutes from RP-HPLC are generally very good, which is an advantage over NP-HPLC. Another advantage of RP-HPLC is that most of the solvents used (e.g. water, acetonitrile, methanol, tetrahydrofuran) are transparent to UV light, so that UV absorbance can be measured at wavelengths as low as 190 or 200 nm. This allows solutes to be detected on the basis of absorption due to unconjugated olefinic double bonds, which are present in nearly all eicosanoids. Selection of the injection medium is not nearly as critical in RP-HPLC as in NP-HPLC. Although the sample must be injected in a solvent which is miscible in water, the question of solubility is not as much of a problem, since most samples are soluble in solvents such as ethanol or isopropanol. Retention times or peak shapes are not usually significantly affected by injection of as much as 30 µl of these solvents. An alternative method of introducing samples on to an RP-HPLC column is to pump or inject relatively large volumes (>1ml) of material dissolved in water or 30% methanol in water directly on to the column (27,29,30). The sample will be completely adsorbed on the top of the column and can then be eluted by the addition of a stronger solvent, such as methanol or acetonitrile, to the mobile phase. One disadvantage of RP-HPLC is that the mobile phase normally contains water, and is thus more difficult to remove than the mobile phases used in NP-HPLC.

Stationary phases of ODS silica with particle sizes of 5 µm give good separations of many eicosanoids. Although 6-oxoPGF$_{1\alpha}$ normally gives a single sharp peak with RP-HPLC, TXB$_2$ is usually observed as a fairly broad peak, as is the case with NP-HPLC. The separation of a mixture of eicosanoids (similar to that shown in Figures 3 and 5) by HPLC on a

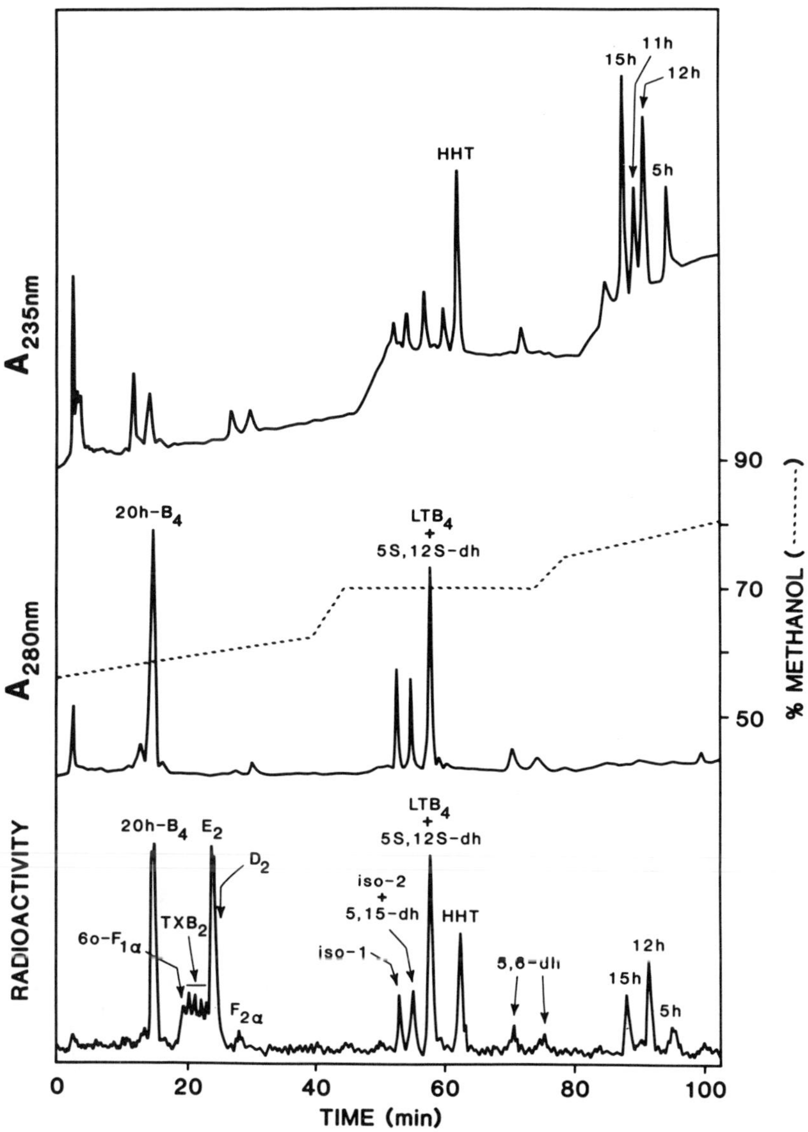

FIGURE 6. Reversed-phase HPLC of the mixture of eicosanoids described in the legend to Figure 5. The mobile phase consisted of a gradient prepared from solvent E (water/acetic acid, 100:0.05) and solvent F (methanol/acetic acid, 100:0.05) as follows: 0 min, 56% F; 40 min, 62% F; 45 min, 70% F; 75 min, 70% F; 80 min, 75% F; 110 min, 82% F. The flow rate was 1.5 ml/min. 5,15dh, 5,15-dh-20:4; 5,6-dh, isomers of 5,6-dh-20:4; 5h, 5-HETE.

column (4.6 x 250 mm) of Ultrasphere ODS (5 um particle size, Beckman Instruments) with methanol/water/acetic acid is shown in Figure 6. This mobile phase gives very good separation of LTB_4 and its two trans isomers, and resolves monohydroxy-20:4 positional isomers from one another (1). It is a very poor system for separating cyclooxygenase products, however, since 6-oxo$PGF_{1\alpha}$, TXB_2, PGE_2, and PGD_2 are not resolved from one another. The peak for LTB_4 with a retention time of 58.5 min also contains the double lipoxygenase product, 5S,12S-dh-20:4, whereas that for the second trans isomer of LTB_4 (isomer-2) with a retention time of

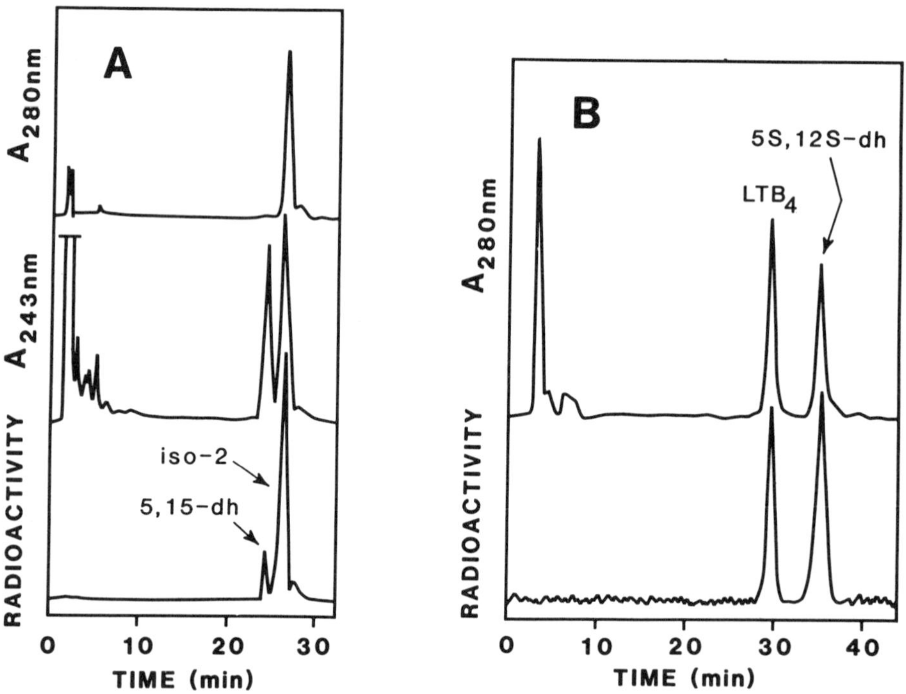

FIGURE 7. Separation by RP-HPLC of the components present in the peaks with retention times of (A) 55.4 min (iso-2 + 5,15-dh) and (B) 58.5 min (LTB_4 + 5S,12S-dh) in the chromatogram illustrated in Figure 6. The mobile phase was water/acetonitrile/acetic acid (60:40:0.05) and the flow rate, 1.5 ml/min.

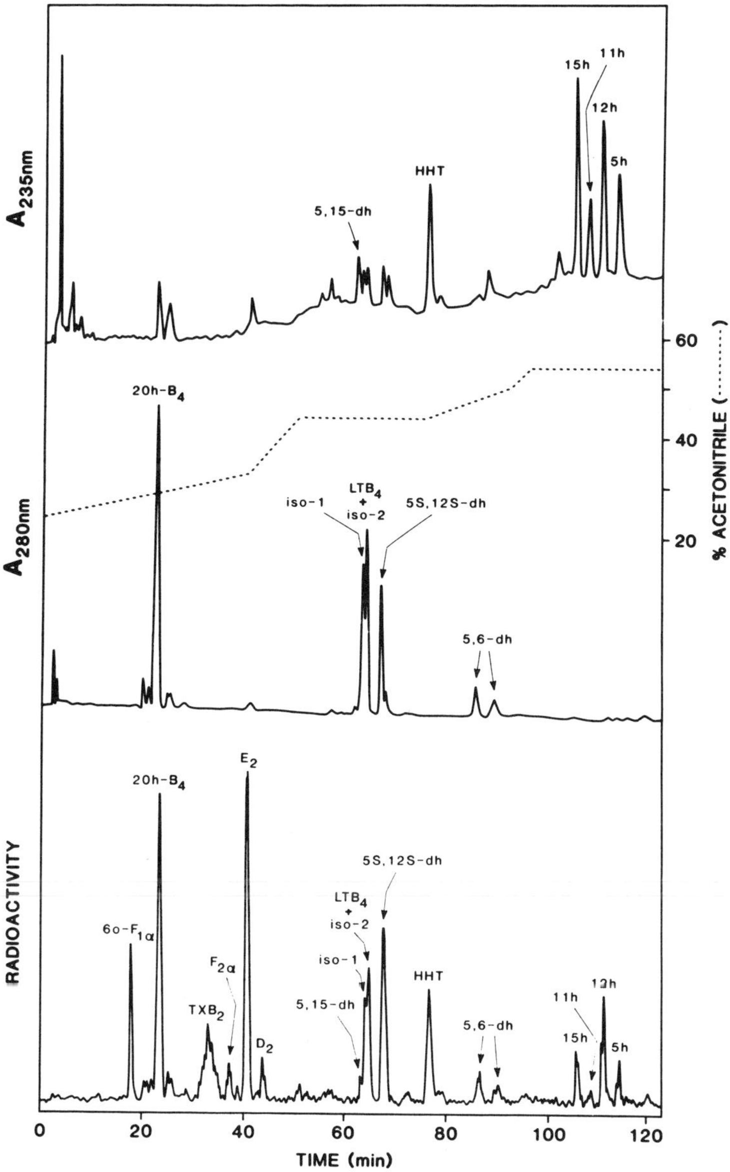

FIGURE 8. Reversed-phase HPLC of the mixture of eicosanoids described in the legend to Figure 5. The mobile phase consisted of a gradient prepared from solvent E (water/acetic acid, 100:0.05) and solvent G (acetonitrile/acetic acid, 100:0.05) as follows: 0 min, 24% G; 40 min, 33% G; 50 min, 44% G; 76 min, 44% G; 92 min, 50% G; 96 min, 54% G. The flow rate was 1.5 ml/min.

55.5 min also contains 5,15-dh-20:4. Fortunately, the selectivity of the column can be changed considerably by using acetonitrile/water/acetic acid as the mobile phase, and this system can be used to completely separate the above two peaks into their two components (Figures 7A and 7B). Unlike isomer-2, 5,15-dh-20:4 does not absorb at 280 nm, since it does not have three conjugated double bonds, but rather two pairs of conjugated double bonds. It can therefore be detected at 243 nm (Fig 7A).

Cyclooxygenase products can be separated much better by RP-HPLC using acetonitrile/water/acetic acid (33,34) as the mobile phase (Figure 8). Although the peak for TXB_2 is still rather broad, this system gives very good separation of 6-oxo$PGF_{1\alpha}$, TXB_2, $PGF_{2\alpha}$, PGE_2, and PGD_2. The acetonitrile system also gives very good resolution of the monohydroxy-20:4 isomers, but unfortunately does not resolve LTB_4 from its two trans isomers. As noted above, there are considerable differences in selectivity between the above two mobile phases. With the acetonitrile system, the retention times of 6-oxo$PGF_{1\alpha}$ and $PGF_{2\alpha}$ are less than those of 20h-LTB_4 and PGE_2, respectively, whereas the opposite is true for the methanol/water/acetic acid system. This is probably due to the ability of methanol to form hydrogen bonds with polar groups on solute molecules.

A mobile phase similar to that illustrated in Figure 8, but using phosphoric acid instead of acetic acid has been reported to give good separation of various eicosanoids (35). We compared the two systems by using conditions identical to those described in Figure 8, except that instead of 0.05% acetic acid, we used 0.02% phosphoric acid (pH 2.5). These two mobile phases gave nearly identical results, so we prefer to use the one containing acetic acid, since it is a volatile solvent, and can be removed much more easily. Mobile phases containing phosphoric acid may have some special applications, however, such as separation of amino acid-containing leukotrienes, where it is necessary to use a lower pH than that which can be obtained with acetic acid.

8. HPLC OF PEPTIDO-LEUKOTRIENES

Leukotrienes containing amino acids (leukotrienes C_4, D_4, E_4, and F_4) are not present in extracts obtained by the procedure illustrated in Figure 2, but must be extracted by other methods as discussed above. These substances behave much differently on HPLC than other eicosanoids, since their retention times are strongly dependent on the pH of the mobile phase. Reversed-phase HPLC can be used to separate these products using mobile phases of different pHs containing 0.02% phosphoric acid (ph 3.0) (29,30), 0.02% acetic acid (pH 4.0) (27) or acetic acid/ammonium hydroxide (pH 5.7 or 6.2) (27,28). At the highest pH, the retention times of peptido-leukotrienes are relatively low, whereas at the lowest pH, they are much longer than those of most other eicosanoids. In contrast, the retention times of other eicosanoids with RP-HPLC are relatively insensitive to the pH of the mobile phase. The chromatographic behavior of peptido-leukotrienes can be adversely affected due to the accumulation of metal ions in the stationary phase. This can be corrected by treatment of the column with EDTA (28,38).

A very useful method for the separation of lipoxygenase products has recently been reported by Borgeat (29,30). In this procedure, media (2 ml) from incubations with leukocytes is acidified with phosphoric acid and immediately injected directly on to an ODS silica column. Trihydroxy, dihydroxy, and monohydroxy metabolites of 20:4 are first eluted using a mobile phase consisting of various proportions of methanol, acetonitrile and water, and containing 0.02% phosphoric acid and small amounts of tetrahydrofuran and dimethylsulfoxide (pH ca. 3.0). Leukotrienes C_4, D_4, and E_4 are then eluted by raising the pH of the mobile phase to 5.5 by the addition of ammonium hydroxide (29,30). The main advantages of this procedure are that it is not necessary to extract the samples, and that there is no overlap between peptido-leukotrienes and other lipoxygenase products which do not contain amino acids.

9. ARGENTATION HPLC

Argentation HPLC of eicosanoids can be carried out on cation exchange columns which have been treated with silver nitrate (37,38). We have used columns of RSiL KAT (4.6 x 350 mm; 5 μm particle size; Alltech Associates) for this purpose. The column is first washed with water (300 ml), and then treated with 1 M silver nitrate (150 ml). It is then washed with water (300 ml), methanol (300 ml), acetone (200 ml), ethyl acetate (200 ml), methylene chloride (200 ml),

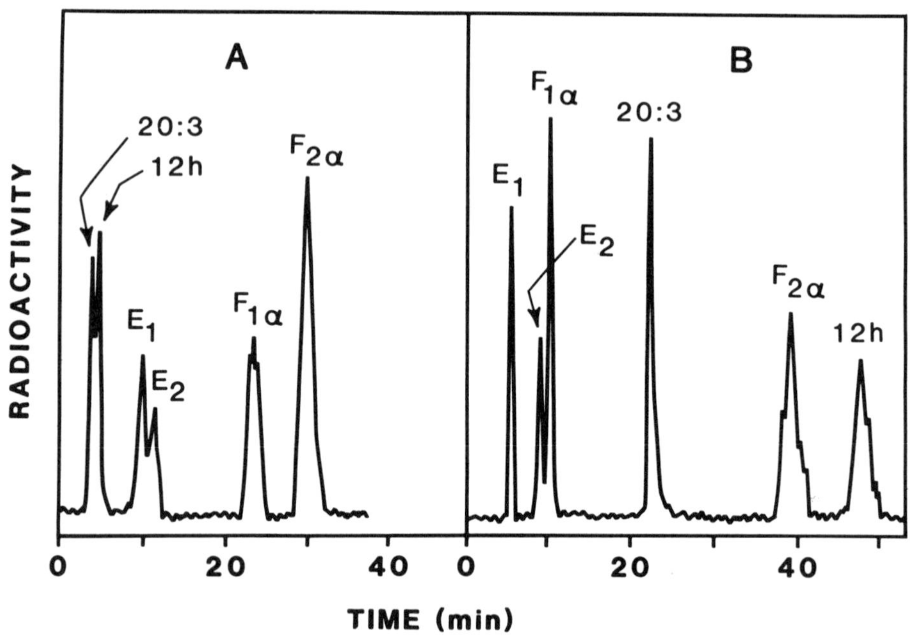

FIGURE 9. Argentation HPLC of a mixture of radioactively-labeled 8,11,14-eicosatrienoic acid (20:3), 12-HETE (12h), PGE_1, PGE_2, $PGF_{1\alpha}$, and $PGF_{2\alpha}$ using different mixtures of solvent H (chloroform/methanol/acetic acid, 79.5:20:0.5) and solvent I (chloroform/acetonitrile/acetic acid, 79.5:20:0.5) at a flow rate of 1.5 ml/min. (A) 90% solvent I; (B) 5% solvent I.

and hexane (200 ml).

This column interacts with solutes by two mechanisms: (i) charge-transfer interactions between silver ions in the stationary phase and olefinic double bonds of the solute, and (ii) interactions between polar groups of the stationary phase and polar groups of the solute. The relative contributions of these two types of interactions can be controlled by altering the composition of the mobile phase. Acetonitrile can be added to reduce silver ion-olefin interactions, and, under appropriate conditions, the selectivity of the silver ion-loaded column can be quite similar to that of a silica column. On the other hand, polar interactions can be reduced by the addition of methanol, allowing silver ion-olefin interactions to predominate. This is illustrated in Figure 9, which shows chromatograms of prostaglandins E_1, E_2, $F_{1\alpha}$, and $F_{2\alpha}$, and 12-HETE (12h) and 8,11,14-eicosatrienoic acid (20:3) with different mixtures of chloroform/methanol/acetic acid (79.5:20:0.5) (solvent H) and chloroform/acetonitrile/acetic acid (79.5:20:0.5) (solvent I). With 90% solvent I (2% methanol; 18% acetonitrile), the order of retention times is similar to that which would be expected with NP-HPLC (i.e. fatty acids < monohydroxy fatty acids < PGE < PGF) (Figure 9A). With this mobile phase, $PGF_{1\alpha}$ and $PGF_{2\alpha}$, but not PGE_1 and PGE_2, are well separated. If the acetonitrile concentration is reduced to 1% and the methanol concentration increased to 19% (i.e. 5% solvent I), the selectivity of the stationary phase is quite different (Figure 9B). With these conditions, the retention times of PGE_1 and $PGF_{1\alpha}$ are considerably shorter, whereas those for 12-HETE and 20:3 are much longer. Thus both PGE_1 and PGE_2, and $PGF_{1\alpha}$ and $PGF_{2\alpha}$ are very well separated from one another. With this mobile phase, the retention time of 20:4 (not shown) is much longer than that of 12-HETE. The contribution of interactions between polar groups and the stationary phase can be suppressed even further by using methanol/acetic acid (99.8:0.2) as the mobile phase (Table 2). In

Table 2. Retention times (tR) of some eicosanoids chromatographed by argentation HPLC with methanol/acetic acid (99.8:0.2) as the mobile phase. Reproduced with permission from the publisher from reference 37.

Compound	tR (min)	Compound	tR (min)
18:0	2.4	15h-20:3	4.8
		20:3	19.0
PGE_1	3.1	15h-20:4	10.9
6-oxo$PGF_{1\alpha}$	3.4	5h-20:4	13.8
$PGF_{1\alpha}$	3.8	11h-20:4	21.4
		9h-20:4	22.4
PGE_2	5.1	8h-20:4	24.0
PGD_2	6.9	12h-20:4	25.1
$PGF_{2\alpha}$	10.8	20:4	65.0

this case, the retention times of prostaglandins are very low, but they can be increased, with a corresponding improvement in resolution, by adding hexane/propanol to the methanol/acetic acid (cf. Figure 10). An advantage of these mobile phases is that they are transparent to UV light, so that solutes can be detected at 200 nm (Figure 10).

Argentation HPLC is generally not very useful for the separation of mixtures containing large numbers of eicosanoids, since solutes which are structurally quite different (e.g. 15-HETE and $PGF_{2\alpha}$, Table 2) can have nearly identical retention times. On the other hand, this method can be very useful for the separation of groups of closely related compounds such as prostaglandins and TXB_2 (39) and monohydroxy-20:4 isomers (Table 2). Obviously, this system is very well suited for separating eicosanoids containing different numbers of double bonds, such as prostaglandins of the 1 ,2, and 3 series, or monohydroxy metabolites of 20:3, 20:4 and

20:5 (Figure 9B).

Another application of argentation HPLC is for the purification of biosynthetically-prepared deuterium- or tritium-labeled 20:4 metabolites. Solutes which contain olefinic deuterium or tritium atoms have longer retention times than the corresponding unlabeled compounds due to the fact that carbon-hydrogen bond lengths decrease in the order $C{-}^{1}H > C{-}^{2}H > C{-}^{3}H$. Thus olefinic groups containing deuterium or tritium atoms can interact more strongly with the silver ions of the stationary phase than olefinic groups containing only protium atoms (39,40). This approach can therefore be used to increase the specific activities of tritium-labeled compounds, which is an important factor in such applications as receptor binding studies and radioimmunoassays. This method can also be used to separate unlabeled eicosanoids from their deuterium-labeled analogs in order to

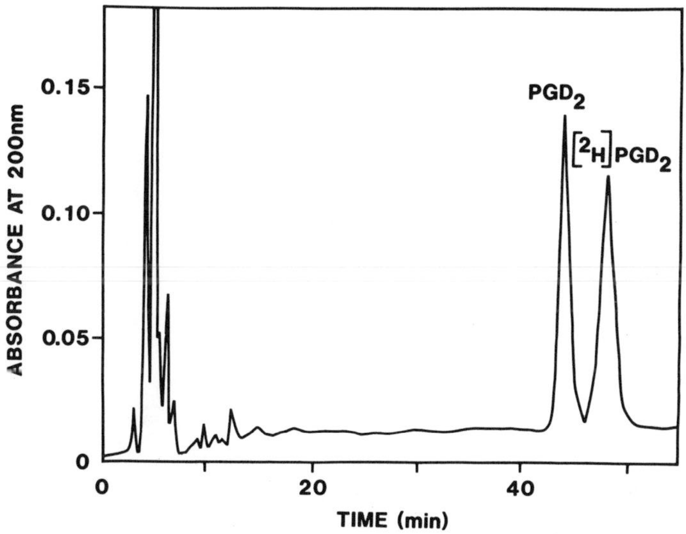

FIGURE 10. Separation of a mixture of PGD_2 and $[5,6,8,9,12,14,15{-}^{2}H]PGD_2$, which was prepared by incubating $[5,6,8,9,11,12,14,15{-}^{2}H]$ 20:4 with a homogenate from rat spleen. The mobile phase was hexane/n-propanol/methanol (50:10:40) and the flow rate, 1 ml/min. Reproduced with the permission of the publisher from reference 40.

obtain deuterium-labeled internal standards with low blank values for analysis by gas chromatography-mass spectrometry with selected ion monitoring. An example of this is shown in Figure 10, which shows the separation of [5,6,8,9,12,14,15-^{2}H]PGD_2 from PGD_2. This method can also be used to separate other prostaglandins, HETEs, and LTB_4 labeled with deuterium from the corresponding unlabeled compounds (39,40).

10. QUANTITATION OF EICOSANOIDS BY HPLC

Eicosanoids can be quantitated by HPLC using a UV or fluorescence detector after derivatization (20-23) as discussed above. Alternatively, monohydroxy metabolites of arachidonic acid and leukotrienes can be quantitated on the basis of their absorbance at 235 nm and 280 nm, respectivly, (1,32). This can be done either by RP-HPLC with PGB_2 as the internal standard (1), or by NP-HPLC with 5S-hydroxy-12-(2-hydroxy)ethoxy-6,8,10,14-eicosatetraenoic acid as internal standard (32). Monohydroxy metabolites of 20:4 can also be quantitated using decanophenone as an internal standard (41).

Radioactive products of enzymatic reactions can also be quantitated by HPLC using either a radioactive monitor or by liquid scintillation counting. In this case, we normally incubate preparations containing cyclooxygenase or lipoxygenase with ^{14}C-labeled substrate and add [^{3}H]$PGF_{1\alpha}$ or [^{3}H]$PGF_{1\beta}$ as an internal standard (9,10).

11. CONCLUSIONS

Both reversed-phase and normal-phase HPLC are powerful methods for separating mixtures of eicosanoids. In general, we prefer to use RP-HPLC on columns of 5 µm ODS silica, because this method gives good recoveries, is relatively insensitive to the injection medium, and gives a very good UV baseline. For lipoxygenase products we normally initially use methanol/water/acetic acid as the mobile phase, whereas for cyclooxygenase products we prefer acetonitrile/water/

acetic acid. Normal-phase HPLC using hexane/toluene/ethyl acetate/acetonitrile/methanol/acetic acid also gives good separation of cyclooxygenase products. In many cases, when all of the products of interest are not resolved in the first chromatography, it is necessary to carry out a second HPLC, either with the same stationary phase and a different mobile phase to alter the selectivity, or with a different stationary phase. Monohydroxy metabolites of 20:4 are very well separated by NP-HPLC with hexane/isopropanol/acetic acid. Argentation HPLC is not usually advantageous for the separation of mixtures containing many different eicosanoids, but has some specific applications, such as the separation of similar eicosanoids differing in their degrees of unsaturation, or for separating unlabeled eicosanoids from their deuterium- or tritium- labeled analogs.

ACKNOWLEDGEMENTS

The excellent technical assistance of F. Gravelle is gratefully acknowledged. This work was supported by grants from the Medical Research Council of Canada and the Quebec Heart Foundation. The author holds a Scientist award from the Medical Research Council.

REFERENCES

1. Borgeat P, Samuelsson B: Arachidonic acid metabolism in polymorphonuclear leukocytes: effects of ionophore A23187. Proc Natl Acad Sci USA (76): 2148-2152, 1979.

2. Murphy RC, Hammmarström S, Samuelsson B: Leukotriene C: a slow-reacting substance from murine mastocytoma cells. Proc Natl Acad Sci USA (76): 4275-4279, 1979.

3. Oliw EH, Guengerich FP, Oates JA: Oxygenation of arachidonic acid by hepatic monooxygenases. Isolation and metabolism of four epoxide intermediates. J Biol Chem (257): 3771-3782, 1982.

4. Chacos N, Falck JR, Wixtrom C, Capdevila J: Novel epoxides formed during the liver cytochrome P-450 oxidation of arachidonic acid. Biochem Biophys Res Commun (104): 916-922, 1982.

5. Powell WS: Rapid extraction of oxygenated metabolites of arachidonic acid from biological samples using octadecylsilyl silica. Prostaglandins (20): 947-957, 1980.

6. Powell WS: Properties of leukotriene B_4 20-hydroxylase from polymorphonuclear leukocytes. J Biol Chem (259): 3082-3089, 1984.

7. Hamberg M, Samuelsson B: On the specificity of the oxygenation of unsaturated fatty acids catalyzed by soybean lipoxidase. J Biol Chem (242): 5329-5335, 1967.

8. Boeynaems JM, Brash AR, Oates JA, Hubbard WC: Preparation and assay of monohydroxyeicosatetraenoic acids. Anal Biochem (104): 259-267, 1980.

9. Powell WS: Formation of 6-oxoprostaglandin $F_{1\alpha}$, 6,15-dioxoprostaglandin $F_{1\alpha}$, and monohydroxyicosatetraenoic acids from arachidonic acid by fetal calf aorta and ductus arteriosus. J Biol Chem (257): 9457-9464, 1982.

10. Powell WS: Formation of trihydroxyheptadecenoic acids, monohydroxyicosatrienoic acids, and prostaglandins from 8,11,14-icosatrienoic acid by adult and fetal aorta. J Biol Chem (257): 9465-9472, 1982.

11. Borgeat P, Samuelsson B: Transformation of arachidonic acid by rabbit polymorphonuclear leukocytes. Formation of a novel dihydroxyeicosatetraenoic acid. J Biol Chem (254): 2643-2646, 1979.

12. Borgeat P, Samuelsson B: Metabolism of arachidonic acid in polymorphonuclear leukocytes. Structural analysis of novel hydroxylated compounds. J Biol Chem (254) 7865-7869, 1979.

13. Borgeat P, Picard S, Vallerand P, Sirois P: Transformation of arachidonic acid in leukocytes. Isolation and structural analysis of a novel dihydroxy derivative. Prostaglandins Med (6): 557-570, 1981.

14. Maas RL, Turk J, Oates JA, Brash AR: Formation of a novel dihydroxy acid from arachidonic acid by lipoxygenase-catalyzed double oxygenation in rat mononuclear cells and human Leukocytes. J Biol Chem (257): 7056-7067, 1982.

15. Orning L, Hammarström S, Samuelsson B: Leukotriene D: a slow reacting substance from rat basophilic leukemia cells. Proc Natl Acad Sci USA (77): 2014-2017, 1980.

16. Lewis RA, Drazen JM, Austen KF, Clark DA, Corey EJ: Identification of the C(6)-S-conjugate of leukotriene A with cysteine as a naturally occurring slow reacting substance of anaphylaxis (SRS-A). Importance of the 11-cis geometry for biological activity. Biochem Biophys Res Commun (96): 271-277, 1980.

17. Borgeat P, Samuelsson B: Transformation of arachidonic acid and homo-γ-linolenic acid by rabbit polymorphonuclear leukocytes. J Biol Chem (251): 7816-7820, 1976.

18. Hamberg M, Samuelsson B: Prostaglandin endoperoxides. Novel tranformations of arachidonic acid in human platelets. Proc Natl Acad Sci USA (71) 3400-3404, 1974.

19. Bergström S, Ryhage R, Samuelsson B, Sjovall J: Prostaglandins and related factors 15. The structure of prostaglandins E_1, $F_{1\alpha}$, and $F_{1\beta}$. J Biol Chem (238): 3555-3564, 1963.

20. Morozowich W, Douglas SL: Resolution of prostaglandin p-nitrophenacyl esters by liquid chromatography and conditions for rapid, quantitative p-nitrophenacylation. Prostaglandins (10): 19-40, 1975.

21. Fitzpatrick F: High performance liquid chromatographic determination of prostaglandins $F_{2\alpha}$, E_2, and D_2 from in vitro enzyme incubations. Anal Chem (48): 499-502, 1976.

22. Turk J, Weiss SJ, Davis JE, Needleman P: Fluorescent derivatives of prostaglandins and thromboxanes for liquid chromatography. Prostaglandins (16): 291-309, 1978.

23. Hatsumi M, Kimata SI, Hirosawa K: 9-anthryldiazomethane derivatives of prostaglandins for high-performance liquid chromatographic analysis. J Chromatogr (253) 271-275, 1982.

24. Gréen K, Hamberg M, Samuelsson B, Frölich J: Extraction and chromatographic procedures for purification of prostaglandins, thromboxanes, prostacyclin, and their metabolites. Adv Prostaglandin Thromboxane Res (5): 15-38, 1978.

25. Powell WS: Rapid extraction of arachidonic acid metabolites from biological samples using octadecylsilyl silica. Methods Enzymol (86): 467-477, 1982.

26. Bennett HPJ, Hudson AM, McMartin C, Purdon GE: Use of octadecylsilyl silica for the extraction and purification of peptides in biological samples. Biochem J (168): 9-13, 1977.

27. Mathews WR, Rokach J, Murphy RC: Analysis of leukotrienes by high-pressure liquid chromatography. Anal Biochem (118): 96-101, 1981.

28. Metz SA, Hall ME, Harper TW, Murphy RC: Rapid extraction of leukotrienes from biological fluids and quantitation by high-performance liquid chromatography. J Chromatogr (233): 193-201, 1982.

29. Borgeat P, Fruteau de Laclos B, Rabinovitch H, Picard S, Braquet P, Hebert J, Laviolette M: Eosinophil-rich human polymorphonuclear leukocyte preparations characteristically release leukotriene C_4 upon ionophore A23187 challenge. J Allergy Clin Immunol, in press.

30. Borgeat P: Reversed-phase high-pressure liquid chromatography profiling and quantitation of lipoxygenase products. Prostaglandins, in press.

31. Porter NA, Logan J, Kontoyiannidou V: Preparaation and purification of arachidonic acid hydroperoxides of biological importance. J Org Chem (44): 3177-3181, 1979.

32. Maclouf J, Fruteau de Laclos B, Borgeat P: Stimulation of leukotriene biosynthesis in human blood leukocytes by platelet-derived 12-hydroperoxyicosatetraenoic acid. Proc Natl Acad Sci USA (79): 6042-6046, 1982.

33. Alam I, Ohuchi I, Levine L: Determination of cyclooxygenase products and prostaglandin metabolites using high-pressure liquid chromatography and radioimmunoassay. Anal Biochem (93): 339-345, 1979.

34. Eling T, Tainer B, Ally A, Warnock R: Separation of arachidonic acid metabolites by high-pressure liquid chromatography. Methods Enzymol (86): 511-517, 1982.

35. Van Rollins M, Aveldano MI, Sprecher HW, Horrocks LA: High-pressure liquid chromatography of underivatized fatty acids, hydroxy acids, and prostanoids having different chain lengths and double-bond positions. Methods Enzymol (86): 518-530, 1982.

36. Metz SA, Hall ME, Harper TW, Murphy RC: Rapid extraction of leukotrienes from biological fluids and quantitation by high-performance liquid chromatography. J Chromatogr (275): 468, 1983.

37. Powell WS: Separation of icosenoic acids, monohydroxyicosenoic acids, and prostaglandins by high-pressure liquid chromatography on a silver ion-loaded cation exchange column. Anal Biochem (115): 267-277, 1981.

38. Powell WS: Argentation-high-pressure liquid chromatography of prostaglandins and monohydroxyeicosatetraenoic acids. Methods Enzymol (86): 530-543, 1982.

39. Powell WS: Separation of unlabeled and isotopically-labeled metabolites of arachidonic acid by argentation high-pressure liquid chromatography. Adv Prostaglanin Thromboxane Leukotriene Res (11): 207-213, 1983.

40. Powell WS: Separation of unlabeled metabolites of arachidonic acid from their deuterium- and tritium-labeled analogs by argentation high-pressure liquid chromatography. Anal Biochem (128): 93-103, 1983.

41. Sun FF, McGuire,JC: Inhibition of human neutrophil arachidonate 5-lipoxygenase by 6,9-deepoxy-6,9-(phenylimino)-$\Delta^{6,8}$-prostaglandin I_1 (U-60257). Prostaglandins (26): 211-221, 1983.

23

RADIOIMMUNOASSAY OF EICOSANOIDS ASSOCIATED WITH TUMOR GROWTH

LAWRENCE LEVINE

1. INTRODUCTION

Although prostaglandin production has been associated with neoplasia, a causal relationship has never been demonstrated. Most mammalian cells have the enzymes that metabolize arachidonic acid, but the capacity to generate the thromboxanes, prostacyclins, the various prostaglandins, mono- and dihydroxyfatty acids and the peptide-containing leukotrienes from endogenous substrate varies among the cells (1-7). The relative levels of these arachidonic acid metabolites produced endogenously (the cell's arachidonic acid metabolite profile) also can vary with the cell's state of differentiation and nutritional history. Examples of how 1) genetic specification (although it should be realized that postreplicative events also are important), 2) differentiation, and 3) nutrition can affect these profiles are given in Table 1, and Fig. 1 and Fig. 2, respectively.

In Table 1 are given the arachidonic acid profiles of several cells that have been cultured under normal nutritional conditions and then stimulated to metabolize arachidonic acid. All of the cells were cultured in monolayer, and at less than confluent cell densities all were stimulated by treatment with the Ca^{2+} ionophore, A-23187, for 60 minutes. The levels of arachidonic acid metabolites were measured by radioimmunoassay of the culture fluids and with most cells the analyses were confirmed by immunochromatography (high performance liquid chromatography and radioimmunoassay). Clearly, the arachidonate metabolite profile varies from cell to cell. For example, greater than 90% of the products generated by endothelial cells and smooth muscle cells freshly isolated from

W.E.M. Lands (ed.), *BIOCHEMISTRY OF ARACHIDONIC ACID METABOLISM*.

Table 1. Arachidonic acid metabolism by cells in culture

		% Metabolic Product in Culture Fluid					
Cell	Source	$6\text{-k-}F_{1\alpha}$	$PGF_{2\alpha}$	PGE_2	TxB_2	iSRS	iHETE
MDCK	Dog kidney	22	58–68	10–20	<0.1	<0.1	<0.1
Smooth muscle	Bovine aorta	75–90	5–15	5	<0.1	<0.1	<0.1
C-9	Rat liver	92	4	4	<0.1	<0.1	<0.1
Endothelial	Bovine aorta	93	3.5	2	1.5	<0.1	<0.1
WI-38	Normal human embryonic lung	3	18	67	12	<0.1	<0.1
D-550	Normal human foreskin	1	19	79	1	<0.1	<0.1
Epithelial	Rat glomerulus	88	8	4	<0.1	<0.1	<0.1
$HSDM_1$	Mouse fibrosarcoma	<0.1	11	89	<0.1	<0.1	<0.1
MC5-5	Methyl-cholanthrene transformed fibroblast	<0.1	11	89	<0.1	<0.1	<0.1
3T3	Mouse fibroblast	3	33	64	<0.1	<0.1	<0.1
WEHI	Mouse lymphoma	0.1	5	5	1	33	56
RBL-1	Rat basophil leukemia	<0.1	0.3	0.3	0.1	58	41
LC-540	Rat testicular tumor	0.3	59	30	0.5	4	6

bovine aorta is prostacyclin. Not all endothelial cells produce primarily prostacyclin (PGI_2), as we have found that in endothelial cells, freshly isolated from bovine adrenal and human umbilicus vein, 80% and 26% respectively of their cyclooxygenase products are PGI_2. Of the cells listed in Table 1, only the two leukocyte cell lines, WEHI-3 and RBL-1, and the Leydig cell line produced hydroxyfatty acids and protein-containing leukotrienes, and among those the ratio of the hydroxyfatty acids to protein-containing-leukotrienes to cyclooxygenase products varied considerably. It should be noted that all of the cells listed in Table 1 produced some measurable prostaglandin (PG)E_2 and $PGF_{2\alpha}$, the arachidonic acid metabolites whose increased levels in biologic fluids of animals with cancer have been used to implicate arachidonic acid metabolism with cancer (8-10).

In Fig. 1 is shown the immunochromatographic analysis for

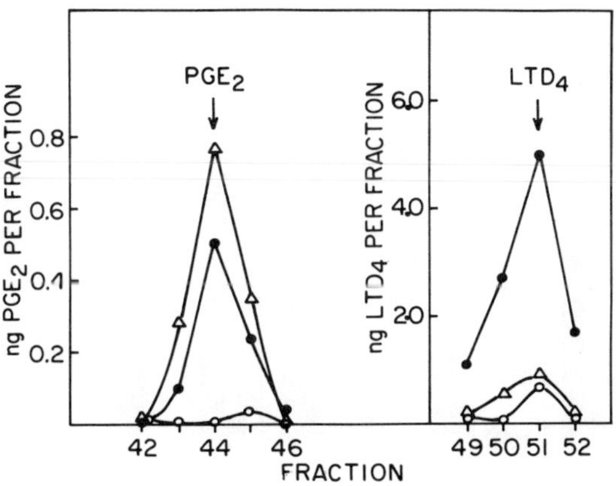

FIGURE 1. Immunochromatographic analysis for PGE_2 and LTD_4 synthesized by HL60 cells before (o) and after differentiation induced by 13-trans retinoic acid (Δ) and DMSO (•).

PGE_2 and leukotriene (LT)D_4 in the culture fluids of A-23187 (0.3 μM) stimulated HL60 cells before and after differentiation induced by incubation of the HL60 cells with 1 μM 13-trans retinoic acid and 1.3% dimethylsulfoxide (DMSO) for 4 days. The levels of PGE_2 produced by the retinoic acid differentiated cells after stimulation by A-23187 are 1.6 times higher than those produced by the DMSO differentiated cells, whereas the levels of LTD_4 are 10 times higher in the DMSO differentiated cells than in the retinoic acid-differentiated cells. Increased levels of LTC_4, LTE_4, LTB_4 and thromboxane (Tx)B_2 also are found in the culture fluids of DMSO-differentiated cells stimulated with A-23187 when compared to those found in the culture fluids of the retinoic acid differentiated cells (L. Levine and T. Breitman, unpublished).

The effect of nutrition on the arachidonate metabolic profile of a cell line that produces both cyclooxygenase and lipoxygenase products is illustrated in Fig. 2. A steroid-secreting cell line isolated from a Leydig cell testicular tumor of the rat (the LC-540 cell line) when stimulated by melittin, a polypeptide from bee venom, synthesizes $PGF_{2\alpha}$, PGE_2, TxA_2 (measured as TxB_2) and also serologically active hydroxyfatty acids (HETE) and peptide-containing LTs (SRS).

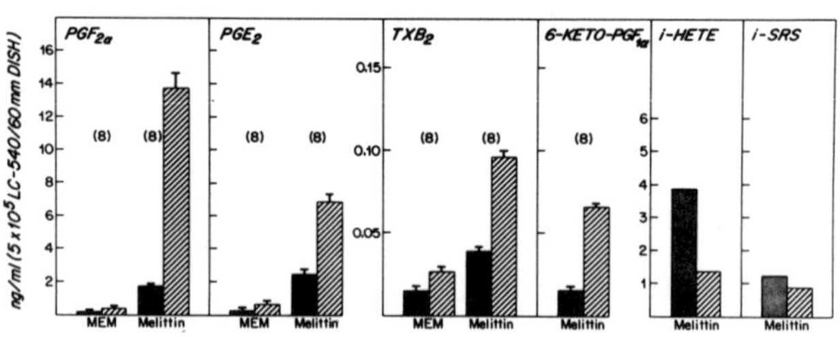

FIGURE 2. Effect of EPA on arachidonic acid metabolism by rat Leydig cells. Cells (5×10^5/60 mm dish) were preloaded with EPA (5 μg/2×10^5 cells) and after washing to remove unincorporated EPA, they were stimulated with melittin (5 μg/ml) for 20 hours. EPA preloaded cells (▨); control cells (▨) (from 11).

Whereas the levels of cyclooxygenase products PGE_2, $PGF_{2\alpha}$, and TxB_2 are reduced in culture media of cells preloaded with eicosapentaenoic acid (EPA), the levels of lipoxygenase products, HETE and SRS, are not; they are increased, suggesting that EPA is affecting generation of the PGG_2 (H_2) endoperoxide, and not the lipoxygenase-generated epoxide (11). Similar reductions of cyclooxygenase generated PGE_2, $PGF_{2\alpha}$ and TxB_2 are found in culture media of the rat basophil leukemia (RBL-1) cells preloaded with EPA that are stimulated by the Ca^{2+} ionophore A-23187, while the levels of lipoxygenase products, SRS and HETE, are not affected significantly or even are increased in the culture fluids of the cells preloaded with EPA.

2. ANALYSES OF ARACHIDONIC ACID METABOLITES IN BIOLOGIC FLUIDS

As recorded above, elevated levels of prostaglandins have been found in the blood and urines of patients with cancers or in animals bearing tumors (10-12). Several analytical procedures were used to generate such data. No attempt will be made in this chapter to compare the specificities and/or sensitivities of the various analytical techniques for the measurement of arachidonic acid metabolites in biologic fluids. Because of the number of samples that can be analyzed, radioiummunoassay has been used most extensively.

Although each cell produces its unique arachidonic acid metabolic profile, PGE_2 and $PGF_{2\alpha}$ have been found as two of its products in every cell, albeit to different extents. In addition, we have found that this unique profile, although it may be affected slightly, is not drastically altered by stimulating that cell to metabolize arachidonic acid with several different agonists. However, analysis for PGE_2 or $PGF_{2\alpha}$ whose stability *in vivo* is low may not be sufficiently sensitive to yield information on its association with pathological conditions. The physiological stability *in vivo* of the 15-OH dehydrogenase and $\Delta 13$ reductase products of the primary prostaglandins E_2 and $F_{2\alpha}$, however, is much greater (12), and analyses for 13,14-dihydro-15-keto-PGE_2 and 13,14-dihydro -15-keto-$PGF_{2\alpha}$ are more likely to give information concerning an association between arachidonic acid metabolism and cancer.

As an example of the greater sensitivity of the analyses of the 13,14-dihydro-15-keto-PGE_2 and 13,14-dihydro-15-keto-$PGF_{2\alpha}$ over that of their parent prostaglandins, PGE_2 and $PGF_{2\alpha}$, their levels in the peripheral plasma of normal rabbits and rabbits bearing the VX_2 carcinoma (13) are shown in Fig. 3. In addition to physiological stability, the chemical stability of the product to be assayed must be considered. Whereas 13,14-dihydro-15-keto-$PGF_{2\alpha}$ is chemically stable in peripheral plasma, 13,14-dihydro-15-keto-PGE_2 is not (14,15); and it has been suggested that analyses for a stable degradation product, 11-deoxy-13,14-dihydro-15-keto-11β,16ξ-cyclo-PGE_2 is more reliable (16). The serologic activity of an antigen depends on the binding energy of the antigen-antibody complex. For example, antibodies to heme-free apomyoglobin have such high binding energy that when they react with metmyoglobin, heme is released from the metmyoglobin (17). To evaluate the serologic stability of 13,14-dihydro-15-keto-PGE_2 with our antiserum, a known

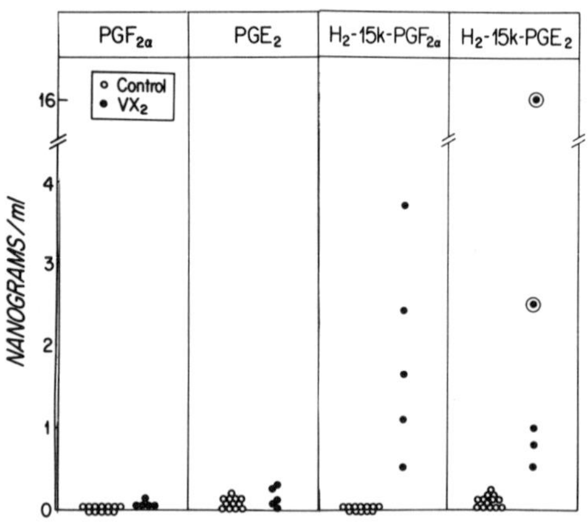

FIGURE 3. Plasma concentrations of $PGF_{2\alpha}$, E_2, 13,14-dihydro-15-keto-$PGF_{2\alpha}$ and 13,14-dihydro-15-keto-PGE_2 in 5 hypercalcemic rabbits bearing VX_2 carcinomas (●) and in a group of normal control rabbits (o). Two of the VX_2 plasma samples (⊙) were assayed for 15-k-PGE_2 and none was detected (from 13).

quantity of 13,14-dihydro-15-keto-PGE_2 dissolved in 70% ethanol (10 mg/ml) was diluted to contain 100 µg/ml of rat plasma or Tris buffer (0.14 M NaCl, 0.01 M Tris, pH 7.4 containing 0.1% gelatin) and incubated at 36°C. At various times, 50 µl was removed, added to 10 ml of iced Tris buffer, and several decrements were assayed by radioimmunoassay (18). During incubation from 0 h to 12.1 h in rat plasma or in Tris buffer, no loss of serologic activity is observed (Fig. 4). Such serologic stability analyses must be done with peripheral plasma of all species in which assay for 13,14-dihydro-15-keto-PGE_2 is to be performed. In human plasma dehydration of 13,14-dihydro-15-keto-PGE_2 is rapid [t 1/2 is approximately 45 min at 37°C (15)]. With our antiserum, the serologic t 1/2 in human plasma is 12 hours. More recently it has been shown that there are much longer-lived metabolites in peripheral plasma, and it has been suggested that these are more reliable indeces of prostaglandin production (12).

For confirmation of serologic analyses, high performance liquid chromatography (HPLC) is a method that has been used

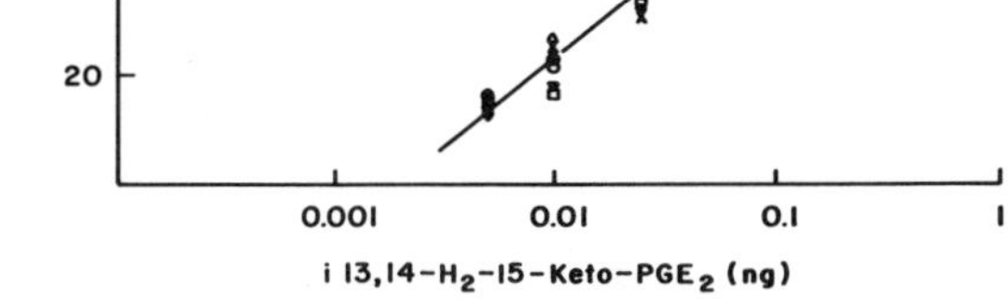

FIGURE 4. The stability of immunoreactive 13,14-dihydro-15-keto-PGE_2 in rat plasma. The serologic activity was unchanged from 0 to 12.1 h (from 18).

(19,20). Such a confirmation is shown in Fig. 5. The chromatographic profile of arachidonate metabolites in the plasma sample from a rabbit 3 weeks after injection with VX_2 tumor cells is compared with the profile of metabolites in normal control rabbit plasma. Increases in 13,14-dihydro-15-keto-$PGF_{2\alpha}$ and 13,14-dihydro-15-keto-PGE_2 in the plasma of the tumor-bearing rabbit were found (13). The HPLC effluent fractions also were analysed for 6-keto-$PGF_{1\alpha}$, TxB_2, $PGF_{2\alpha}$, PGE_2, 13,14-dihydro-PGE_2, and 15-keto-PGE_2; none was found (13).

3. PRECAUTIONS IN INTERPRETING SEROLOGICALLY DETERMINED LEVELS OF ARACHIDONIC ACID METABOLITES

It must be emphasized that although radioimmunoassay (RIA) is a remarkably specific technique whose sensitivity equals or surpasses that of the available chromatographic and spectral methods, antibody specificity with respect to arachidonic acid

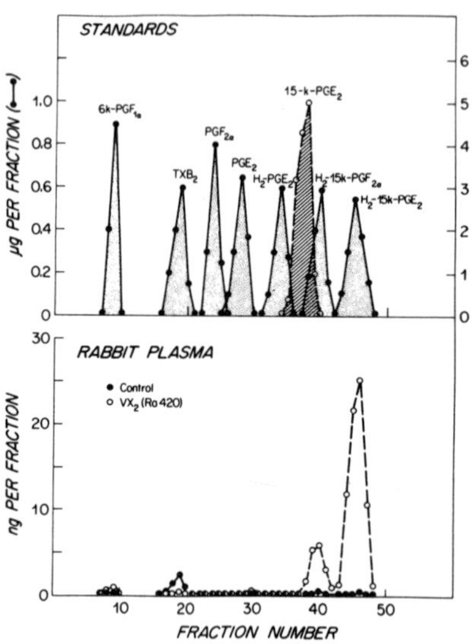

FIGURE 5. The elution patterns of 8 cyclooxygenase metabolites of arachidonic acid separated by HPLC and analyzed serologically. The top panel shows the profiles for standard metabolites, and the bottom panel gives the patterns in a control (•) and a tumor-bearing rabbit plasma (o) (from 13).

metabolites is directed toward functions distal to the common carboxyl function used to synthesize the immunogen (21). The fatty acid chain proximate to that carboxyl function adds little to the considerable binding energy of the eicosanoid-antibody complex. The areas in PGE_2 and $PGF_{2\alpha}$ which contribute most to the binding energy of the PGE_2 anti-PGE_2 and $PGF_{2\alpha}$ anti-$PGF_{2\alpha}$ complexes are shown in Fig. 6. Reaction with anti-12-HETE identifies the immunodominant hydroxyl groups as well as the fatty acid conformation (20); and reactivity with anti-SRS identifies both the 6-sulfidopeptide-containing function and the overall conformation of fatty acid (6). These antibodies,

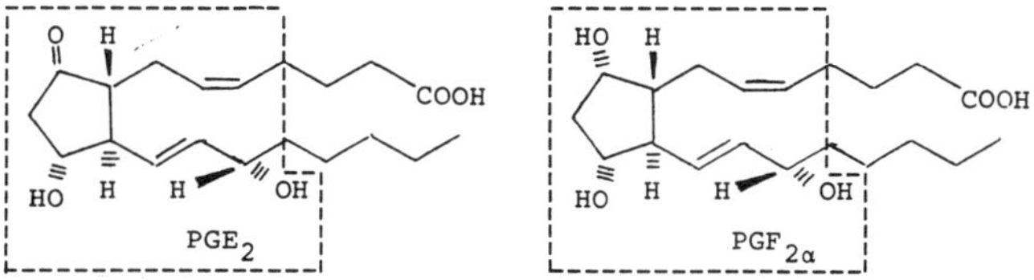

FIGURE 6. PGE_2 and $PGF_{2\alpha}$. The enclosed areas (-----) represent structures that contribute >90% to the binding energy of the homologous prostaglandin-antibody complexes (from 25).

because of the limits of the size of the antibody combining site, do not specify the chain length of the eicosanoid. Thus, the immunoreactive compounds in extracts of some invertebrates have some or all of the structures recognized by the appropriate antibodies even though they may be derivatives of arachidonic acid metabolites or products synthesized from different polyunsaturated fatty acid substrates. Because of this antigen-antibody specificity and the sensitivity of the RIA, several molecules related to PGE_2, $PGF_{2\alpha}$, TxB_2, 6-keto-$F_{1\alpha}$ and HETE have been detected in extracts of these invertebrates. They probably are derivatives of these eicosanoids, such as have been identified in coelenterates (22) (epi derivatives, methyl esters) as well as products of polyunsaturated fatty acids other than arachidonic acid. But only analyses by mass spectrometry can identify them.

Serologic heterogeneity is more striking among the invertebrates when assaying for lipoxygenase rather than cyclooxygenase products. When a vesicular gland microsomal fraction is used as the cyclooxygenase, only the (n-6) polyenoic acids (C_{18} to C_{20} acids) are substrates (23); the presence of the terminal (n-3) ethylenic bond inhibits oxygenation of the (n-6) acids. The C_{20}:3(n-6) is a better substrate than the C_{20}:4(n-6). The serologic homogeneity with respect to cyclooxygenase products of mammalian tissue extracts after HPLC in contrast to the serologic heterogeneity found in invertebrate extracts is consistent with these considerations. Lipoxygenases do not have the rigid substrate specificities of the cyclooxygenases (25), and their products are even more serologically heterogeneous than cyclooxygenase products in both the vertebrates and invertebrates.

Immunoassays for prostaglandins and their metabolites are being used by many investigators in studies of arachidonic acid metabolism in mammalian tissues. In most mammalian tissues, but not all, greater than 90% of the polyunsaturated fatty acid substrate is C_{20}:4(n-6) so that in those tissues in which arachidonic acid predominates, RIAs of biological fluids with the use of some antisera without identification by a second intrinsic property of the eicosanoid is probably valid. However, some products of oxygenation of a longer carbon chain polyunsaturated fatty acid, e.g. C_{22}:5(n-6) (high in rat testes) or the polyunsaturated fatty acids of invertebrates may be recognized serologically; and in invertebrates, in which the polyunsaturated fatty acids may vary both qualitatively and quantitatively, quantitative values obtained by RIA can only be made after identification of the eicosanoid by a second property.

ACKNOWLEDGEMENTS

The work done in the author's laboratory was supported by Grants GM 27256 and CA 17309 (National Cancer Institute) from the National Institutes of Health. LL is an American Cancer Society Research Professor of Biochemistry (award PRP-21).

REFERENCES

1. Levine L, Alam A: Deacylation of cellular lipids and arachidonid acid metabolism. Prog in Lipid Res (20): 81-88, 1982.
2. Levine L, Moskowitz, MA: Alpha- and beta-adrenergic stimulation of arachidonic acid metabolism by cells in culture. Proc Natl Acad Sci USA (76): 6632-6636, 1979.
3. Coughlin SR, Moskowitz MA, Zetter BR, Antoniades HN, Levine L: Stimulation of prostacyclin synthesis by platelet-derived growth factor. Nature (288): 600-602, 1980.
4. Coughlin SR, Moskowitz MA, Antoniades HN, Levine L: Serotonin receptor-mediated stimulation of bovine smooth muscle cell prostacyclin synthesis and its modulation by platelet-derived growth factor. Proc Natl Acad Sci USA (78): 7134-7138, 1981.
5. Kreisberg JI, Karnovsky MJ, Levine L: Prostaglandin production by homogeneous cultures of rat glomerular epithelial and mesangial cells. Kidney Internat (22): 355-359, 1982.
6. Levine L, Morgan RA, Lewis RA, Austen KF, Clark DA, Marfat A, Corey EJ: Radioimmunoassay of the leukotrienes of slow reacting substance of anaphylaxis. Proc Natl Acad Sci USA (78): 7692-7696, 1981.
7. Snoek GT, Levine L: Requirements for protein synthesis and calcium for stimulation of prostaglandin synthesis in cultured rat liver cells by tumor promoters. Cancer Res (43): 4743-4746, 1983.
8. Bennett A: Prostaglandins and cancer. In: Karim SM (ed) Practical applications of prostaglandins and their synthesis inhibitors. MTP Press Ltd., Lancaster, 1979, pp. 149-188.
9. Karmali RA: Prostaglandins and cancer. Prostaglandins Med (5): 11-28, 1980.
10. Levine L: Arachidonic acid transformation and tumor production. Adv Cancer Res (35): 49-79, 1981.
11. Levine L, Worth N: Eicosapentaenoic acid: Its effects on arachidonic acid metabolism by cells in culture. J Allergy Clin Immunol. (In press).
12. Oliw E, Granstrom E, Anggard E: The prostaglandins and essential fatty acids. In: Pace-Asciak C, Granstrom E (eds) Prostaglandins and related substances. Elsevier, 1983, pp. 1-11.
13. Alam I, Voelkel EF, Tashjian AH Jr, Levine L: Analysis by high performance liquid chromatography of cyclooxygenase products of arachidonic acid metabolism in plasma of rabbits bearing the VX_2 carcinoma. Prostaglandins Med (4): 227-237, 1980.
14. Fitzpatrick FA, Aguirre R, Pike JE, Lincoln FH: The stability of 13,14-dihydro-15-keto-PGE_2. Prostaglandins (19): 917-913, 1980.
15. Granstrom E, Hamberg M, Hansson G, Kindahl H: Chemical instability of 15-keto-13,14-dihydro-PGE_2: The reason for low assay reliability. Prostaglandins (19): 933-957, 1980.

16. Granstrom E, Kindahl H: Radioimmunologic determination of 15-keto-13,14-dihydro-PGE_2: A method for its stable degradation product, 11-deoxy-15-keto-13,14-dihydro-11β,16-cyclo-PGE_2. Adv Prostaglandin Thromboxane Res (6): 181-182, 1980.
17. Crumpton MJ: Conformational changes in sperm-whale metmyoglobin due to combination with antibodies to apomyoglobin. Biochem J (100): 223-232, 1965.
18. Axelrod L, Levine L: Plasma prostaglandin levels in rats with diabetes mellitus and diabetic ketoacidosis. Diabetes (31): 994-1001, 1982.
19. Alam I, Ohuchi K, Levine L: Determination of cyclooxygenase products and prostaglandin metabolites using high pressure liquid chromatography and radioimmunoassay. Anal Biochem (93): 339-345, 1979.
20. Rigas B, Lewis RA, Austen RA, Corey EJ, Levine L: Identification and quantitation of arachidonic acid metabolic products in rabbit, rat and human saliva. Archs oral Biol (28): 1031-1035, 1983.
21. Levine L, Cernosek RMG, Van Vunakis H: Specificities of prostaglandins B_1, $F_{1\alpha}$, and $F_{2\alpha}$ antigen-antibody reactions. J Biol Chem (246): 6782-6785, 1971.
22. Light RJ, Samuelsson B: Identification of prostaglandins in the gorgonian, Plexaura homomalla. Eur J Biochem (28): 232-240, 1972.
23. Lands WEM, LeTellier PR, Rome LH, Vanderhoek JY: Inhibition of prostaglandin biosynthesis. In: Bergstrom S (ed) Advances in the biosciences (9). 1973, pp. 15-28.
24. Nugteren DH: Arachidonate lipoxygenase in blood platelets. Biochim Biophys Acta (380): 299-307, 1975.
25. Levine L, Kobayashi T: Detection of compounds immunologically related to arachidonic acid transformation products in extracts of invertebrates. Prostaglandins Leukotrienes Med (12): 357-369, 1983.

24

MASS SPECTROMETRY AND EICOSANOID ANALYSIS

ROBERT C. MURPHY AND TIMOTHY W. HARPER

1. INTRODUCTION

Mass spectrometry and its associated techniques have been among the most valuable and powerful analytical techniques applied to the study of arachidonic acid metabolism. In fact, many of the advances made in the discipline of mass spectrometry have been made by individuals trying to further their understanding of prostaglandin biosynthesis. Much of the value of mass spectrometry lies in the rather specific information one can gain concerning a molecule such as a derivatized eicosanoid from rather small quantities of material, e.g., a few nanomoles. The information is typically qualitative in nature; however, in the properly designed mass spectrometry experiment, quantitative information can be obtained. In fact, the combined degree of precision, specificity, accuracy and sensitivity attainable is unmatched by any other quantitative analytical technique currently available.

In this brief chapter, the basic concepts in methodology, instrumentation and interpretation of data will be presented. No attempt will be made to be complete, but reference to arachidonic acid metabolites will be made. The student interested in furthering his study of mass spectrometry is directed to books on interpretation (1), instrumentation (2) and special techniques (3). Recent reviews have dealt specifically with quantitative analysis by combined gas chromatography - mass spectrometry (4,5) and the reader is directed to these for details of the use of stable isotope internal standards and selected ion recording.

2. THE MASS SPECTRUM

The fundamental data generated by the mass spectrometer is a representation of the relative abundance of ions with a certain mass-to-charge ratio (m/z) that have been generated by an ionization process. In order to obtain such data, one must introduce a sample into a vacuum, generate ions from the gaseous molecules, separate ions according to mass, detect separated ions and record ion abundances.

2.1 Sample introduction

Until a few years ago, the major limitation of mass spectrometry as applied to biomedical research was the absolute necessity for a molecule to have sufficient vapor pressure, typically with a partial pressure of 10^{-6} to 10^{-7} torr (1 torr equals 1 mm Hg), within the ion source region of the mass spectrometer. For permanent gases and low boiling liquids or solids, a simple capillary leak from a heated reservoir served as an excellent inlet. For prostaglandins, which are highly polar molecules that associate with themselves in the condensed phase by intermolecular bonds, vapor pressure is a limiting feature. Placing a sample of low volatility directly in the ion source and heating it to 200° to 300° (heated direct insertion probe) may permit one to achieve adequate volatility for only the least polar of the eicosanoids, such as free arachidonic acid itself.

Perhaps the most powerful means to introduce a prostanoid into the mass spectrometer is as the effluent directly from a gas chromatograph. Molecules which can successfully transit a gas chromatograph have, by definition, sufficient vapor pressure for mass spectrometry. Typically, packed GC columns having flow rates of 30 ml min^{-1} are connected to the mass spectrometer ion source by means of a carrier gas separator. The separator, such as an all-glass jet separator, can very efficiently concentrate the GC effluent by removing a very light carrier gas such as helium (6). In more recent years, combined capillary gas chromatography - mass spectrometry has become much more popular due to the enhanced chromatographic resolving power and ease of working with small diameter fused silica columns. Capillary GC flow rates of 1-3 ml/min can be rather easily accommodated with large vacuum pumps attached to the ion source of the mass spectrometers and, therefore, no separator is needed to interface capillary GC to a mass spectrometer. This feature~simplifies the GC/MS system and reduces the unavoidable and sometimes serious losses that occur while molecules traverse a carrier gas separator.

2.2 Derivative formation

Conversion of eicosanoids into compounds with significantly higher vapor pressure has been an important part of analysis by mass spectrometry. This is an absolute requirement for analysis of eicosanoids by GC/MS techniques. Since prostaglandins and leukotrienes are both biologically produced in small quantities and are somewhat unstable chemically, very efficient chemical derivatization reactions, which require relatively mild conditions, are employed. Side products or multiple derivatives are quite undesirable. Table 1 lists the typical derivatives which are formed with eicosanoids for GC/MS analysis. These derivatives enhance volatility by converting polar moieties with active hydrogen atoms (e.g., -O-H) into esters or ethers. Blau (7) has very adequately reviewed details of derivative formation.

Eicosanoids all contain the carboxyl moiety which can be easily esterified by diazomethane (CH_2N_2) or CH_3OH/HCl solutions. Numerous reagents are available to rapidly form trimethylsilyl (TMS) ethers and esters and such derivatives are quite popular for GC/MS analysis (8,9,10). The t-butyldimethylsilyl (TBDMS) derivative has been found to possess some favorable mass spectral properties as compared to TMS derivatives (11,12). Finally, the keto group in PGE_2, PGD_2 and PG-6-keto-$F_{1\alpha}$ is usually derivatized to reduce potential hydrogen-bonding during GC/MS analysis. The reaction of the carbonyl group with methoxylamine hydrochloride to form a Schiff base leads to the methoxime derivative. Unfortunately, two isomers (syn and anti) can form with keto-prostaglandins and this can complicate the GC/MS analysis (13,14). The formation of mixed derivatives is quite typical, for example, methoxime, trimethylsilyl ether, methyl ester of PG-6-keto-$F_{1\alpha}$ (5). The pentafluorobenzyl ester derivatives have been found to be quite useful for negative ion chemical ionization of prostaglandins (16,17) and leukotrienes (18,19). It should be stressed that the derivative chosen greatly influences the mass spectral behavior of the eicosanoid.

Table 1:
Typical Derivatives Employed to Enhance the
Volatility of Prostaglandins and Leukotrienes for GC/MS Analysis

Functional Group	Derivative
Carboxyl, –COOH	Methyl ester, $-COOCH_3$
	Trimethylsilyl ester, $-COO-Si(CH_3)_3$
	t-Butyldimethylsilyl ester, $-COO-Si(CH_3)_2C(CH_3)_3$
	Pentafluorobenzyl ester, $-COOCH_2C_6F_5$
Hydroxyl, R–OH	Trimethylsilyl ether, $-O-Si(CH_3)_3$
	Acetate, $-O-OCCH_3$
	t-Butyldimethylsilyl ether, $-O-Si(CH_3)_2C(CH_3)_3$
Ketone, C=O	Methoxime, $C = N-OCH_3$[1]

[1] Two isomers can be formed, syn and anti

3. ION FORMATION AND SEPARATION

3.1 Electron impact ionization

Conceptually, the most straightforward mode of ion formation is the electron impact mode. Figure 1 outlines schematically the events which take place in the ion source region of the mass spectrometer. The derivatized prostaglandin (M) enters the ion source (heated to prevent condensation on the walls) into which electrons froma heated filament are accelerated to typically 70 eV, for reproducibility considerations. Thus, the electrons have energies corresponding to 2000 kcal/mole which is more than enough to break all covalent bonds. When an electron passes near the molecular orbital clouds of the prostaglandin, energy can be transferred to form electronically excited M*, indicted as enlarged M in Figure 1. This hypothetical intermediate expells an electron (in fact it is promoted to an infinite molecular orbital) to form the less energetic positive ion $M^{+\cdot}$, the molecular ion. It is felt that $M^{+\cdot}$ retains all of the covalent character of M but still has excess energy and is deficient one electron. This excess energy can be dissipated through rupture of single bonds to form fragment ions (as well as neutrals or radicals which cannot be detected), such as F_1^+, F_2^+, F_3^+ which are quite straightforward portions of the original prostaglandin M. There still may be sufficient energy for these fragments (e.g., F_1^+) to further decompose to smaller fragments F_4^+, F_5^+ and F_6^+. Another important route of decomposition is through rearrangement processes in which different portions of the molecule form new intramolecular bonds (R^+). These rearrangement ions are extremely useful for structure elucidation and qualitative analysis since severe steric criteria come into effect for efficient molecular rearrangement. However, they can also be the source of some confusion.

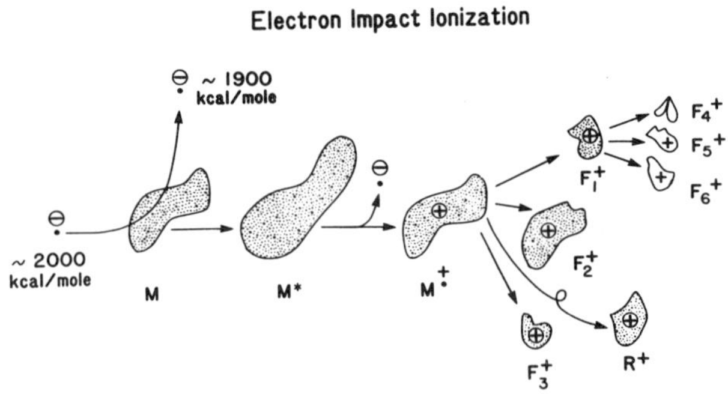

FIGURE 1. Schematic representation of ionization by electron impact (70 eV).

Overall, the electron impact ionization process leads to a population of several types of ions inside the ion source within 10^{-6} to 10^{-12} sec. The EI mass spectrum is, simply stated, a representation of the reaction of an energetic electron with a molecule through a presentation of the positive ion products of that reaction. Figure 2A is the EI mass spectrum of LTB_4 (TMS, methyl ester) as such an example.

3.2 Chemical ionization

A less energetic mode of ion formation is through ion-molecule reactions, termed chemical ionization. With this technique, the ion source is maintained at relatively high pressure (approximately 1 torr) with a gas such as methane, isobutane or ammonia. This reagent gas is ionized by a beam of electrons by the same process of electron impact described in Figure 1. However, due to the high pressure, primary ions arising from the reagent gas, react with neutral molecules to form secondary products as follows:

$CH_4 + e^- \longrightarrow CH_4^{+\cdot} + 2e^-$	Electron impact ionization
$CH_4^{+\cdot} + CH_4 \longrightarrow CH_5^{+} + CH_3$	Ion molecule reaction
$CH_5^{+} + M \longrightarrow MH^{+} + CH_4$	Chemical ionization (proton transfer)

When a small amount of sample (M) is introduced into this chemical ionization plasma, it is more likely that the sample becomes ionized by an ion such as CH_5^+ or $C_2H_5^+$ than by the electron. Since the ion-molecule reaction is a rather slow equilibrium process, chemical ionization yields an ion population which is much different from that obtained with electron impact. Typically, protonated molecular ions (MH^+) or molecular ions having lost a hydrogen atom $(M-1)^+$ are quite abundant when compared to electron impact ionization. The chemistry of this ionization process has been recently reviewed (20).

Since the abundance of molecular ions of eicosanoids produced by electron impact ionization is very small (Figure 2A), chemical ionization offers the advantage of being able to usually observe this very important ion. Yet, methane and isobutane chemical ionization have not been widely used in the prostaglandin field. In spite of the simplified mass spectra obtained (21), molecular ions are still not particularly intense. Ammonia has some advantage as a reagent gas for the chemical ionization of eicosanoids since prominent molecular ions are usually observed (22). Figure 2B is the CI (CH_4) mass spectrum of LTB_4 (TMS, methyl ester) and ions due to the addition of a proton (MH^+) and loss of hydride $(M-H)^+$ can now be seen.

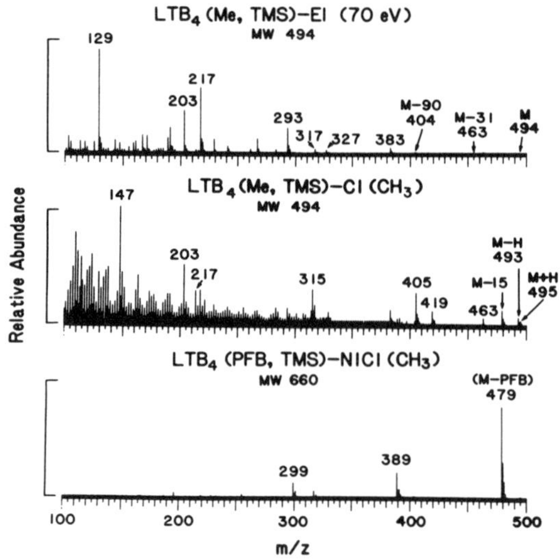

FIGURE 2. Mass spectra of leukotriene B_4 (as the methyl ester, trimethylsilyl ether derivative) using different ionization processes. A: Electron impact. B: Chemical ionization using methane as reagent gas. C: Pentafluorobenzyl ester, trimethylsilyl ether derivative and negative ion chemical ionization techniques.

3.3 Negative ion chemical ionization

In the ionization processes described above, emphasis was placed on the formation of positive ion species. However, under certain conditions, negative ions can be produced in high abundance. Even though mass spectrometers, in theory, can analyze negative ions quite well, the potential for negative ion formation has not been widely appreciated until recently. In particular, the formation of negative ions by the electron capture process under chemical ionization conditions can be quite efficient with certain eicosanoid derivatives such as pentafluorobenzyl esters.

Figure 3 is a representation of the events leading to the formation of negative ions by the process of electron capture. As with chemical ionization (3.2), energetic electrons (2000 kcal/mole) react with methane at 1 torr. If we now consider the fate of the electrons, it is obvious that at these high pressures, numerous collisions/encounters take place to reduce the energy (and, therefore, velocity) of the electrons. In effect, they quickly are reduced to the thermal energy of the ion source or become "thermalized". At this point, a molecule (M) which has high electrophilicity can "capture" the electron, incorporate it into its molecular orbitals and, thereby, become a negative ion (M^-). There is some excess energy imparted to the molecule and further fragmentation may take place leading to fragments (A^-).

In general, eicosanoids do not avidly accept electrons in NCI mass spectrometry. Recently, it has been found that the pentafluorobenzyl esters of prostaglandins (16,17) and leukotrienes (18,19) are efficiently ionized. Rather simple mass spectra result (Figure 2C), dominated by the loss of the pentafluorobenzyl radical to form the stabilized caboxylate anion. This mode of ionization forms the basis for some of the most sensitive quantitative assays for eicosanoids yet developed (23).

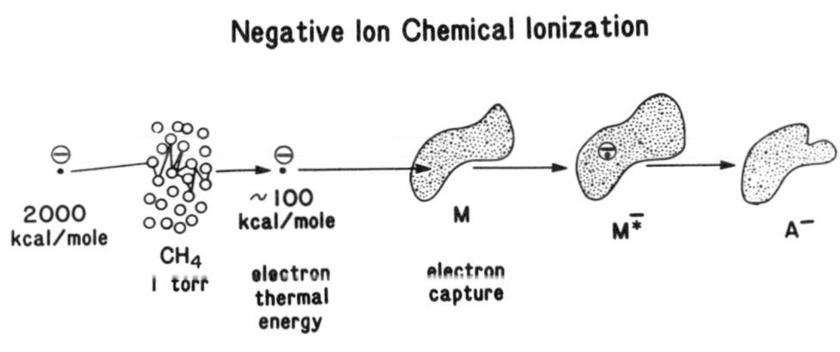

FIGURE 3. Schematic representation of negative ion formation by electron capture under chemical ionization conditions.

3.4 Ion separation

Once ions are formed by processes discussed above, they are accelerated out of the ion source compartment of the mass spectrometer by an electric field of the same polarity. These accelerated ions can then be separated according to their mass-to-charge ratio by numerous physical principles, but most commonly either by magnetic sector based instruments or quadrupole electric field instruments.

In the magnetic sector instrument, ions are accelerated out of the ion source by high potentials (e.g., 4000 - 8000 volts, $\underline{V}$) and passed through a strong magnetic field (B). A force experienced at right angles to the magnetic field and direction of motion, causes the ion to begin to move in a circular path with radius $\underline{r}$. The resultant ion trajectory must satisfy the equation:

$$m/z = B^2r^2 / 2V$$

where m/z is the ion's mass-to-charge ratio. Since the radius is constant in most mass spectrometers, one can vary either $\underline{B}$ or $\underline{V}$ to focus each ion at the detector. The electron multiplier is usually used to detect ions because of its very high sensitivity and low noise. The magnetic sector instrument has been quite popular due to unique advantages of high-mass sensitivity and resolving power.

In the quadrupole mass spectrometer, ions are injected into the middle of a square array of four circular or hyperbolic metallic rods. Non-adjacent rod pairs are connected to either a DC power supply or RF power supply which generates a complex electrical field within the rod array. Ions interact with this field to alter their trajectory in such a manner that only ions with a specific m/z value can travel the entire rod set length (stable path) depending on the voltage potentials and RF frequency. Since voltage potentials are rather easily altered, ions with different m/z values can be permitted to traverse the rod set and be detected by the electron multiplier. This mass spectrometer or mass filter is quite well suited for computer control and data acquisition. Furthermore selected ion recording experiments are easily implemented.

4. THE MASS SPECTRUM

The mass spectrum as typically reported is a histogram of the ions produced in the ion source with m/z values on the abscissa and the relative abundance (intensity) of each ion on the ordinate. By convention, the abundances of all ions are expressed as percentages of the most abundant ion, or the base peak. It has become the practice of some laboratories to report mass spectral data of prostanoids only above a certain mass, e.g., above m/z 100, 150 or 200. This leads to a distortion of the magnitudes of the actual abundances of ions emanating from the prostaglandin since the highly abundant ions at low mass are thereby ignored. Another representation of the abundance data is the percentage of the total ion current (Σ) which an ion carries. The percent sigma scale is often abbreviated "% Σ_{50}" which means the total ionization of the molecule from m/z 50 to the molecular ion was summed.

4.1 Types of ions

4.1.1 Molecular ion. This important ion carries the entire mass of the molecule, but for many eicosanoids this ion may be totally absent. Molecules that do not contain nitrogen have molecular ions at even m/z values. However, methoxime derivatives of PGE_2, PG-6-keto-$F_{1\alpha}$ and TxB_2 have molecular ions at odd m/z values since a single nitrogen atom is present. Figure 4 is the molecular ion region of 5-HETE (methyl ester, TMS ether) which reveals ions higher in mass than m/z 406. These are due to the natural presence of heavy stable isotopes of carbon, hydrogen, oxygen and silicon. Table 2 lists the exact masses and natural abundances of the various isotopes for these elements. The ion at m/z 407 is predo}inantly due to a single carbon-13 or silicon-29 present in the 5-HETE which has an elemental composition of $C_{24}H_{42}O_3Si$. Since there are 24 carbon atoms, each of which could be replaced with carbon-13 at the natural abundance frequency, one would predict m/z 407 to be 26.4% (0.011 x 24) as abundant as m/z 406 due to carbon-13 alone. The ion at m/z 408 arises from species with oxygen-18, silicon-30, or two carbon-13 atoms. The precise measurement of abundances of isotope peaks is the basis of quantitative mass spectrometry when stable isotope-labeled internal standards are used (5).

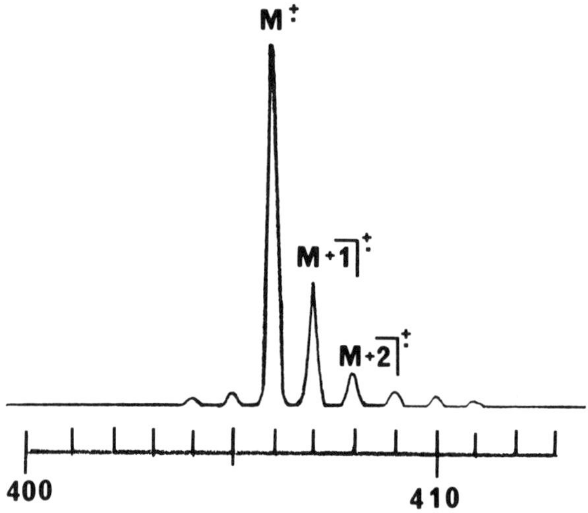

FIGURE 4. Molecular ion region of 5-HETE (methyl ester, trimethylsilyl ether) obtained by electron impact ionization (70 eV).

Table 2:
Exact Masses and Stable Isotope Abundances of Elements in Eicosanoid Derivatives for GC/MS
(Taken from Reference 1)

Isotope	Exact Mass	Natural Abundance
^{1}H	1.00783	99.9985
^{2}H	2.01410	0.015
^{12}C	12.00000	98.9
^{13}C	13.00335	1.108
^{14}N	14.00307	99.6
^{15}N	15.00011	0.4
^{16}O	15.99491	99.8
^{18}O	17.99916	0.2
^{19}F	18.99841	100
^{28}Si	27.97693	92.2
^{29}Si	28.97649	4.7
^{30}Si	29.97376	3.1

4.1.2 Fragment ions. At the risk of oversimplification, many of the major ions in the mass spectra of eicosanoid derivatives are straightforward fragment ions, apparently involving single bond cleavage reactions or loss of neutral species such as methanol (-32 daltons) or trimethylsilanol (-90 daltons). Typically, fragmentation most commonly occurs at points of branching with charge retention on the most stable carbonium atom ($3^{o} > 2^{o} > 1^{o}$). Another highly favored cleavage site is adjacent to ether oxygen atoms with stabilization of the cation provided by the non-bonded electrons of oxygen. Table 3 indicates some of the studies of the mass spectral behavior of derivatized eicosanoids where detailed fragmentation pathways are discussed.

While it is certainly impossible to fully understand the origin of each observed ion in a complex molecule such as a prostaglandin or leukotriene, it is important to understand the origin of all major ions in the mass spectra and even the less abundant ions at high mass. Odd electron ions (those at even mass numbers in compounds not containing nitrogen) are often important rearrangement ions which provide additional valuable information.

Table 3:
Studies of the Mass Spectometric
Fragmentation of Derivatized Eicosanoids

Eicosanoid	Derivative[1]			Ionization	Reference
	COOH	C=O	OH		
PGA	Me	MOX	TMS, tBDMS	EI	10
	Me	MOX	TMS	CI	22
	TMS	MOX	TMS	EI	28
PGB	ME	-	TMS	EI	10
	TMS	MOX	TMS	EI	14
	ME	MOX	TMS	EI	22
PGD	TMS	-	TMS	EI	9
PGE	ME	MOX	TMS	EI	13
	ME	MOX	TMS	EI	22
	TMS	MOX	TMS	EI	29
	ME	-	TMS	EI	30
	ME	-	-	EI	31
PGF	ME		TMS, TBDMS	EI	11
	ME	-	TBDMS	EI	12
	ME	-	R-BO, TMS	CI	21
	ME		TMS	CI	22
	ME	-	-	EI	32
TxB	ME	MOX	TBDMS		12
LTA	ME	-	-	EI	33
LTB	ME		TMS	EI	34

1 Abbreviations: ME, methyl ester; MOX, methoxime; TMS, trimethylsilyl; TBDMS, tert butyldimethylsilyl; R-BO, alkyl cyclic boronate.

There are numerous books available describing the empirical rules employed in mass spectral interpretation. Recent advances in metastable techniques (linked scans), isotope labeling (^{13}C, ^{18}O, ^{2}H) and high resolution mass spectrometry need to be employed for such detailed studies of fragmentation.

5. FAST ATOM BOMBARDMENT

Recently a new mass spectrometric technique has been developed which obviates the requirement of any sample vapor pressure in order to obtain a mass spectrum (25). This technique, termed fast atom bombardment (FAB) or liquid matrix assisted, particle-induced desorption, has already had substantial impact on eicosanoid analysis (26,27). Importantly, it is the only technique with which mass spectral data from leukotriene C_4 has been obtained (26).

The process of FAB mass spectrometry is schematically outlined in Figure 5. This is a surface ionization technique where an intense beam of rapidly moving atoms, such as xenon or argon, obliquely strikes the surface of glycerol (or another suitable matrix) which contains the eicosanoid either dissolved or suspended. Glycerol is a particularly useful matrix since many amphoteric substances, such as phospholipids and leukotrienes, orient themselves on the surface. At the surface, a complex collisional cascade takes place which results in some surface molecules being expelled from the surface. Perhaps this can be better explained by an analogy to the game, pool, where there is a high probability that a pool ball will come directly toward you when you shoot away from you a cue ball (at high velocity) into the triangular rack of fifteen pool balls at rest. By such a process, desorption of nonvolatile molecules from the surface is effected.

If the molecule is already ionized (as are all non-derivatized eicosanoids at pH 7), this desorption process leads to ions in the gaseous state even though they have no equilibrium vapor pressure. Ions can then be accelerated into a magnetic field, separated and detected. Figure 6 is the mass spectra of LTC_4, LTD_4 and LTE_4 obtained by fast atom bombardment conditions. As can be seen, strong molecular ions (MH^+) at 626, 497 and 440 are observed for LTC_4, LTD_4 and LTE_4, respectively. Some fragmentation of these molecules does take place as seen by the ions at m/z 301, 319 and 333, which arise from the arachidonic acid side chain following cleavage of the C-5 bond. The ion at m/z 308 in LTC_4 is the protonated glutathionyl cation.

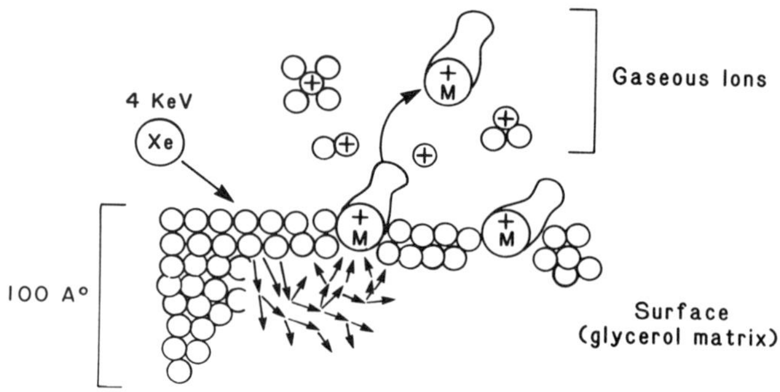

FIGURE 5. Schematic representation of the formation of gaseous ions from nonvolatile molecules (M) in a glycerol matrix (O) by fast atom bombardment (Xe).

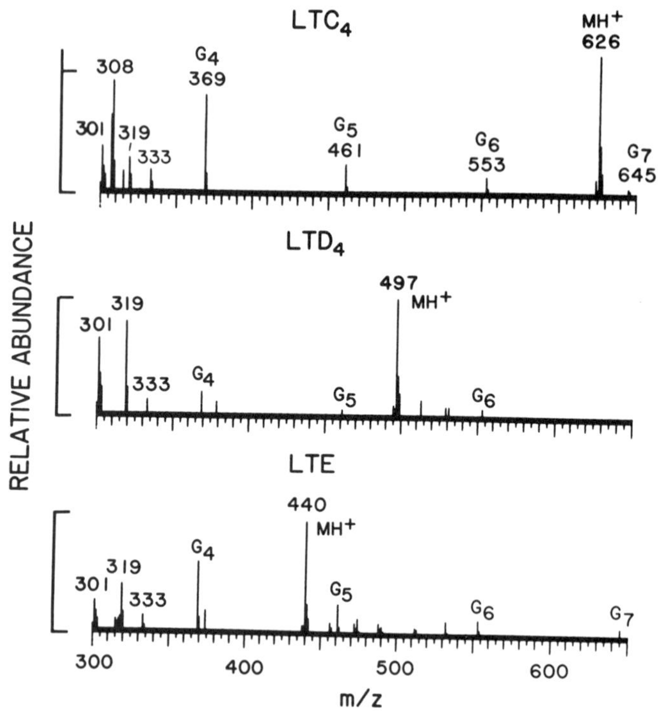

FIGURE 6. Fast atom bombardment (FAB) mass spectra of leukotrienes LTC_4, LTD_4 and LTE_4. Gycerol cluster ions as 92 dalton intervals at indicated by Gn, where n $\doteq$ 4-7.

Figure 7 is the FAB mass spectrum of LTB_4, both positive and negative ions. It should be noted that FAB is currently substantially less sensitive than other mass spectrometric techniques.

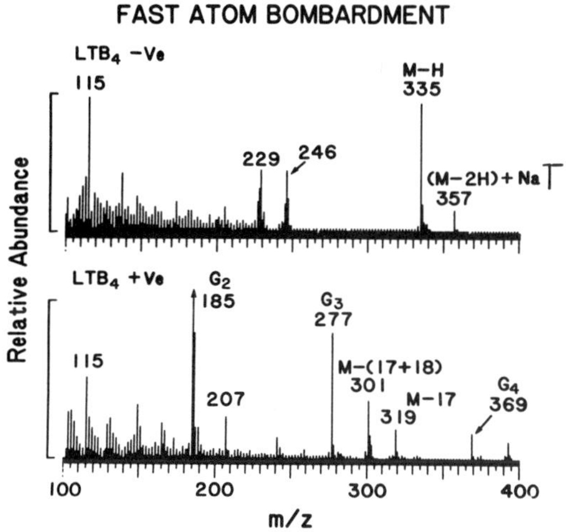

FIGURE 7. Positive and negative ions formed by FAB analysis of leukotriene B_4.

6. CONCLUSION

Mass spectrometry, using electron impact ionization, has had substantial application to the study of arachidonic acid metabolism. The information content in a complete mass spectrum is typically more than can be fully understood at this time. Nonetheless, sufficient information is usually available to greatly assist in structure elucidation and identification by pattern comparison. Newer techniques of ionization, such as negative ion chemical ionization, suggest even greater sensitivities in analysis will become available for biological studies. Fast atom bombardment mass spectrometry has now opened entirely new avenues of eicosanoid studies. No longer is there an absolute limitation in vapor pressure for mass spectral analysis and all eicosanoids are now amenable for qualitative mass spectrometric analysis.

ACKNOWLEDGEMENT

This work was supported in part by grants from the National Institutes of Health (NS 09199 and RR 01152).

REFERENCES

1. McLafferty FW: Interpretation of mass spectra. 3rd Edition, University Science Books, Mill Valley, CA, 1980.
2. Rose ME, Johnstone RAW: Mass spectrometry for chemists and biochemists. Cambridge University Press, Cambridge, 1982.
3. Cooks, RG, Beynon JH, Caprioli RM, Lester GR: Metastable ions. Elsevier, Amsterdam, 1973.
4. Green K, Hamberg M, Samuelsson B, Smigley M, Froelich JC: Measurement of prostaglandins, thromboxanes, prostacyclin and their metabolites by gas liquid chromatography-mass spectrometry. Adv Prostaglandin Thromboxane Res (5):39-94, 1978.
5. Fischer C, Frolich JC: Analysis of prostanoids by GC/MS measurement. Adv Lipid Res (19):185-202, 1982.
6. McFadden WH: Techniques of combined gas chromatography/mass spectrometry. Wiley-Interscience, 1973.
7. Blau K, King G: Handbook of derivatives for chromatography. Heyden, London, 1978.
8. Middleditch BS, Desiderio DM: Mass spectra of prostaglandins. IV. Trimethylsilyl derivatives of prostaglandins of the F series. Anal Biochem (55):509-520, 1973.
9. Rosello J, Gelpi E, Regand M, Breton JC: Mass spectrometric identification of new derivatives of prostaglandins D_1 and D_2. Adv Mass Spectrum (8B):1329-1333, 1980.
10. Watson JT, Sweetman BJ: A comparative vapor-phase study of various derivatives of prostaglandins A and B. Org Mass Spectrom (9):39-52, 1974.
11. Brash AR, Baille TA: A comparison of t-butyldimethylsilyl and trimethylsilyl ether derivatives for the characterization of urinary metabolites of prostaglandin $F_{2\alpha}$ by gas chromatography-mass spectrometry. Biomed Mass Spectrom (5):1346-1356, 1978.
12. Mai J, German B, Kinsella JE: Prostaglandins and thromboxanes: tert-butyldimethylsilyl ethers as derivatives for their analysis by gas chromatography-mass spectrometry selected ion monitoring. J Chromatog (254):91-107, 1983.
13. Green K: Gas chromatography-mass spectrometry of O-methyloxime derivatives of prostaglandins. Chem Phys Lipids (3):254-272, 1969.
14. Middleditch BS, Desiderio DM: Mass spectra of prostaglandins: II. Trimethylsilyl and alkyloxime-trimethylsilyl derivatives of prostaglandins B_1 and B_2. Lipids (8):267-270, 1973.
15. Goldyne ME, Hammarstrom S: Quantitative determination of prostaglandins E_1 and F_1 alpha by multiple ion analysis. Anal Biochem (88):675-681, 1978.
16. Waddell KA, Blair IA, Wellby J: Combined capillary column gas chromatography negative ion chemical ionization mass spectrometry of prostanoids. Biomed Mass Spectrom (10):83-88, 1983.

17. Waddell KA, Robinson C, Orchard MA, Borrow SE, Dollery CT, Blair IA: Quantitative analysis of arachidonic acid metabolites in complex biological fluids using capillary gas chromatography/negative ion chemical ionization mass spectrometry. Int J Mass Spectrom (48):233-236, 1983.
18. Strife R, Murphy RC: Preparation of pentafluorobenzyl esters of arachidonic acid lipoxygenase metabolites. Analysis by gas chromatography and negative ion chemical ionization mass spectrometry. J Chromatog (305): 3-12,1984
19. Strife R, Murphy RC: Stable isotope labeled 5-lipoxygenase metabolites of arachidonic acid. Analysis by negative ion chemical ionization mass spectrometry. Leukotrienes Prostaglandins Med. (13):1-8, 1983
20. Harrison AG: Chemical Ionization Mass Spectrometry, CRC Press, Boca Raton, 1983
21. Smith AG, Brooks CJW: Gas chromatography chemical ionization mass spectrometry of prostaglandin $F_{2\alpha}$ cyclic boronate derivatives. Biomed Mass Spectrom (4):258-264, 1977.
22. Ariga T, Suzuki M, Morita I, Murota S-I, Mujatake T: Chemical ionization mass spectrometry of trimethylsilyl derivatives of various prostaglandins. Anal Biochem (90):174-182, 1978.
23. Blair, IA, Barrow SE, Waddel KA, Lewis PJ, Dollery CT: Prostacyclin is not a circulating hormone in man. Prostaglandins (23):579-589, 1983.
24. McCloskey JA: Mass spectrometry of fatty acid derivatives. In: Gunstone FD (ed) Topics in lipid chemistry. Logos Press, Glasgow, 1970, pp 369-440.
25. Barber M, Bordeli RS, Sedwick RD, Tyler AN: Fast atom bombardment of solids as an ion source in mass spectrometry. Nature (293):270-275, 1981.
26. Murphy RC, Mathews WR, Rokach J, Fenselau C: Comparison of biologically-derived and synthetic leukotriene C_4 by fast atom bombardment mass spectrometry. Prostaglandins (23):201-206, 1982.
27. Taylor GW, Morris HR, Beaubien B, Clinton PM: Fast atom bombardment mass spectrometry of eicosanoids. In: Piper P (ed) Prostaglandins ser., Leukotrienes and other lipoxygenase products. John Wiley, Chichester, 1983, pp 277-282.
28. Middleditch BS, Desiderio DM: Mass spectra of prostaglandins. I. Trimethylsilyl and alkyloxime-trimethylsilyl derivatives of prostaglandin A_1. Prostaglandins (4):31-41, 1973.
29. Middleditch BS, Desiderio DM: Mass spectra of prostaglandins. III. Trimethylsilyl and alkyloxime-trimethylsilyl derivatives of prostaglandins of the E-series. J Org Chem (38):2204-2209, 1973.
30. Mel'nikova VI, Pivnitsky KK: The mass spectra and the structure of proctaglandins E 9-enol trimethylsilyl ethers. Biomed Mass Spectrum (10):399-43, 1983.
31. Horvath G: Mass spectral studies on prostaglandins. Biomed Mass Spectrom (3):4-13, 1976.
32. Horvath G: Mass spectral studies on prostaglandins. VI. Prostaglandin $F_{1\alpha}$ and 6-keto-prostaglandin $F_{1\alpha}$. Adv Mass Spectrom (8B):1334-1349, 1980.
33. McKay SW, Mallen DNB, Shrubsill PR, Smith JM, Baker SR, Jamieson WB, Ross WJ, Morgan SE, Rockhom DR: Semi-preparative high performance liquid chromatography and spectroscopic characterization of eight geometric isomers of leukotriene A methyl ester. J Chromatog (214):249-256, 1981.
34. Harper TW, Murphy RC: Mass spectrometric characterization of leukotriene hydroxy fatty acids. Proc Ann Conf Mass Spectrom (3):127-128, 1982.

INDEX